S0-ATH-870

Cellulose Chemistry and Technology

Cellulose Chemistry and Technology

Jett C. Arthur, Jr., EDITOR
Southern Regional Research Center, USDA

A symposium sponsored by the Cellulose, Paper and Textile Division at the 171st Meeting of the American Chemical Society, New York, N.Y., April 5–9, 1976.

ACS SYMPOSIUM SERIES **48**

AMERICAN CHEMICAL SOCIETY
WASHINGTON, D. C. 1977

Library of Congress CIP Data

Cellulose chemistry and technology.
(ACS symposium series; 48 ISSN 0097-6156)

Includes bibliographical references and index.

1. Cellulose—Congresses.
I. Arthur, Jett C., 1918– . II. American Chemical Society. Cellulose, Paper and Textile Division. III. Title. IV. Series: American Chemical Society. ACS symposium series; 48.

TS933.C4S92 1976 676'01'54 77-6649
ISBN 0-8412-0374-1 ACSMC8 48 1–397

PRINTED IN THE UNITED STATES OF AMERICA

ACS Symposium Series

FOREWORD

The ACS Symposium Series was founded in 1974 to provide a medium for publishing symposia quickly in book form. The format of the Series parallels that of the continuing Advances in Chemistry Series except that in order to save time the papers are not typeset but are reproduced as they are submitted by the authors in camera-ready form. As a further means of saving time, the papers are not edited or reviewed except by the symposium chairman, who becomes editor of the book. Papers published in the ACS Symposium Series are original contributions not published elsewhere in whole or major part and include reports of research as well as reviews since symposia may embrace both types of presentation.

CONTENTS

PREFACE

The centennial anniversary meeting of the American Chemical Society gave the Cellulose, Paper and Textile Division the opportunity to present a timely Symposium on International Developments in Cellulose, Paper, and Textiles. Research scientists from academia, industry, and government, representing more than sixteen countries, cooperated in presenting significant research accomplishments in paper, wood, and cellulose chemistry and in cotton, wool, and textile fiber chemistry.

In this volume research advancements on structure and on properties and reactions in cellulose chemistry have been contributed by investigators from Australia, Canada, Germany, Japan, Norway, the United Kingdom, the Union of Soviet Socialist Republics, and the United States. Two additional volumes, "Textile and Paper Chemistry and Technology" and "Cellulose and Fiber Science Developments: A World View," will include other contributed symposium manuscripts.

I would like to thank the participants, presiding chairmen, and particularly P. Albersheim, D. F. Durso, C. T. Handy, B. Leopold, A. Sarko, L. Segal, and A. M. Sookne whose leadership made the 22 sessions of the symposium truly international in scope. In addition, Herman Mark kindly made significant remarks to open the symposium.

New Orleans, LA
March 1, 1977

JETT C. ARTHUR, JR.
Organizing Chairman

Structure

1

Crystal Structures of Oligocellulose Acetates and Cellulose Acetate II

R. H. MARCHESSAULT

Department of Chemistry, Université de Montréal, Montreal, Québec, Canada

H. CHANZY

Centre de recherches sur les macromolécules végétales, CNRS, Grenoble, France

The crystal structure of a polysaccharide can be obtained by first determining the crystal structures of its oligosaccharide single crystals. By extrapolation and comparison with the fiber diagram, the structure of the polysaccharide is obtained. This is the classical approach for polymer structures and some efforts along these lines using powder diffraction data (1) and a single crystal study on cellotetraose (2) have been reported for the cellulose system. Historically, it was extensive work on the cellulose oligosaccharides by Freudenberg (hydrolysis, optical rotation) which provided a proof of structure for cellulose (3) by showing that the behaviour of the oligomer molecules, for which the structure could be established rigorously, extrapolated to the observed properties of cellulose. The present approach aims at the same objective in the area of conformation and packing of the repeating anhydroglucose triacetate in the crystal structure of cellulose triacetate (4).

The acetate oligosaccharides of cellulose which relate to the crystal structure of cellulose triacetate (CTA), were chosen for the following reasons:

- pure oligosaccharide fractions are readily available;
- single crystals of suitable size and perfection can be easily obtained from the acetylated oligomers compared to the unacetylated;
- the absence of hydrogen bonding in the system was expected to lead to oligomers which have a conformation close to that of the polymer;
- the oligomer crystal structures can be compared with data from a fiber diagram of outstanding quality: cellulose triacetate II;
- lamellar single crystals of CTA II yield electron diffraction data which can be used to define the base plane projection of the CTA II crystal.

The last argument may be a decisive one in all crystal structure work on oligosaccharides in the future. Ordinary fiber diagrams yield at best 5 to 10 equatorial (hk0) diffraction spots while the electron diffractogram from the lamellar single crystals show at least 5 times as many hk0 diffraction data. Since hk0 diffraction corresponds to the Fourier transform of the unit cell projected down the fiber axis, electron diffractograms such as these will help in solving the packing problem for CTA II (4). Furthermore from this additional information we may expect much greater certainty in the space group selection and better precision in the cell dimensions. The method is applicable also in cases where the polysaccharide is hydrated since it is possible to observe diffraction from the hydrated crystal even in the vacuum of the electron microscope by using a cooling stage (6).

The CTA crystal structure bears a close relation to native cellulose morphology. Heterogeneous acetylation (7) leads to a preservation of the native cellulose microfibrils and a unit cell referred to as CTA I. Simple heat treatment of the isolated microfibrils leads to a transformation into CTA II microfibrils. It has been observed (5) that for the isolated microfibril this transformation remains intramicellar and is accompanied by development of a shish-kebab structure. In crystallographic terms it is to be expected that such a transformation implies an antiparallel chain arrangement in CTA II. Since previous studies on this system have been inconclusive (4) in this respect, one of our objectives is to settle the chain polarity question.

Finally, it has been shown in a recent study (9) that conformational analysis of polysaccharide chains is most reliable when chain coordinates are derived from a single crystal structure of the directly related dimer. The present approach will provide such data. It should be appreciated however that cellotriose acetate ($C_{40}O_{27}H_{54}$) has a molecular weight of 966 which makes it the largest oligosaccharide structure so far undertaken. Furthermore, it would appear desirable to solve even larger oligomers in the series if our objectives are to be fully attained.

Experimental. The crystal structure determination of β-cellobiose octaacetate (G_2) has been reported (10).

Samples of β-cellotriose undecaacetate were obtained from D. Horton (Ohio State University, Columbus, Ohio) and are part of the material prepared by Dickey and Wolfrom (11). They were dissolved in hot ethanol with about 5% water. After slow evaporation of the solvent (about two weeks), long needles suitable for x-ray study were obtained. These crystals were similar in shape to those used for the cellobiose octaacetate study. In detail, the crystals are parallelepipeds and the two longest axes of the parallelepiped are perpendicular to the unique axis.

The unit cell and space group data for β-cellobiose acetate and β-cellotriose acetate were derived from Weissenberg and precession photographs. Three dimensional diffraction was recorded for β-cellobiose acetate, 3013 diffraction data were recorded and the structure was resolved by the direct method (10) followed by a three dimensional refinement.

For β-cellotriose acetate (G_3) the 3396 recorded diffraction data allowed solution by the direct method but the latter was modified (12) to take advantage of the known conformation of β-cellobiose acetate and the obvious relation between the values of the molecular length of the two molecules as derived from the c dimension of the unit cell (cf. below). This approach involved the generation of coordinates for β-cellotriose acetate by using the coordinates of β-cellobiose acetate and adding another unit to the non-reducing end with the same geometry as that of the non-reducing unit in β-cellobiose acetate. In this way the direct method was biased by the accumulated stereochemical knowledge which we have for this system and the crystallographic solution of a larger structure by the direct method was helped by solution of the previous structure of the homologous series.

Densities were measured by the classical floatation method.

Results. Since polymer X-ray fiber data follows the convention of making the c axis parallel to the chain axis we will report our oligomer data in a slightly non-conventional way from a crystallographic point of view. This ultimately will permit an easier comparison between the single crystal data on oligomers and the fiber X-ray data.

The needle shaped crystals of the three oligomers seem to have a similar morphology and molecular orientation in the crystal. Remembering that c is the unique axis in the polymer system, the relative shape and orientation of the molecular (c) axis in the needles and lamellar polymer single crystals are shown in Fig. 1. These data were derived from the single crystal studies of β-cellobiose acetate (10), β-cellotriose acetate (12), from electron diffraction of CTA II single crystals (8) and fiber diffraction data (14). Actually, the molecular axis for G_2 is not quite parallel to c and data for β-cellotriose acetate are not sufficient yet to establish whether or not there is a perfect parallelity. In fact the difference between the c dimensions for G_2 and G_3 is 5.4 A which is significantly greater than the value of 5.2 - 5.3 A derived from fiber diffraction hence it appears that only for the structure of G_3 or G_4 will one probably have a suitable equivalence between oligomer and polymer.

The pertinent crystal structure data are summarized in Table I. The c axis increases in going from G_2 to G_3 by an amount related to the increment for one added monomer, but it is slightly too high compared to the fiber repeat. The b dimension is relatively constant and from the crystal structure of G_2 it is known that this is the thickness of the ribbon-like G_2 molecule. The a dimension is related to the width of the ribbon-like molecule.

The theoretical density for G_3 is close to what is observed for the polymer hence it may be assumed that the same sort of interchain forces are present in the oligomer and polymer. The space group for G_3 is $P2_1$ and there are two molecules per unit

TABLE I

UNIT CELL PARAMETERS OF OLIGOCELLULOSE ACETATES AND CELLULOSE TRIACETATE II

sample	*a* A	*b* A	*c* A	γ^{o}	unit cell	space group	molecs. per cell	density g/cc.	F.W.	Emp. form.
Cellobiose Ac.	19.414	5.614	31.814	90	ortho.	$P2_12_12_1$	4	1.26	678	$C_{28}H_{38}O_{19}$
Cello-triose Ac.	11.709	5.675	37.216	94.3	mono.	$P2_1$	2	1.28	966	$C_{40}H_{54}O_{27}$
Cellulose Acetate II	24.36	11.42 (2 x 5.71)	10.54 (fiber axis)	90	ortho.	$P2_12_12_1$	4	1.29	288	$(C_{12}O_8H_{16})$

cell hence a "parallel" chain structure is operative in this case. By contrast the polymer has a $P2_12_12_1$ space group* and 4 cellobiose triacetate units per cell. This space group necessarily implies antiparallel chains, Dulmage (4) had suggested in his original study of this system that pairs of parallel chains formed the asymmetric unit but did not conclude, because of space group ambiguity, that the unit cell had equal members of parallel and antiparallel chains.

One of the sought for results from the single crystal data was the value of τ, the C(1)-O-C(4') bond angle, and the conformational angles ϕ and ψ at the glycosidic linkage (cf. Fig. 2). The data available to date are limited but Table II leaves one to hope that a trend may develop from G_3 onward wherein τ and ϕ, ψ converge toward the values which have been derived for the polymer on the basis of conformational analysis using the virtual bond approach (13).

An example of the sort of extrapolation approach which will be used in going from oligomer data to the polymer is the following. The $\underline{c}$ axis of the oligomers should lead eventually to the value observed for the polymer and in fact the oligomers may provide the most correct value for the polymer (e.g. the fiber repeat proposed by Dulmage was 10.43 A while a least squares refinement of the fiber data yields 10.54 A). A linear plot of $\underline{c}_n$ vs (n-2), where $\underline{n}$ is the degree of polymerization, is predicted when in the oligomers:

- the molecular axis is truly parallel to $\underline{c}$
- the angles τ, ϕ, ψ have attained their limiting values.

The proposed plot is based on a "superposition" model wherein the $\underline{c}$ axis of the unit cell is made up of a contribution from each end monomer (a + e) and the middle monomers contributes a value $\underline{\ell}$. Deviations from the above assumptions may make the plot non-linear for the lower oligomers of the series;

$$\underline{c}_3 = a + e + \ell$$

$$\underline{c}_n = a + e + (n - 2)\,\ell$$

$$\frac{\underline{c}_n}{n - 2} = \frac{a + e}{n - 2} + \ell$$

An important feature of the crystal structure of β-cellobiose acetate is the poor definition of the C(6) acetate group at the reducing end. This same problem occurs in the β-cellotriose acetate structure (12) nevertheless the conformation is clearly

---

* based on systematic absences in the fiber diagram and the systematic absences in the two principal directions of the electron diffractogram of the polymer single crystal, it was concluded that the most likely space group was $P2_12_12_1$(14).

TABLE II

VALUES FOR THE GLYCOSIDIC BOND ANGLE (τ) AND THE GLYCOSIDIC TORSIONAL ANGLES (ϕ,ψ) IN DEGREES

sample		τ	ϕ	ψ
β-Cellobiose Ac.		116.8	44	16
β-Cellotriose Ac.	reducing	115.5	46	12
	non-red.	117.0	24	-20
CTA II (conformational anal.)		119	6	-12

gt i.e. O(5)-C(5)-C(6)-O(6) = ~ 60°. However, the conformation of the C(6) acetate for the other groups is well defined and corresponds to the gg conformation, i.e. the torsional angle O(5)-C(5)-C(6)-O(6) = ~ -60°. Thus while the absolute values of the torsional angles defining the acetate groups are different in G_2 and G_3 the pattern is the same i.e. the planar C(2) and C(3) acetates are arranged so that the carbonyl nearly eclipses the axial hydrogen at the corresponding ring carbon and the C(6) acetates at the reducing end are different conformationally from those in the middle and end units. Thus the gg conformation would appear to be the likely choice for the C(6) acetate in CTA.

The electron diffraction data on single crystals of the CTA II structure and the X-ray fiber data on CTA II fibers are being used in combination with a classical conformational analysis approach based on the virtual bond method (13) to refine the structure. The coordinates of the non-reducing end of β-cellobiose acetate were used to obtain the τ, ϕ, ψ values (cf. Fig. 2) listed in Table II based on the known repeat of 10.5 A and the assumption that a twofold screw axis was along the chain direction when the acetate groups at C(2) and C(3) only, are considered. The minimum energy position of the rotatable groups was established by working with a tetramer and using an iterative procedure (15) to find the minimum energy position of these groups. The two middle residues were then considered to correspond to the crystallographic repeat of CTA II. A projection of the repeating cellobiose acetate residue is shown in Fig. 2. The twofold screw symmetry is not maintained since the C(6) acetate groups are shown in the two different positions found in the oligomer structures. The fact that the conformational energy undergoes little variation for a broad range of rotations of the various bonds in the C(6) acetate group makes the solution to the polymer structure quite difficult. In the present approach, conformations for the C(6) acetate group are tested by comparison between intensities predicted for trial structures and observed hk0 intensities.

As to the chain packing aspects, this is being deduced from a stereochemical analysis based on energy minimization. The CTA II structure so far evolved bears a great similarity to what was proposed by Dulmage (4) i.e. two pairs of CTA chains with the same polarity (related by 2_1 symmetry) are arranged in antiparallel fashion in an orthorhombic 4 chain unit cell. However the exact position of the chains in the unit cell is somewhat different in the present work.

Conclusions. The approach is breaking new ground in the methodology of oligosaccharide crystallography. The parallel chain arrangement in the triose seems non-characteristic of CTA II packing and suggests that only at higher molecular weights does the chain adopt the characteristic CTA II structure. The oligomer structures provide information on the arrangement of the C(6) acetates which will be compared with results from electron dif-

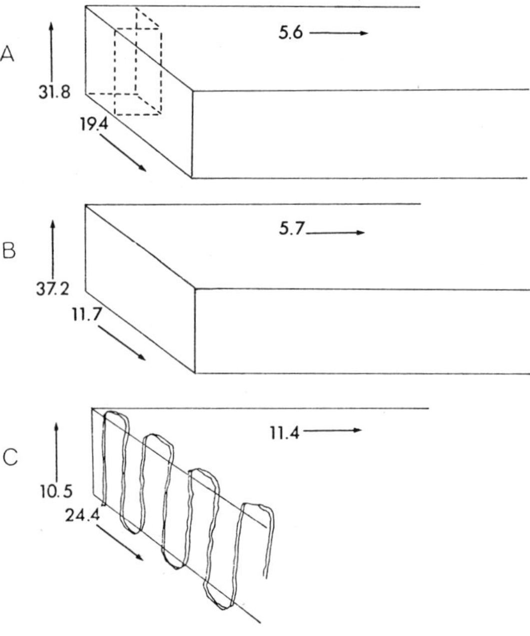

Figure 1

Figure 2

fraction for the single crystals of CTA II. The importance and usefulness of this new input of information would seem to be confirmed by our results so far.

Acknowledgements. The work reported herein is a continuing project involving collaborators from our respective laboratories at Montreal and Grenoble. The work is supported in part by the National Research Council of Canada and the Echanges France-Québec. Collaborators who are actively contributing to this program in Montreal are: F. Brisse, S. Pérez, and P.R. Sundararajan; at Grenoble: E. Roche; and from the Laboratoire de minéralogie at the Université de Lyon: Micheline Boudeulle.

Literature Cited

(1) Williams, D.G., J. Pol. Sci., (1970), 8, Part A-2, 637.
(2) Poppleton, J. and Mathieson, A.M.L., Nat. (London), (1968), 219, 1046.
(3) Purves, C.B., "Cellulose and Cellulose Derivatives", Chap. IIA, E. Ott, Editor, Interscience Publ. Inc., New York (1943)
(4) Dulmage, W.J., J. Pol. Sci., (1957), 26, 277.
(5) Chanzy, H., "Proceedings of the Eighth Cellulose Conf.", Appl. Polymer Symposia, Syracuse, New York, No. 28 (in press)
(6) Taylor, K., Chanzy, H. and Marchessault, R.H., J. Mol. Biol. (1975), 92, 165.
(7) Sprague, B.S., Riley, J.L. and Noether, H.D., Text. Res. J., (1958), 28, 275.
(8) Roche, E. and Chanzy, H., J. Pol. Sci., Part A-2 (in press).
(9) Marchessault, R.H. and Sundararajan, P.R., Pure and Appl. Chem., (1975), 42, 399.
(10) Leung, F., Chanzy, H., Pérez, S. and Marchessault, R.H., Can. J. Chem., (1976), 54, 1365.
(11) Dickey, E.F. and Wolfrom, M.L., J. Am. Chem. Soc., (1949), 71, 825.
(12) Pérez, S. and Brisse, F., Acta Cryst. (in press).
(13) Sundararajan, P.R. and Marchessault, R.H., Can. J. Chem., (1975), 53, 3563.
(14) Chanzy, H., Roche, E., Boudeulle, M., Marchessault, R.H. and Sundararajan, P.R., Third European Crystallography Conference, Zurich, Sept. (1976).
(15) Sundararajan, P.R. and Marchessault, R.H., Biopolymers, (1972), 11, 829.

2

A Virtual Bond Modeling Study of Cellulose I

ALFRED D. FRENCH

Southern Regional Research Center, U.S. Dept. of Agriculture, ARS, P.O. Box 19687, New Orleans, LA 70197

VINCENT G. MURPHY

Department of Chemical Engineering, Iowa State University, Ames, IA 50011

Use of a molecular model in solving fiber diffraction patterns for the positions of atoms is necessitated by the small number of diffraction data in comparison to the large number of atoms that must be located in each of three dimensions. Thus, it is current practice to construct a stereochemically reasonable polymer model and then use the diffraction intensity information to locate the chain within the unit cell and to determine the correct placement of any side groups. However, when the proposed structure with lowest X-ray error, R, is compared with competing models, the differences in R are often small and statistical tests must be used to determine the level of confidence one may have in a specific model. For example, Gardner and Blackwell (1) recently reported with high confidence that native cellulose from *Valonia* is composed of parallel *up* chains as opposed to parallel *down* or antiparallel chains. These conclusions were drawn for models that gave R values of 0.179, 0.202, and 0.207 based on observed reflections only.

The confidence tests (2), essentially the F-tests for analysis of variance, specifically assume that errors in the data are not systematic. This assumption is not rigorously true for fiber photography because of the manner in which corrections are applied to the photographs and the usual exclusion of meridonal data. The neglect of intensity contributions from the 8-chain unit cell is another small but systematic error for *Valonia*. A modest increase in the probability of error is one possibility if systematic errors are suspected. The major factors in the probability of error, however, are the ratio of the R values for the competing models and the number of variables used in the study. Small differences in the R values are significant only when the data/parameter ratio is large. Because the number of variable parameters is drastically reduced by the polymer modeling process, the *apparent* confidence is considerably enhanced for structures favored by only small differences in R.

Are the small differences in R, such as have been reported for cellulose, really significant when determined using rigid chain models? Is there some other cellulose model, equally reasonable but with somewhat different internal parameters, that would favor, for example, a parallel down structure? These questions were addressed in the present study by examining the effects of changes in chain geometry upon the structural determination of native cellulose from Valonia. The virtual bond method was employed because it provides a computationally simple procedure that allows some degree of flexibility at the glycosidic linkages in the model refining process (3,4).

Computer Modeling Techniques

A major assumption in the construction of polysaccharide models is that the geometry of the actual monomer units is similar to that of saccharide residues from simple carbohydrate materials. Table I contains structural data from the β-glucose residues of a number of crystalline substances (5-8). The bond lengths and bond angles are relatively constant; however, the individual ring conformation angles show a variation of as much as 12°. These angles, characteristic of the shapes of the rings, do not vary as much as those in a considerably larger number of α-glucose rings (3). (The individual torsion angles of currently available α-rings vary as much as 20°.) Faced with a number of possible monomers, a common choice is to use an averaged residue geometry for polymer modeling. Arnott and Scott describe both α- and β-residues averaged from many pyranoses (9). Sarko and Muggli (10) report another β-ring, obtained by averaging residue coordinates from crystalline β-D-glucose (5) and an early study of β-cellobiose (11). Table I shows the parameters of both average residues to be similar. In addition to providing models of pyranose rings, surveys of single crystal studies indicate the glycosidic angle to be about 117° for 1,4 linkages and provide information on permissible inter-residue atomic contacts and hydrogen bond lengths.

Once a monomeric geometry is selected, that coordinate set may be repeated, generating a model for the polymer structure. A common approach is to choose some value for the glycosidic bond angle (τ) and then to examine the stereochemical acceptability of the model as a function of the residue rotations about the C(1)--O and O--C(4') bonds. The linkage torsion angle variation method is sometimes referred to as the Φ-Ψ method after the names assigned to the torsion angles that measure the residue rotations (12). The degree of stereochemical acceptability at given increments of Φ-Ψ is usually determined by the number of hard-sphere contacts in the model (13) or by its total potential energy (10,12). As an example, the Φ-Ψ map for cellulose with O(6) gt, as obtained by Sarko and Muggli (10) using their average residue and a glycosidic angle of 116°, is reproduced as Figure 1.

TABLE I
PARAMETERS FOR β-D-GLUCOSE RINGS

		β-D GLUCOSE	MALTOSE MONOHYDRATE	β-METHYL MALTOSIDE	CELLOBIOSE NONREDUCING	CELLOBIOSE REDUCING	METHYL CELLOBIOSIDE METHANOL, NONREDUCING	METHYL CELLOBIOSIDE METHANOL, REDUCING	ARNOTT-SCOTT AVERAGE	SARKO-MUGGLI AVERAGE
Virtual Bond Lengths	O4-O1	5.454	5.507	5.448	5.452	5.444	5.558	5.513	5.468	5.438
	O3-O1	4.735	4.795	4.774	4.820	4.726	4.875	4.715	4.766	4.754
	O2-O1	2.848	2.821	2.869	2.754	2.889	2.894	2.854	2.807	2.807
Ring Bond Lengths	C1-C2	1.526	1.515	1.511	1.525	1.514	1.527	1.537	1.523	1.507
	C2-C3	1.519	1.540	1.535	1.520	1.520	1.561	1.540	1.521	1.513
	C3-C4	1.512	1.549	1.514	1.534	1.531	1.544	1.538	1.523	1.489
	C4-C5	1.529	1.552	1.512	1.532	1.528	1.535	1.552	1.525	1.550
	C5-O5	1.437	1.435	1.428	1.436	1.436	1.453	1.442	1.436	1.430
	O5-C1	1.433	1.442	1.424	1.425	1.435	1.451	1.442	1.429	1.424
exo-Ring Bond Lengths	C1-O1	1.384	1.421	1.377	1.397	1.380	1.413	1.390	1.389	1.384
	C2-O2	1.429	1.409	1.419	1.416	1.423	1.433	1.450	1.423	1.402
	C3-O3	1.433	1.433	1.430	1.427	1.409	1.433	1.449	1.429	1.421
	C4-O4	1.419	1.447	1.437	1.419	1.446	1.433	1.445	1.426	1.426
	C5-C6	1.513	1.524	1.511	1.519	1.500	1.530	1.511	1.514	1.505
Ring Bond Angles	O5-C1-C2	108.5	111.2	110.4	108.3	109.3	106.6	108.4	109.3	108.9
	C1-C2-C3	112.1	108.6	109.7	108.3	110.0	109.2	114.5	110.5	110.5
	C2-C3-C4	110.5	110.1	110.6	109.5	111.8	114.1	111.1	110.3	111.1
	C3-C4-C5	109.8	108.6	111.0	111.0	112.3	107.6	108.8	110.2	110.8
	C4-C5-O5	107.6	107.3	108.2	110.6	109.2	110.9	109.0	110.2	108.4
	C5-O5-C1	112.7	112.9	111.4	112.4	113.5	111.1	110.4	112.3	113.3
exo-Ring Bond Angles	O5-C1-O1	107.0	107.4	107.1	107.5	106.9	107.8	107.4	107.3	107.6
	C2-C1-O1	108.2	107.2	108.7	109.0	110.3	110.8	107.8	108.4	109.8
	C1-C2-O2	108.5	108.9	111.0	110.1	110.6	110.2	106.8	109.3	109.8
	C3-C2-O2	109.8	109.6	107.6	113.6	106.4	109.3	109.6	110.8	110.5
	C2-C3-O3	108.7	110.6	111.2	112.0	107.2	110.7	105.9	109.6	109.3
	C4-C3-O3	109.1	106.7	107.4	111.5	112.5	108.3	112.0	109.7	110.7
	C3-C4-O4	111.1	108.0	107.8	108.1	109.1	110.1	110.6	110.4	109.8
	C5-C4-O4	108.2	109.0	111.2	109.4	106.4	108.4	107.4	108.6	106.7
	C4-C5-C6	114.9	113.5	115.2	110.9	113.6	111.1	114.8	112.7	113.9
	O5-C5-C6	107.1	110.9	108.4	105.4	106.4	106.8	107.3	106.9	106.5
Ring Torsion Angles[a]	O5-C1-C2-C3 (I)	53.7	56.1	55.8	63.5	57.8	58.7	51.6	57.5	57.5
	C1-C2-C3-C4 (II)	-50.8	-54.4	-50.0	-57.3	-50.7	-51.6	-45.4	-53.2	-53.5
	C2-C3-C4-C5 (III)	53.4	57.1	52.0	52.0	48.0	47.7	48.0	52.2	51.9
	C3-C4-C5-O5 (IV)	-59.8	-59.9	-58.3	-51.9	-51.1	-54.0	-60.7	-56.0	-54.7
	C4-C5-O5-C1 (V)	66.3	63.8	65.2	59.8	60.9	68.0	70.5	62.4	61.8
	C5-O5-C1-C2 (VI)	-62.8	-63.4	-65.0	-65.7	-65.1	-68.9	-64.0	-62.8	-63.9
Primary[a] Alcohol conformation	C4-C5-C6-O6	59.1	60.1	53.5	168.4	-169.3	173.2	65.9	61.2	-59.3
	O5-C5-C6-O6	-60.4	-60.8	-67.8	48.7	70.5	52.1	-55.4	-60.0	-178.8
	TORSION ANGLE INDEX	120.4	120.7	125.7	142.4	135.3	145.5	122.8	127.7	130.1

[a] Torsion angle A-B-C-D is the clockwise angle from bond A-B to bond C-D when the four atoms are projected onto a plane perpendicular to bond B-C and atom A is closest to viewer.

[b] The torsion angle index is the sum of absolute values of torsion angle I, II, V and VI minus the sum of the absolute values of torsion angles III and IV.

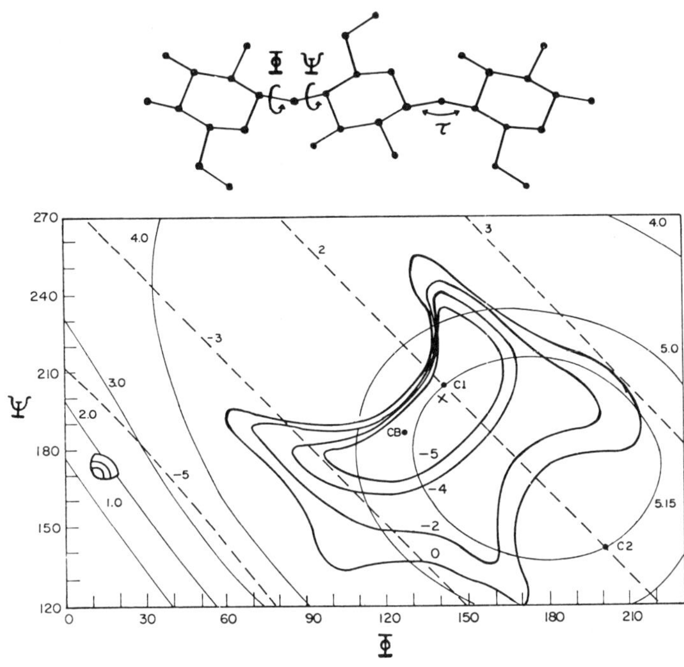

Macromolecules

Figure 1. Conformational energy map for 1–4 linked, β-D-glucose chains with O(6) in a gt position. The torsion angles Φ and Ψ that correspond to geometrically possible cellulose structures are at C1 and C2. The minimum in potential energy (X) occurs very near C1. (10)

In Figure 1, certain combinations of Φ and Ψ result in chains with special features. For instance, the innermost "circular" line denotes the Φ - Ψ combinations yielding helical models with a rise per residue of 5.15 Å, and the "straight" line more or less bisecting this "circle" indicates the Φ - Ψ combinations yielding helical models with two residues per turn. Also, the conformation possessing minimum energy (denoted by an X in the figure) occurs very near the intersection of the h = 5.15 Å and n = 2 lines. If we consider only models with an integral number of residues per turn and the observed 10.3 Å fiber-repeat spacing, then no other Φ - Ψ combination is nearly as attractive. As noted previously, (3) however, the Φ - Ψ values determined in this manner (and even the feasibility of the model) are often highly dependent upon the choice of monomer geometry and glycosidic angle.

An alternate and older technique is the virtual bond method. Jones (14,15) was perhaps the first to use it to full advantage. To be used efficiently, this method requires that the fiber repeat distance be known and the number of residues per turn be proposed from some outside source such as a diffraction pattern or a Φ - Ψ map. However, the only models that are built are those that meet these two requirements. A further advantage is that effects of changes in residue geometry can readily be seen.

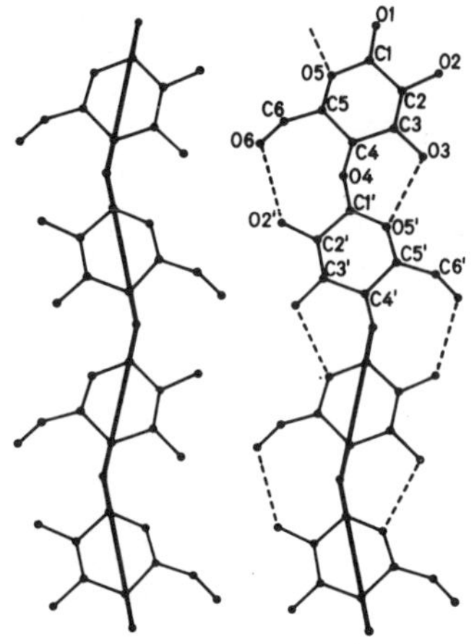

Figure 2. Virtual bond model of cellulose I (projection on ac *plane). Atomic designations are given, with likely hydrogen bonds indicated by dotted lines. The virtual bonds are shown as heavy black lines connecting the linkage atoms, O(1) and O(4). The coordinates used in this figure are from a model composed of nonreducing β-cellobiose residues and with $\tau = 115.7°$.*

As shown in Figure 2, cellulose is considered to be a collection of vectors (or virtual bonds) linked at the glycosidic oxygens. The end points of these vectors are fixed by the previously determined fiber repeat distance and number of residues per turn. Varying the residue rotations (θ) about the virtual bond changes both the magnitude of the glycosidic angle and the relative positions of atoms on successive residues. We have adopted the convention that a positive rotation appears clockwise when viewed from O(4) to O(1) and $\theta = 0°$ when the plane formed by O(4)—O(1)—C(6) is parallel to the fiber axis. For a structure with screw symmetry, all residues have the same value of θ; therefore, once the initial residue is positioned, the rest of the model can be generated by a simple screw operation. Further details on the mechanics of these procedures are contained in earlier reports (3,4).

For the purposes of rapidly generating, screening, and drawing a large number of potential models, we have developed a versatile computer program, HELIX, based on the virtual bond method. Starting with initial residue coordinates in any arbitrary orientation, HELIX generates a stereochemical survey, such as shown in Figure 3, in about 15 minutes on our relatively slow, batch-mode computer (CDC 1700). Although Figure 3 only shows plots of glycosidic angle and O(3)--O(5') distance vs. residue orientation, this particular option of HELIX computes τ , Φ , Ψ and all inter-residue atomic distances at each value of θ. In the course of the distance calculations, any value less than fully allowed, as defined by Rees and Skerrett (13), is output along with the identity of the offending atomic pair.

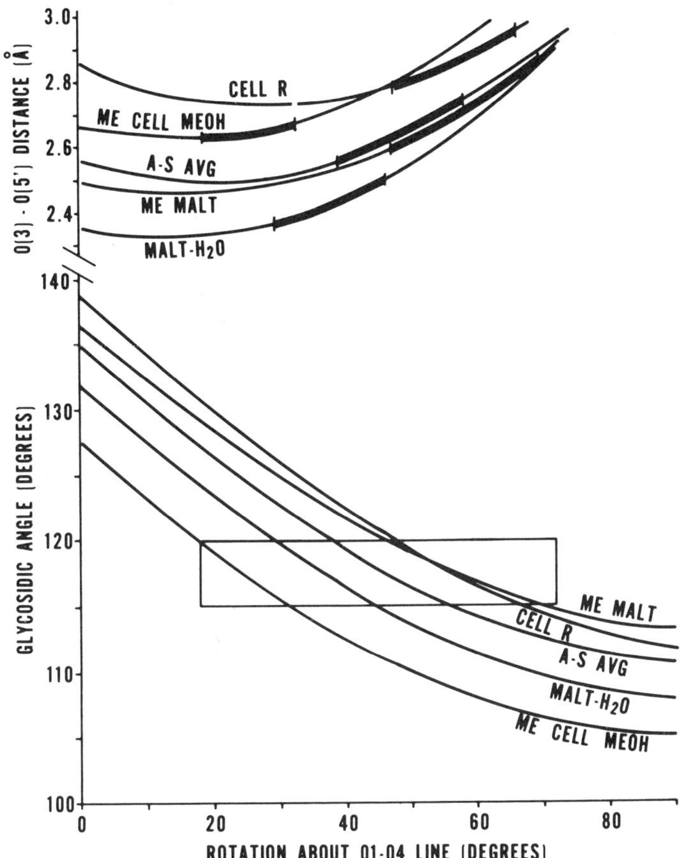

Figure 3. Glycosidic angle and O(3)–O(5′) distance for five monomer geometries as a function of θ, the rotation about the O(4)–O(1) virtual bond. The five glucose residues are: Me Malt, (β-anomeric residue from methyl β-maltoside monohydrate (7)), A–S Avg (Arnott-Scott average residue (9)), Malt–H_2O (β-anomeric residue from maltose monohydrate (6)), Cell R (reducing residue from β-cellobiose (5)), and Me Cell MeOH (reducing residue from methyl β-cellobioside methanol complex (8)).

residue from maltose monohydrate (6)), Cell R (reducing residue from β-cellobiose (5)), and Me Cell MeOH (reducing residue from methyl β-cellobioside methanol complex (8)).

The horizontal lines at 115 and 120° of glycosidic angle depict the range considered reasonable, and the vertical lines at 18 and 72° of rotation about the virtual bond show the limits of θ for production of geometrically feasible cellulose models having a 10.38 Å repeat. Heavy areas on the curves of O(3)--O(5') distance indicate the regions corresponding to acceptable glycosidic angles.

Because the exact value of the glycosidic angle in cellulose is unknown, we considered a range of values, $115° \leqslant \tau \leqslant 120°$, as indicated by the horizontal lines in Figure 3. Thus, depending upon the residue, θ values anywhere from 18 to 72° yield models with acceptable glycosidic angles. Over this range of θ, the linkage torsion angles Φ and Ψ vary about 80° and 45°, respectively. Bearing in mind that all of these models have the correct rise per residue and the correct number of residues per turn, we see that a large number of models can satisfy the rudimentary requirements of the cellulose chain. This number, however, can be reduced by rejecting the models that incorporate overly short inter-residue contacts.

Although all possible contacts must be considered, the most important one for cellulose is the 0(3)--0(5') distance. As noted by Gardner and Blackwell (1), if a glycosidic angle of 117° is used with the Arnott-Scott average residue (9), the 0(3)--0(5') distance is only 2.60 Å. From Figure 3, it is clear that a glycosidic angle slightly less than 115° is required if this distance is to be a more acceptable value of 2.75 Å. Two other residues offer some promise of providing models more in conformity with accepted convention. The reducing residue from β-cellobiose (5) leads to models with 0(3)--0(5') distances of 2.8--3.0 Å when τ is in the 115--120° range, and the corresponding values for models derived from the β anomer of methyl β-maltoside (7) are 2.6--2.9 Å. Of course, we cannot completely discount the possibility that cellulose actually has a small glycosidic angle or an unusually short hydrogen bond between 0(3) and 0(5').

An important result of the calculations summarized in Figure 3 is that none of the available residues yielded models with 0(3)--0(5') distances too long for hydrogen bonding when the glycosidic angle was in the most probable range. Because all available residues were tested on this point, and they have a wide range of internal variations, it is likely that this contact is a consequence of the $^4C_1$ monomer geometry and the fiber-repeat distance. In the absence of such results, however, we feel that it would be inappropriate to limit consideration to models that contain this (assumed) feature.

As noted in the Introduction, the virtual bond method was selected for this study because it facilitates the introduction of chain geometry as a variable in the structural determination. Under more favorable conditions (i.e., larger data/variable ratio), the linked atom method (9,16), would be the technique of choice since it permits even greater chain flexibility during the refinement process. With this method, the individual atom positions may be allowed progressively increasing freedom, ultimately subject only to the constraints imposed by predetermined bond lengths, the very parameters in which we have the most confidence (cf. Table I). The full potential of this method, however, has not been exploited in polysaccharide studies due to the insufficient number and poor quality of the X-ray data. A few structural studies employing a hybrid of the virtual bond and linked atom methods have been reported (17,18).

Procedures for X-ray Diffraction Calculations

The observed only data of Gardner and Blackwell (1) were selected for comparison of observed and calculated intensity square roots. We recognize that there are substantial differences between this data set and that of Mann *et al.* (19) for ramie and that of Sarko and Muggli (10) for *Valonia*. On one hand, ramie and *Valonia* diffraction patterns are different, with no evidence of eight-chain unit cells in ramie (20); on the other hand, the Gardner-Blackwell and Sarko-Muggli data vary due to differing interpretations of the same photograph. Nevertheless, the Gardner-Blackwell data set has resulted in one of the best diffraction intensity agreements reported in a polysaccharide modeling study.

Cellulose chains, constructed from various residues by the virtual bond method, as discussed above and in more detail earlier (3), were positioned at the corner (0,0) and center (1/2,1/2) of the two-chain, monoclinic unit cell described by Gardner and Blackwell (1). In computing intensity square roots, only contributions from the carbon and oxygen atoms were included, using standard scattering factors (21) and an isotropic temperature factor of 2.5 (1). In attempting to obtain the best match between experimental and calculated values, variations in the following parameters were considered: residue rotation about the virtual bond (θ), rotation of the primary alcohol group (χ), rotation of the corner and center chains about their axes (ROT1 and ROT2), relative translation of the center chain along its axis (SHIFT), and packing mode (parallel *up*, parallel *down* and antiparallel).

Instead of a least squares minimization technique such as that employed by Gardner and Blackwell (1), we used a direct systematic search of the variable space similar to that described in previous reports (10,22,23). This method requires much more computer time but is readily adapted to modest-sized computers and can better avoid false (local) minima.

Results and Discussion

Coarse-grid Searches Using Models Derived from Arnott-Scott and Reducing β-Cellobiose Residues. Coarse-grid tests were run prior to fine-grid refinement in order to determine the optimal value for θ and the approximate positions of the chains in the unit cell. Table II shows results obtained in a coarse-grid search of the variable space using models constructed from the Arnott-Scott average residue (9). A feature of these calculations is that R values less than 0.21 were found only for parallel chain models with O(6) in the *tg* position. Also, R is sensitive to the internal variations in the chain caused by changing θ. A graph of best R *vs.* θ predicts that a minimum in R would be found very near to

TABLE II
COARSE-GRID* SEARCH FOR MODELS MADE FROM ARNOTT-SCOTT AVERAGE RESIDUE

θ	O(6)	R	ROT1	ROT2	SHIFT	τ
			PARALLEL UP			
40°	tg	.21	65°	60°	.27	119.4°
	gt	.22	65	65	.26	
50	tg	.17	55	50	.27	116.7
	gt	.23	50	55	.26	
60	tg	.16	45	40	.27	114.4
	gt	.23	45	40	.26	
70	tg	.20	35	35	.27	112.6
	gt	.25	30	35	.26	
			ANTIPARALLEL			
40	tg	.25	55	50	.31	119.4
	gt	.28	60	60	.31	
50	tg	.22	50	45	.31	116.7
	gt	.25	40	40	.31	
60	tg	.21	45	40	.31	114.4
	gt	.23	30	30	.31	
70	tg	.21	35	30	.31	112.6
	gt	.23	25	25	.31	

*Values of ROT1 and ROT2 and SHIFT are those yielding the lowest value of R for given θ and O(6) conformation. Increments of variation were: θ (10°), ROT (5°), χ (10°), SHIFT (0.02). Even values of SHIFT denote that adjacent odd shifts had equal values of R.

θ = 57.9°, the θ value for the Gardner-Blackwell model. This suggests that their model happens to have the optimum θ for models based on the Arnott-Scott average residue.

Table III contains results from models made with the reducing glucose residue from β-cellobiose (5). Again, the parallel models with O(6) *tg* gave better agreement than did the others, although in this case, the distinctions were not as clear. The calculations definitely indicated that models with O(6) in a *gg* conformation were inferior, but they showed only a marginal preference for *tg* over *gt* and for parallel over antiparallel. Once again, values of θ in the range of 50--60° were favored; however, this time the preferred glycosidic angle and O(3)--O(5') distances were both somewhat larger, 116.5° and 2.92 Å respectively.

Effect of Rotation About the Virtual Bond on Diffraction Intensity. From the viewpoint of computer model-building, changing the rotation of the residue about its virtual bond is equivalent to changing the glycosidic angle and finding the new values of Φ and Ψ that correspond to the required rise per residue and required number of residues per turn. At the level of resolution provided by most fiber X-ray diagrams, however, R values should be minimal at the rotation of a roughly planar mass of electron density that best matches the "true" structure, regardless of deficiencies in internal detail such as an unreasonable glycosidic angle or nonoptimal values of Φ and Ψ.

Using the β-maltose monohydrate residue, we tested more thoroughly the validity of using θ as a variable in fiber X-ray analysis. As shown in Figure 3, this residue does not yield feasible cellulose models; hence, the calculation of R at different values of θ should show whether R minimizes at feasible glycosidic angles, at reasonable O(3)--O(5') distances or simply when the atoms are rotated about the virtual bond in the approximately correct amount. As shown in Table IV, the minimum R occurred at θ = 55°, roughly equal to the values found for the other residues. At this point, the glycosidic angle is quite small and the O(3)--O(5') distance less than optimal.

Tables II, III, and IV show that the optimal values of ROT1 and ROT2 vary as the residues assume different orientations about the virtual bond. This is due to the fact that cellulose is an extended chain and the virtual bond is nearly parallel to the chain axis. A similar study with V amylose (23), a helix with a rise per residue one-fourth that of cellulose, showed very little such effect. There again, however, the lowest R values were attained at similar values of θ for all residue geometries studied.

TABLE III

COARSE-GRID SEARCH FOR MODELS MADE FROM REDUCING RESIDUE OF β-CELLOBIOSE*

θ	O(6)	R	ROT1	ROT2	SHIFT	τ
				PARALLEL UP		
40	tg	.20	70	65	.27	122.3
	gg	.32	65	60	.25	
	gt	.22	65	70	.26	
50	tg	.19	60	55	.27	119.2
	gg	.30	60	50	.25	
	gt	.21	55	60	.26	
60	tg	.19	50	45	.24	116.5
	gg	.29	50	40	.25	
	gt	.22	50	40	.25	
70	tg	.22	35	40	.26	114.4
	gg	.29	35	35	.27	
	gt	.23	40	30	.25	
				ANTIPARALLEL		
40	tg	.23	60	55	.31	122.3
	gg	.32	60	65	.28/.34	
	gt	.23	70	65	.35	
50	tg	.21	50	50	.33	119.2
	gg	.29	50	45	.25	
	gt	.22	55	50	.35	
60	tg	.20	40	40	.32	116.5
	gg	.26	40	40	.27	
	gt	.22	50	45	.35	
70	tg	.23	30	30	.31	114.4
	gg	.27	35	30	.33	
	gt	.22	35	30	.32	

* See footnote to Table II.

TABLE IV

R vs. θ FOR STEREOCHEMICALLY POOR MODELS MADE FROM THE β -RESIDUE FROM MALTOSE MONOHYDRATE*

R	θ	τ	O(3)--O(5') DISTANCE	ROT1	ROT2
0.24	25°	122°	2.35 Å	75°	75°
0.23	35	118	2.41	65	65
0.21	45	115	2.51	55	55
0.19	55	112	2.64	45	45
0.23	65	110	2.79	35	35
0.32	75	109	2.90	25	25

* All models were parallel up and O(6) and SHIFT were fixed. In addition, values of ROT2 were required to equal ROT1 and were given 10° increments.

Refinements of Models Made with Several Different Residues. Models derived from five different residues were selected for fine-grid refinement following preliminary tests such as those described above. On the basis of these calculations, models with O(6) in gt positions were eliminated from consideration after they showed no evidence that R values substantially lower than on the coarse-grid searches would be found. Parallel up, parallel down, and anti-parallel models were all considered, and the results in terms of packing mode are given in Table V. Included in this table are R" values, as defined by Hamilton (2), a different definition than that given by Gardner and Blackwell (1). These results show that in each instance the best (lowest R) parallel up model is superior to the best parallel down which, in turn, is better than the best antiparallel model. Although no attempt was made (at this time) to find the best R" values, the rankings based on R" are the same. If one allots a single degree of freedom for variations in residue geometry and applies Hamilton's test to the R values, which is permissible for rough work (2), the same high level of confidence in the parallel up arrangement as reported by Gardner and Blackwell (1) is obtained. Alternatively, one might consider the results presented in Table V as being analogous to tossing a 3-sided coin five times and obtaining "heads" each time. This constitutes a simple test, then, for whether a feature determined by the diffraction analysis probably applies to the true structure.

Another indication from these fine-grid searches is that residue geometry has an effect on R comparable to that of other modeling parameters such as packing mode and O(6) conformation. The differences in R for the best models derived from the various

TABLE V

EFFECTS OF PACKING MODE ON R VALUE*

RESIDUE GEOMETRY	PACKING MODE	R-VALUE R	R-VALUE R"	ROT1	ROT2	-X	SHIFT	θ	τ	O(3)--O(5')
Arnott-Scott (9)	UP	.157	.165	45°	45°	81°	.265	57.9°	114.8	2.75Å
	DOWN	.174	.179	138	135	73	.265			
	ANTI	.200	.215	45	43	69	.310			
Sarko-Muggli (10)	UP	.156	.174	52	47	79	.265	57.9	115.4	2.76
	DOWN	.168	.185	136	134	74	.265	55	116.0	2.72
	ANTI	.193	.205	43	41	69	.310			
β-Cellobiose (5) Reducing	UP	.161	.181	50	46	80	.265	60	116.5	2.92
	DOWN	.180	.197	138	130	74	.265			
	ANTI	.187	.198	44	42	75	.310			
Me-β-Maltoside (7)	UP	.176	.189	45	45	74	.265	55	118.0	2.70
	DOWN	.193	.203	140	134	68	.265			
	ANTI	.223	.227	40	38	54	.310			
β -Maltose-H_2O (6)	UP	.179	.202	43	45	75	.265	55	112.0	2.51
	DOWN	.205	.216	140	136	75	.265			
	ANTI	.230	.241	41	39	55	.320			

* The variables ROT1, ROT2, and SHIFT are all at optimum values for models with a given residue geometry and packing mode. "UP" refers to models in the cellulose I unit cell as described by Gardner and Blackwell that have the z coordinate of O(5) greater than the z coordinate of C(5).

residues are similar to those observed in an earlier study of V amylose (23). Also, models with the poorest stereochemical features, those from the β -maltose monohydrate residue (6), yielded the highest values of R and, in general, the results seem to favor models with glycosidic angles and 0(3)—0(5') distances toward the lower end of the allowable ranges. The latter results were somewhat unexpected since, as noted above, models for which these parameters are near their average values were possible with either the reducing residue from β-cellobiose (5) or the β-residue from methyl β-maltoside (7).

Finally, the 0.168 value of R obtained for the best parallel down model based on the Sarko-Muggli residue illustrates the extent of the differences between the Sarko-Muggli (10) and the Gardner-Blackwell data sets. In the Sarko-Muggli study, an R value of 0.29 was reported for a model comparable to that used here.

Models from the Nonreducing β-Cellobiose Residue. In addition to the calculations summarized above, we also tested models derived from two other monomeric geometries. Those based on the residue from crystalline β-D-glucose (5) were not promising; however, models constructed from the nonreducing residue of β-cellobiose (5) yielded the best agreement with the diffraction data and incorporated no intrachain stereochemical disadvantages. With this residue, the minimum R was 0.144 and the minimum in R" 0.155. The slightly different structures resulting in these minima had R" = 0.165 and R = 0.155, respectively. The model having the minimum R" value is somewhat more attractive stereochemically, since it results in a 2.81 Å distance for the 0(6)—0(2') hydrogen bond as opposed to 2.66 Å for the model with the minimum in R.

Table VI compares characteristics of the Gardner-Blackwell (1) model with those of the best model derived from the nonreducing β-cellobiose residue. Not shown are the fractional coordinates. The largest difference between the two models was found in the fractional coordinates of the 0(2) atoms, which differed by 0.5 Å. The advantages of the present model are a larger glycosidic angle, 115.7°, and the possibility of a somewhat stronger 0(6)--0(2') hydrogen bond. Both models have 0(3)--0(5') distances close to 2.77 Å, which is the average of the values reported for the two cellobiose structures (5,8) and three lactose structures (24-26). Cellobiose and lactose have the same environment as cellulose at the glycosidic linkage. For the cellobiose structures (5,8) glycosidic angles are 115.8° and 116.1°, and for the lactoses (24-26), they are 115.7°, 116.0°, and 117.1°, all quite close to the present value of 115.7°.

Fries et al. (26) have proposed that low values of the glycosidic angle are characteristic of the β-1,4 linkage and large values typical of the α-1,4 linkage. Interestingly, studies of V amylose (22,23) that combined diffraction and stereochemical information in a manner similar to the present study showed

TABLE VI

COMPARISON OF GARDNER-BLACKWELL AND NONREDUCING β -CELLOBIOSE MODELS

PARAMETER	GARDNER-BLACKWELL	NONREDUCING β-CELLOBIOSE
θ	57.9°	68.0°
τ	114.8°	115.7°
Packing Mode	Parallel _up_	Parallel _up_
SHIFT	0.266	0.266
ROT1	45°	39°
ROT2	45°	42.5°
Φ 1/	22.7°	17.2°
Ψ 1/	-23.7°	-16.2°
χ	-80.3°	-88.6°
O(3)—O(5')	2.75 Å	2.78 Å
O(6)—O(2')	2.87 Å	2.81 Å
O(6)—O(3")2/	2.79 Å	2.64 Å
R	0.157 3/	0.155
R"	0.165 4/	0.155

1/ Computed from our copy of the Gardner-Blackwell model.
2/ Interchain hydrogen bond between two chains with the same shift.
3/ The corresponding value computed from Gradner and Blackwell's structure factor table is 0.167. This value is different from the above because there are small differences (0.003 _b_ maximum) between their coordinates and those of our "copy" and because we used a cosine table instead of the time-consuming Fortran algorithm in computing structure factors.
4/ The corresponding value computed from Gardner and Blackwell's structure factor table is 0.180. Hamilton's (2) definition of R" is used instead of that given by Gardner-Blackwell. See also footnote 3, above.

that models with glycosidic angles of 118.4° for V hydrate and 118.7° for V dehydrate were nearly optimal. Also, on stereochemical considerations alone, Winter and Sarko reported angles of 120.0° and 120.5° for the α-1,4 linked V dehydrate (27) and V dimethyl sulfoxide (28) structures, respectively.

From an intuitive chemical view, it seems reasonable that one of the nonreducing residues should be the best model for the cellulose monomer. All residues in cellulose (except one per molecule) have had their chemically most active atom, O(1), removed and C(1) is substituted with another glucose ring, analogous to the selected monomer in the parent crystal structure. This substitution appears to affect the geometry of the ring in the neighborhood of C(1) because (see Table I) torsion angles I,II,V, and VI are all at or near the extremes of their ranges in the nonmethyl β-cellobiosic reducing β-cellobiose residue. However, a comparison with the nonreducing residue from the methyl β-cellobioside methanol complex suggests that nonbonded, intermolecular forces can have a dominant influence on ring torsion angles.

Although the R value for the structure favored by this study is not much lower than that of the Gardner-Blackwell structure (see Table VI), R" is, indicating that a somewhat better model has been obtained and that the limit of resolution for the observed only data has been reached. It is probable that the use of individual atomic temperature factors, inclusion of hydrogen atom contribution to diffraction intensity, and removal of the rigid residue constraint would result in a further reduction in R; however, the data are insufficient to warrant such increases in the number of variables.

Conclusions

This work has tested the hypothesis that the results of a polymer structure determination depend upon assumptions incorporated into the rigid chain model. The results showed that monomeric geometry and the assumed glycosidic angle are the key factors in determining the stereochemical acceptability of models for *Valonia* cellulose. Diffraction R values for these models were also found to depend to some extent upon the choice of monomeric geometry. However, extensive calculations for models derived from five different residues resulted in uniform indications of parallel *up* packing mode and *tg* O(6) position as well as similar orientation about the virtual bond.

The virtual bond method readily provided computer models composed of different monomers, permitting selection of a geometry that allowed satisfactory stereochemistry and lowest diffraction R value. The consistent results from refinements of several models indicated that the same packing mode and O(6) position should be characteristic of the "true" structure. Similarly, stereochemical examination of models composed of various residues indicated that the O(3)-O(5) hydrogen bond was characteristic of all celluloses repeating in 10.3 Å.

The evidence presented herein might appear to contraindicate any future need to try alternate chain geometry before reaching conclusions regarding packing mode, etc. It seems preferable, however, when differences in R are small, to enhance the confidence in the conclusions by refinements using several different chain models.

Acknowledgments

The authors thank John Tallant, Biometrician, Southern Region, ARS, for helpful discussions regarding Hamilton's tests. We also thank Professor R. H. Marchessault for suggesting the intuitive chemical suitability of the nonreducing cellobiose residue.

Abstract

A survey shows that ring torsion angles vary in β-D-glucopyranose residues, describing differing glucose geometries within the constraints of the $^4C_1$ conformation. Cellulose chain models composed of monomers with different geometries were generated and examined against intrachain, hard-sphere stereochemical criteria. An O(3)—O(5') contact distance, either appropriate or too short for a hydrogen bond, is found regardless of monomeric geometry. Such a suitable contact distance is thus likely to be characteristic of all cellulose models repeating in 10.3 Å. These models were then tested with the *Valonia* X-ray data of Gardner and Blackwell. The results reinforce conclusions that *Valonia* cellulose I is composed of parallel *up* chains and has O(6) in a *tg* position. Stereochemically attractive models derived from the nonreducing residue of β-cellobiose yielded the lowest R value.

Literature Cited

1. Gardner, K. H. and Blackwell, J. Biopolymers (1974) 13, 1975-2001.
2. Hamilton, W. C. Acta Crystallogr. (1965) 18, 502-510.
3. French, A. D. and Murphy, V. G. Carbohydr. Res. (1973) 27, 391-406.
4. Sundararajan, P. R. and Marchessault, R. H. Can. J. Chem. (1975) 3, 3563-3566.
5. Chu, S. S. C. and Jeffrey, G. A. Acta Crystallogr. (1968) B24, 830-838.
6. Quigley, G. J., Sarko, A., and Marchessault, R. H. J. Amer. Chem. Soc. (1970) 92, 5834-5839.
7. Chu, S. S. C. and Jeffrey, G. A. Acta Crystallogr. (1967) 23, 1038-1049.

8. Ham, J. T. and Williams, D. G. Acta Crystallogr.. (1970) B26, 1373-1383.
9. Arnott, S. and Scott, W. E. J. Chem. Soc., Perkin Trans. II, (1972) 1972, 324-335.
10. Sarko, A. and Muggli, R. Macromolecules (1974) 7, 486-494.
11. Jacobson, R. A., Wunderlich, J. A., and Lipscomb, W. N. Acta Crystallogr. (1961) 14, 598-607.
12. Brant, D. A. Ann. Rev. Biophys., Bioengr. (1972) 1, 369-408.
13. Rees, D. A. and Skerrett, R. J. Carbohydr. Res. (1968) 7, 334-348.
14. Jones, D. W. J. Polym. Sci. (1958) 32, 371-394.
15. Jones, D. W. Biopolymers (1968) 6, 771-773.
16. Arnott, S. and Wonacott, A. J. Polymer (1966) 7, 157-166.
17. Zugenmaier, P. and Sarko, A. Biopolymers (1973) 12, 435-444.
18. Zugenmaier, P., Kuppel, A., and Husemann, E. Presented at A.C.S. Meeting, New York (April, 1976).
19. Mann, J., Roldan-Gonzales, L., and Wellard, H. J. J. Polym. Sci. (1960) 42, 165-171.
20. Hebert, J. J. and Muller, L. L. J. Appl. Polym. Sci. (1974) 18, 3373-3377.
21. Cromer, D. T. and Waber, J. T. Acta Crystallogr. (1965) 18, 104-109.
22. Zaslow, B., Murphy, V. G., and French, A. D. Biopolymers (1974) 13, 779-790.
23. Murphy, V. G., Zaslow, B., and French, A. D. Biopolymers (1975) 14, 1487-1501.
24. Bugg, C. E. J. Amer. Chem. Soc. (1973) 95, 908-913.
25. Cook, W. J. and Bugg, C. E. Acta Crystallogr. (1973) B29, 907-909.
26. Fries, D. C., Rao, S. T., and Sundaralingam, M. Acta Crystallogr. (1971) B27, 994-1005.
27. Winter, W. T. and Sarko, A. Biopolymers (1974) 13, 1447-1460.
28. Winter, W. T. and Sarko, A. Biopolymers (1974) 13, 1461-1482.

3

Studies on Polymorphy in Cellulose

Cellulose IV and Some Effects of Temperature

RAJAI H. ATALLA, BRUCE E. DIMICK, and SEELYE C. NAGEL

The Institute of Paper Chemistry, Appleton, WI 54911

In the first of these studies of polymorphy in cellulose (1, 2) it was proposed that cellulose chains in ordered regions can exist in either of two stable conformations, one of which is predominant in the native (I) form, the other in the mercerized or regenerated (II) form. It had also been established that regeneration of cellulose at elevated temperatures results in recovery in forms other than the cellulose II usually recovered by precipitation at room temperature (3). In the present report the effects of temperature on polymorphic form are examined in greater detail, and the results are interpreted in terms of the proposals concerning the two stable conformations of cellulose.

In the previous report (3) concerned with effects of temperature, attention was focussed on conditions which led to recovery of cellulose in the native lattice upon precipitation from solutions in phosphoric acid. The studies have been expanded to examine regeneration from phosphoric acid under other conditions, as well as regeneration from the recently developed dimethylsulfoxide-paraformaldehyde (DMSO-PF) solvent system (4).

EXPERIMENTAL

Regeneration from phosphoric acid has been explored under a variety of conditions and for a number of different celluloses. The results presented here are confined to solutions prepared with Whatman CF-1 powder, shown by microscopic examination to consist of fragments of cotton fibers; viscosity measurements indicate a DP of approximately 600. Experiments on regeneration from the DMSO-PF system (4) were carried out entirely on the CF-1 powder. The regeneration procedures will be set forth in some detail as previous reports have led to a number of requests for more complete descriptions.

Regeneration from H_3PO_4

Two series of regenerations were carried out with this

system: (a) a series based on regeneration at different temperatures after relatively short aging of the solutions, and (b) a series based on regeneration at different temperatures after aging of the solutions for three to four weeks in order to reduce the DP.

Solutions. Approximately 10 grams of cellulose powder were wet with 7 grams of water and allowed to equilibrate for an hour; 330 grams of 85% phosphoric acid were stirred in a little at a time. The solutions were stored away from strong light for the appropriate period with occasional stirring. They were filtered through "F" glass filters prior to regeneration.

Regeneration. 350 Ml of glycerol were saturated with prepurified nitrogen in an open three neck flask fitted with a mechanically driven agitator and a thermometer. The temperature was raised to the value chosen for regeneration and maintained during the precipitation. The solution, also nitrogen protected, was added in a small stream, near the continuously bubbling nitrogen, over a 25 minute period. Stirring at the temperature was continued for 10 minutes, then 800 ml of boiling water were added, very cautiously at first, and eventually fast enough to complete the addition in 5 minutes. The precipitate, now at approximately 100°C, was centrifuge washed with near boiling distilled water. The washing was repeated 5 times, the third wash including pH adjustment with aqueous ammonia. After the final wash the precipitate was freeze dried.

DMSO-PF

Regeneration from the DMSO-PF system was confined to the undegraded Whatman CF-1 powder. Studies by Swenson (5) have established the nondegrading nature of this solvent system.

Solutions. Since activity of this solvent system is inhibited by moisture the cellulose powder and the paraformaldehyde were first dried in a vacuum oven and the DMSO was dried over molecular sieve. Fifteen grams of cellulose in 500 ml of DMSO were heated to 120°C in a closed container fitted with a stirrer and condenser. At 120°C four to five grams of PF were added. Rapid decomposition of the PF was followed by dissolution of the cellulose to give a clear viscous liquid.

Regeneration. Precipitation of the cellulose from the DMSO-PF solvent system was carried out by addition of the solution in a dropwise manner to the precipitating medium at the temperature of interest. The media were as follows: -2°C, 1:15 methanol-water; 20°C, water; 60°C, water; 100°C, water; 127-130°C, 1:9 water-glycerol under N_2; 165-170°C, glycerol under N_2. Needless to say, at the higher temperatures caution is exercised. The precipitated cellulose was then washed and freeze dried.

Characterization

The samples were characterized by x-ray diffractometry and Raman spectroscopy. The freeze dried powders were pressed into pellets at approximately 2000 psi for this purpose. Some of the pellets pressed for x-ray diffractometry had TiO_2 added as an internal standard.

The x-ray scattering measurements were made with a Norelco diffractometer utilizing nickel filtered copper K-α radiation in the reflecting mode. The Raman spectra were recorded on a Spex Raman system using the 5145 A line of a coherent radiation 52-A (Argon) laser for excitation. The spectra were recorded using the backscattering (180°) mode. In most instances the laser-excited fluorescence (6) decayed to acceptable levels in approximately 30 minutes. Some of the samples regenerated from H_3PO_4 at elevated temperatures were bleached in chlorine dioxide prior to freeze drying and subsequent characterization; it has been established that this bleaching procedure does not in any way influence polymorphic form.

RESULTS

The diversity of polymorphic forms induced by the variation in temperatures are reflected in the x-ray diffractograms of the precipitated celluloses. Figure 1 shows the diffractograms of the celluloses precipitated from the DMSO-PF system. The diffractograms for the series regenerated from H_3PO_4 after aging of the solutions for periods less than one week indicate essentially the same range of polymorphic variation. Comparison of these diffractograms with published diffractograms for the different polymorphic forms (7,8) indicate that the samples precipitated at the lower temperatures are essentially in the cellulose II form. As the temperature of regeneration is increased (20 and 60°C) the celluloses become more crystalline, though the form remains clearly a cellulose II. The first departure from this pattern is noted at 100°C where the (002) peak becomes the most intense feature, and indications of a new feature in the 14-16° region appear. At 127-130°C the diffractogram is typical of low crystallinity cellulose IV, and at 165-170°C, it is clearly a cellulose IV diffractogram.

The diffractograms in Fig. 2, which were recorded for the celluloses regenerated from phosphoric acid after aging for three to four weeks, show an even wider range of polymorphic variation over approximately the same temperature range. At room temperature a highly crystalline cellulose II is formed. At 100°C, substantial conversion to the IV form is noted, although a residual amount of the II form is also indicated. The features characteristic of the II form are further diminished at 120°C and are essentially eliminated at 140°C. At 140°C, however, the feature at approximtely 16° possesses a somewhat complex structure, perhaps suggestive of the beginnings of its resolution into two distinct peaks as clearly occurs at 160°C, where the diffractogram is typical of a high crystallinity cellulose I. The recovery of the cellulose in the I lattice under these conditions has been reported elsewhere (3).

The Raman spectra of the series of regenerated samples represented in Fig. 1 and 2 are shown, respectively, in Fig. 3 and 4. Though the spectra reflect the variation in polymorphic forms, the pattern of the spectral changes does not provide a simple correlation with the x-ray diffractometric indices of polymorphic form. In order to place the Raman spectra of the present samples in perspective x-ray diffractograms and Raman spectra of a series of celluloses mercerized to varying degrees (9) are reproduced in Fig. 5 and 6. The diffractograms in Fig. 5 are consistent with prior observations on systems of this type (8). The Raman spectra of similar partially converted samples have been reported and discussed previously (2). The essential pattern of variation in these spectra is that, particularly in the low frequency region (<600 cm^{-1}), cellulose I and cellulose II have distinctly different spectra. The partially converted samples have spectra with features which occur in the spectra of both the I and II polymorphs.

The most remarkable observation revealed in a comparison of Fig. 3, 4, and 6, is that the spectra of samples which produce distinct cellulose IV diffractograms are essentially similar in most respects to the spectra of the partially converted celluloses which are mixtures of the I and II polymorphs as indicated in the diffractograms in Fig. 5. Indeed this observation is representative of many others wherein spectra of cellulose IV samples appear almost identical to the spectra of the mixed I and II samples obtained by incomplete mercerization. The differences which do occur are usually of the type which arise from variations in the degree of crystallinity.

It should perhaps be noted that although the primary determinant of polymorphic form appears to be the temperature of regeneration, both the DP of the cellulose and the mechanics of the regeneration process are important factors. It is generally true that the lower the DP the higher the degree of order measured in terms of the width-at-half-height of the peaks in the diffractograms. This effect of low DP appears to be operative under all conditions of regeneration.

The effects of regeneration mechanics, though somewhat more complex, are perhaps best characterized in terms of the time allowed for separation from solution. For example, the regeneration from the DMSO-PF system by the procedure described above gave a cellulose of low order at 127-130°C. In contrast, in a related experiment wherein the solution was held at 135°C until the cellulose separated due to the slow thermal dissociation of the

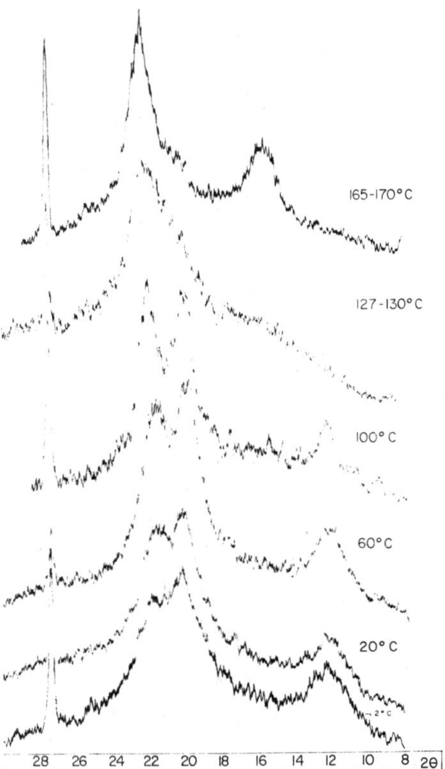

Figure 1. X-Ray diffractograms of celluloses regenerated from the DMSO–PF system (the peak above $2\theta = 27°$ is caused by TiO_2)

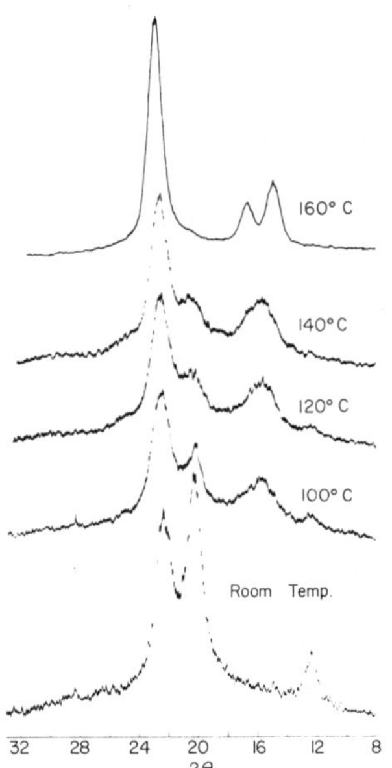

Figure 2. X-Ray diffractograms of low DP celluloses regenerated from phosphoric acid

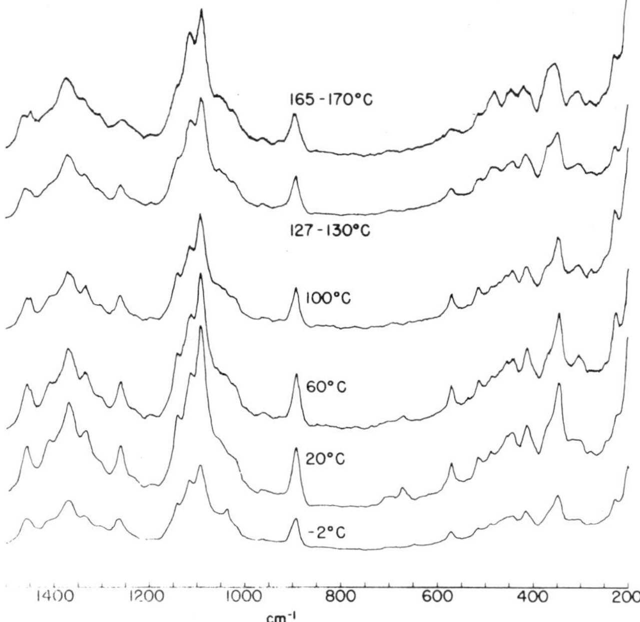

Figure 3. Raman spectra of samples identified in Figure 1

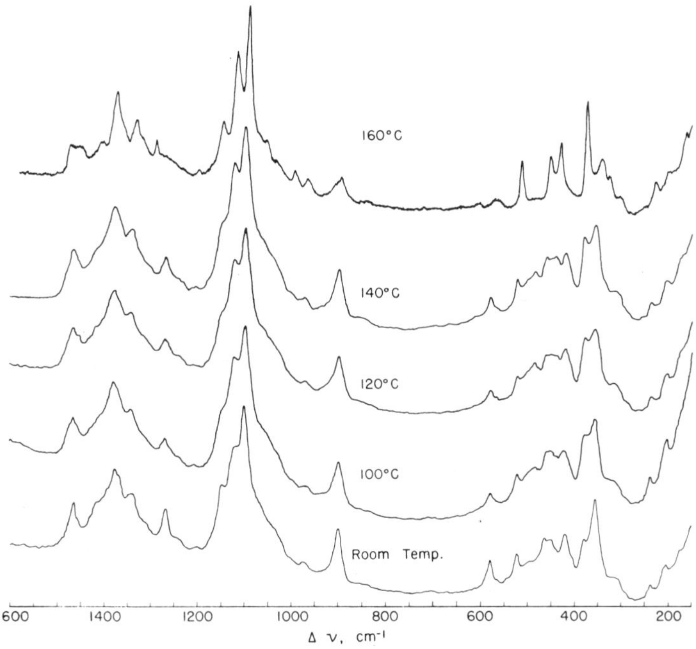

Figure 4. Raman spectra of samples identified in Figure 2

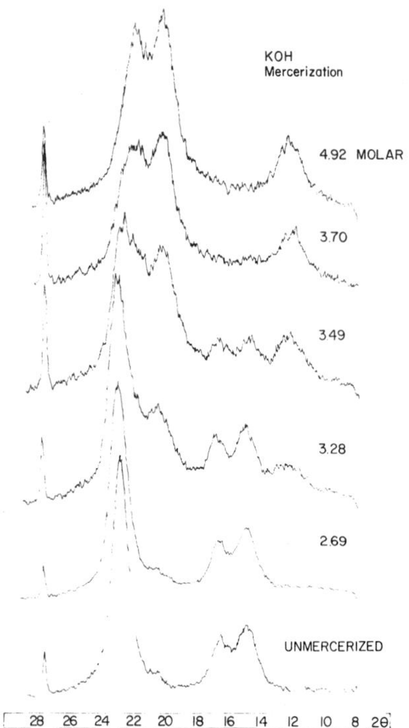

Figure 5. X-Ray diffractogram of celluloses mercerized to different degrees

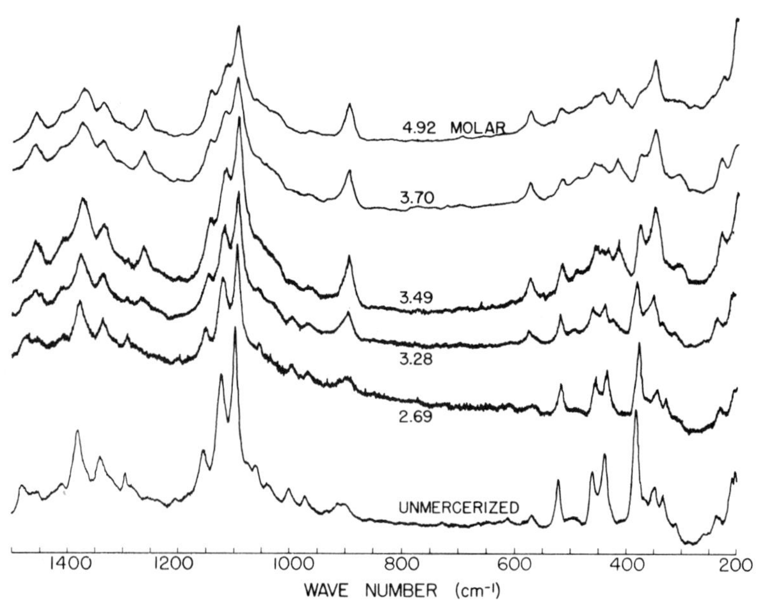

Figure 6. Raman spectra of samples identified in Figure 5

methylol substituent the precipitated cellulose was more ordered, as indicated by the diffractogram shown in Fig. 7 which is typical of highly ordered cellulose IV samples.

In this connection it should also be noted that preparation of cellulose IV by regeneration from solution is novel. Previous reports of its preparation have been based on treatment of celluloses I or II at temperatures in the range 180-250°C in glycerol (7).

DISCUSSION

The first investigation in this series of studies on polymorphy in cellulose focussed on interpretation of the differences between the Raman spectra of celluloses I and II. Comparison of the differences with the spectra of model compounds, and a theoretical analysis of the influence of conformation on vibrational spectra, led to the proposal that the cellulose chains possess different conformations in the two polymorphic forms. Taken together with published mappings of the potential energy (10) and the constraint of a repeat distance of approximately 5.15 A per anhydroglucose unit, the spectra suggested that only two conformations of the cellulose chain are likely to be stable. These have been represented as relatively small left- and right-handed departures from the twofold helix structure. For convenience in the discussion of other polymorphs the two conformations have been identified with the polymorphs in which they are predominant. Thus conformation I is the one dominant in native cellulose while II is the conformation in the mercerized form.

It was noted in (2) that the interpretation set forth in terms of only two stable conformations required, for consistency, that the other polymorphic forms be constituted of these same two conformations. The spectra of the cellulose IV samples shown in Fig. 3 and 4 are in accord with this requirement for, as noted above, they appear to be superimpositions of the spectra of forms I and II. Since this observation is true of many different samples of celluose IV prepared in the course of these studies it seems clear that conformations I and II coexist in the cellulose IV polymorph. When the distinctive x-ray diffractograms of the cellulose IV samples are considered as well, the only interpretation that remains plausible is that the cellulose IV polymorph represents a mixed crystalline habit in which both conformations coexist.

Though the effects of temperature are in part modified by the mechanics of regeneration and the DP of the cellulose, certain clear patterns emerge. The most obvious feature is that as the temperature of regeneration is raised above a certain level, which is determined in part by DP, an increasing proportion of the cellulose chains separate in the I conformation. These seem to coprecipitate with molecular chains separating in the II conformation, giving rise to a fraction of the total precipitate in the

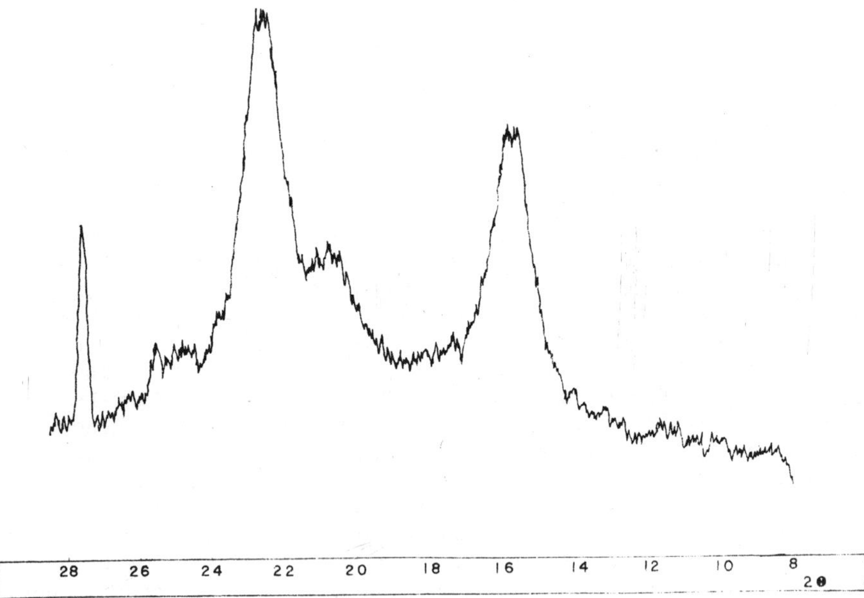

Figure 7. X-Ray diffractogram of sample regenerated from DMSO–PF by slow precipitation at 135°C

IV form, while the majority of the precipitate remains the II polymorph; thus, for example, the 100°C sample in Fig. 1 and 3. As the temperature is increased further, or the DP lowered, a much larger proportion of the molecular chains separate as a co-precipitate of conformations I and II in the IV polymorph; thus for example the 127-130°C sample in Fig. 1 and 3 and the 100°C sample in Fig. 2 and 4. At higher temperatures still, for example the 165-170°C sample in Fig. 1 and 3 and the 140°C sample in Fig. 2 and 4, the IV polymorph dominates and there is no evidence for separate precipitation of the II polymorph. Ultimately, for a low enough DP and 160°C the molecular chains separate entirely in the I conformation and the crystalline form is that of the native cellulose I state, Fig. 2 and 4.

The interpretation set forth above for the present observations concerning the preparation of cellulose IV by precipitation from solution can also be reconciled with the results of past preparations of cellulose IV. The best established procedures previously reported (7,11) are based on treatment of cellulose II in glycerol or water under pressure at temperatures in the range 180-250°C. When this temperature range is compared with results cocerning the glass transition temperature of cellulose (12), as well as the possible plasticizing effects of glycerol (13), it is apparent that the conditions prerequisite for formation of cellulose IV by direct transformation of cellulose II are precisely those which promote conformational mobility. It would seem, therefore, quite likely that a change in conformation is involved in this transformation. Studies of some effects of temperature on mercerization of cellulose (9) appear to confirm this interpretation.

The effect of DP on crystalline form has already been noted above. In Fig. 2 it is clear that the lower DP of the samples aged for 3 to 4 weeks in H_3PO_4 has led to initiation of the conversion to conformation I at lower temperatures, in addition to generally resulting in more ordered precipitates. Ultimately the regeneration in the cellulose I form has so far been possible only for the low DP samples. This may reflect an inherent limitation on the nature of the I conformation, though this possibility has not yet been fully explored.

The influence of time is not unrelated to the effects of DP. In general it appears that for samples of high DP it is necessary that they be held for longer periods under conditions of high molecular mobility if a highly ordered structure is desired; thus the contrast between the 127-130°C sample in Fig. 1, where the precipitation from the DMSO-PF system was rapid, and the sample in Fig. 7, where the separation from solution was much slower. The DP was the same in both instances.

CONCLUSIONS

A number of interrelated conclusions have been developed from the present studies. The Raman spectra of celluloses recovered under a wide range of conditions show features characteristic of one or the other or both conformations identified with celluloses I and II, and thus are consistent with the proposal that only two conformations of the chain are stable in ordered domains in cellulose. The spectra of the cellulose IV samples when taken together with the distinctive x-ray diffractograms, suggest that the IV polymorph is a mixed lattice in which conformations I and II coexist.

Elevation of the temperature of regeneration appears to promote conversion of cellulose chains from the II conformation to the I conformation, the degree of conversion depending on the DP and, to a less clearly defined extent, on the mechanics of the regeneration process. In general, a lower DP permits greater conversion from the II to the I forms; the mechanics of regeneration become important when the time interval allowed for molecular ordering is limited.

On the basis of the above interpretations of the present observations it has been possible to design procedures for the regeneration of cellulose IV under a far wider range of conditions than heretofore reported.

ACKNOWLEDGMENTS

The authors wish to acknowledge many valuable discussions with Dr. Kyle Ward. Ms. Rebecca Whitmore assisted in some of the experimental phases of the work. Support from institutional funds of The Institute of Paper Chemistry is gratefully acknowledged.

ABSTRACT

The supermolecular structure of cellulose precipitated from solution was found quite sensitive to the temperature of regeneration and the degree of polymerization (DP) in studies of regeneration from solutions in phosphoric acid as well as from solutions in the dimethylsulfoxide-paraformaldehyde (DMSO-PF) solvent system. X-ray diffractometry indicated that the samples regenerated at or below room temperature are predominantly of the cellulose II form. As the temperature of regeneration is elevated increasing proportions of cellulose IV are indicated. At temperatures between 100 and 150°C the IV polymorph is dominant form, the degree of order varying with DP and the time scale of the precipitation process. Ultimately at 160°C and for relatively low DP, the regenerated cellulose is in the native or cellulose I polymorph. Raman spectral studies of the same samples show them all to contain varying proportions of cellulose molecules in the conformations typical of celluloses I and II. Conformation II is dominant in samples precipitated at or below room temperature. As the temperature of precipitation increases, a greater proportion of the cellulose I conformation is indicated. The samples precipitated above 100°C, which by x-ray diffractometry appear to be cellulose IV, have Raman spectra showing similar proportions of conformations I and II. Thus it appears that these two conformations form a mixed lattice in the cellulose IV polymorph.

LITERATURE CITED

1. Atalla, R. H. and Dimick, B. E., Carbohydrate Res. 39, C1-3 (1975).
2. Atalla, R. H., Proceedings of the Eighth Cellulose Conference, May, 1975. In press.
3. Atalla, R. H. and Nagel, S. C., Science 158, 522(1974).
4. Johnson, D. C., Nicholson, M. D., and Haigh, F. C. Proceedings of the Eighth Cellulose Conference, May, 1975. In press.
5. Swenson, H. A., Proceedings of the Eighth Cellulose Conference, May, 1975. In press.
6. Atalla, R. H. and Nagel, S. C., Chem. Comm., 1049(1972).
7. Ellefsen, O. and Tonnesen, B. A., in N. M. Bikales and L. Segal, eds., Cellulose and Cellulose Derivatives, (High Polymers, Vol. 5, Part IV), Interscience, New York, 1971. p. 151.
8. Tripp, V. W. and Conrad, C. M., in Robert T. O'Connor, ed., Instrumental Analysis of Cotton Cellulose and Modified Cotton Cellulose, Marcel Dekker, New York, 1972. p. 339.
9. Dimick, B. E., Doctoral Dissertation, The Institute of Paper Chemistry, Appleton, WI, 1976.
10. Rees, D. A. and Skerrett, R. J., Carbohydrate Res. 7, 334 (1968).
11. Howsmon, J. A. and Sisson, W. A., in E. Ott, H. M. Spurlin and M. W. Grafflin, eds., Cellulose and Cellulose Derivatives (High Polymers, Vol. 5, Part I), Interscience, New York, 1954. p. 231.
12. Alfthan, E., de Ruvo, A. and Brown, W., Polymer 14, 329 (1973).
13. Atalla, R. H. and Nagel, S. C., Polymer Letters 12, 565 (1974).

4

Structures of Native and Regenerated Celluloses

JOHN BLACKWELL, FRANCIS J. KOLPAK, and KENNCORWIN H. GARDNER

Department of Macromolecular Science, Case Western Reserve University, Cleveland, OH 44106

We have investigated the crystal structures of cellulose I and II by x-ray methods, based on intensity data from fiber diagrams of *Valonia* and *rayon* celluloses. This has been a long standing problem dating from the 1920s. From the work of Meyer and Misch(1) the native structure was shown to have a monoclinic unit cell with dimensions a = 8.35Å, b = 7.9Å, c = 10.3Å (fiber axis) and γ = 96°. Similarly, from the data collected by Wellard(2), the unit cell for cellulose II is also monoclinic with dimensions a = 7.93Å, b = 9.18Å, c = 10.34Å and γ = 117.3° (average values over a variety of preparations). The unit cells for both forms contain cellobiose residues of two chains, and the space groups are generally thought to be $P2_1$. The cellulose chains are thought to possess two-fold screw symmetry, and to be with their axes through the origin (0,0) and center (1/2,1/2) of the a b projections. The important consequence of this is that the space group symmetries are satisfied whether the two molecules passing through the unit cell have the same (parallel) of opposite (antiparallel) sense.

Considerable effort has been directed to determine the chain polarities for the two forms and thence to elucidate their hydrogen bonding networks. This information is clearly necessary in order to understand the process of cellulose biosynthesis. However, previous workers (notably Jones)(3,4) have generally reported that models containing chains with both the same or alternating polarities could be built for both forms and could not be distinguished in terms of agreement with the x-ray data. In the absense of definite evidence, most researchers favored antiparallel chain structures. Probably the most detailed proposed model is that for native cellulose due to Liang and Marchessault.(5). This model was based on polarized infrared spectra and contained the hydrogen bonding proposed by Frey Wyssling(6) and Hermans *et al*.(7).

Our reinvestigation of the structures of the two polymorphs used rigid body least squares refinement methods. From model building and conformational analysis(8) it was clear that the chain backbone conformation must be at least close to that proposed by Hermann et al.(7), i.e. a 2_1 helix with intramolecular (O(3)-H•••O5') hydrogen bonding. Models could therefore be constructed containing two such chains per unit cell, and the positions of these chains could be refined by least squares methods, using the methods developed for biopolymers by Arnott and Wonacott(9). This work is summarized below, after which the structures determined are described and discussed. More extensive descriptions of the refinement process for the two structures are given elsewhere(10,11).

Experimental

Cellulose I. Samples of the cell walls of the alga *Valonia ventricosa* were purified by boiling in a large excess of 1% aqueous NaOH for 6 hrs, with a change of alkal solution after 3 hrs. They were then allowed to stand overnight in 0.05N HCl at room temperature, rinsed and stored in distilled water. Specimens for x-ray work were prepared as bundles of parallel fibers drawn from the purified cell walls.

Cellulose II. Samples of regenerated cellulose (rayon fibers) were obtained from Celanese Fibers Company. Small individual fibrils were drawn from the larger rayon fibers and prepared as a parallel bundle for x-ray work.

X-ray fiber diagrams were recorded on Kodak No-screen film using CuKα radiation and a flat plate vacuum camera, the collimator consisted of two 400μ pinholes separated by 12cm. The d-spacings were calibrated using inorganic salts. The intensities for cellulose I were measured using a Joyce Loebl densitometer. The areas under radial scans through each reflection were corrected in the manner described by Cella et al.(12). For cellulose II a new method was devised using a Photometrics densitometer which produces an optical density map of the entire x-ray photograph. Contours of equal optical density are drawn from which the integrated intensity of the reflection is determined following subtraction of the background. A few very weak reflections were estimated by eye in each case. Unobserved reflections predicted to occur within the scattering angle covered by the x-ray data were assigned a value such that the structure amplitude was 2/3 of a judged threshold value for the weakest reflection which could be detected in that region of the x-ray photograph.

Unit Cell Parameters

The x-ray fiber diagrams for cellulose I and II are shown in Figures 1 and 2 respectively. Least squares refinement of the unit cell for *Valonia* cellulose I resulted in dimensions a = 16.34Å, b = 15.72Å, c = 10.38Å and γ = 97.0°. From

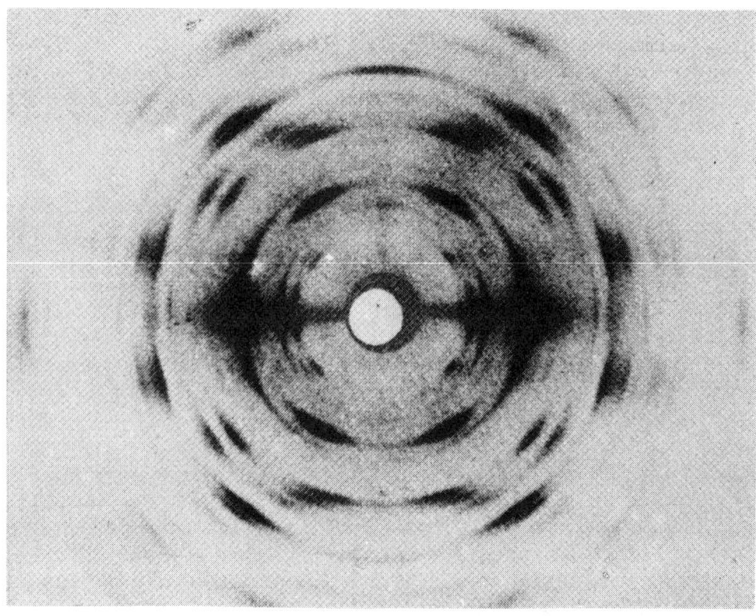

Figure 1. X-Ray diffraction pattern of a bundle of oriented fibers of Valonia cellulose

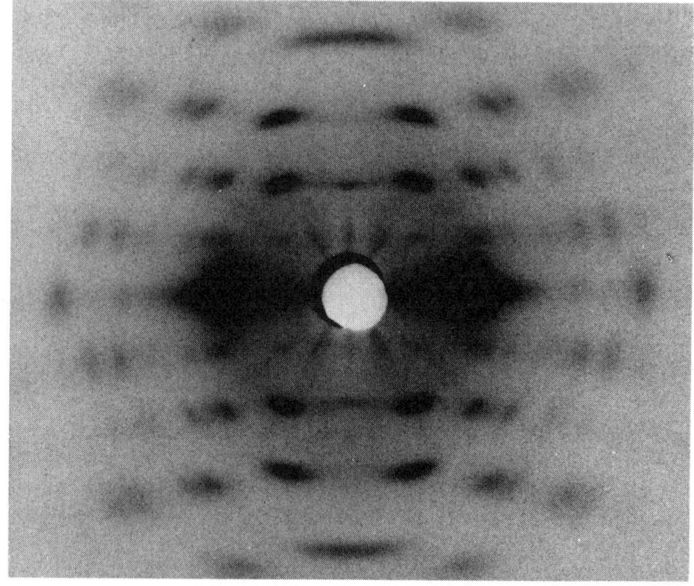

Figure 2. X-Ray diffraction pattern of a bundle of rayon fibers

density measurements, this unit cell contains sections eight chains. Of the 39 observed non meridional reflections, 36 can be indexed by planes with even h and k indices, i.e. they are indexed satisfactorily by a two chain (Meyer and Misch) unit cell with dimensions a = 8.17Å, b = 7.86Å, c = 10.38Å and γ = 97.0°. The remaining three reflections are those necessitating doubling of the a and b axes to give the eight chain cell, first reported by Honjo and Watanabe (13) for Valonia. These three reflections are very weak; furthermore, a large number of reflections with odd h or k are predicted within the range of the x-ray photograph but are too weak to be detected. Hence the differences between the four two-chain units making up the eight-chain cell must be very small, and the two chain Meyer and Misch unit cell is an adequate approximation to the cellulose I structure.

For cellulose II the refined unit cell is monoclinic with dimensions a = 8.01Å, b = 9.04Å, c = 10.36Å and γ = 117.1°. The experimental error for the dimensions for both cellulose I and II is ±0.05Å, and the difference in fiber repeats between the two forms is not significant. The only systematic absenses in each case are for odd order 00ℓ reflections and the space groups are both $P2_1$.

Molecular Model

Consistant with the $P2_1$ symmetries, models were constructed for the cellulose chain as two fold helices with the appropriate repeats. Standard bond lengths and bond angles for pyranose ring structures (14) were used, and the model contained an O(3)-H···O(5') intramolecular hydrogen bond. Standard numbering of the atoms on the cellobiose residue is shown in Figure 3. The C-O-C glycosidic bond angle was 114.8°. The intramolecular hydrogen bond length was 2.75Å and 2.69Å in cellulose I and II, due to the slightly different fiber repeats and glycosidic torsion angles.

The cellulose chain is completely rigid except for possible rotation of the CH_2OH side chain. This rotation is defined by the dihedral angle χ, which is zero when C(6)-O(6) is cis to C(4)-C(5). This orientation is also defined relative to the C(4)-C(5) and C(5)-O(5) bonds: gg guache to both C(5)-O(5) and C(4)-C(5) (χ = -60°); gt, gauche to C(5)-O(5) and trans to C(4)-C(5) (χ = 180°), and tg, trans to C(5)-O(5) and gauche to C(4)-C(5) (χ = +60°). (15).

Chain Packing

The symmetry requirements of the $P2_1$ space group are satisfied by placing two independent chains so that their screw axes are parallel to c and pass through (0,0) and (1/2,1/2) in the a b plane. Two chain unit cells were constructed for both forms, containing parallel and antiparallel chains; each model was then

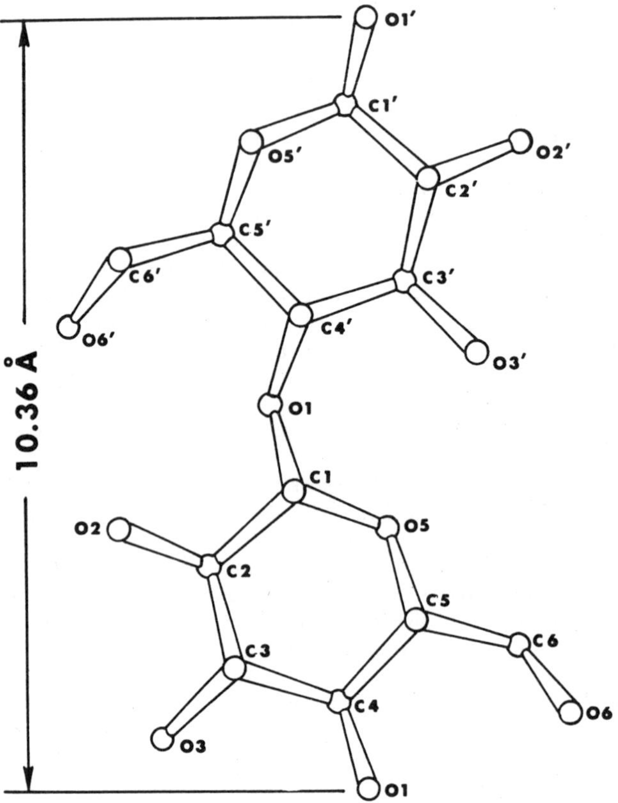

Figure 3. Molecular model for the cellobiose residue showing the numbering of the atoms

refined separately against the x-ray data. The positions of the rigid cellulose chains are completely defined by three packing parameters: the shift of one chain along $\underline{c}$ with respect to the other chain, and two parameters defining the independent rotations of the two chains about their helix axes. A survey of possible packing followed by comparison of the observed and calculated structure amplitudes indicated that four basic two chain models need to be considered. In each case, the chain through (0,0) has the glycosidic oxygen O(1') at z = 0 and "shift" describes the $\underline{c}$ axis displacement of the second chain through (1/2, 1/2): The chain sense is defined as "up" when $z_{O(5)} > z_{C(5)}$. The four models are:

P_1-parallel chains oriented "up" with a shift of $\sim c/4$;
P_2-parallel chains oriented "down" with a shift of $\sim c/4$;
a_1-antiparallel chains with an "up" chain at (0,0) and a down chain at (1/2,1/2), with a shift of $\sim -c/4$; and
a_2-antiparallel chains as in a_1, but with a shift of $\sim +c/4$.

Refinement

Each of the above structures is defined by parameters determining the position and orientation of the rigid chains and their pendant $-CH_2OH$ groups. The least squares procedure refines these parameters to give the best agreement between the observed and calculated structure amplitudes. The seven refinable parameters for each model are:

1) SHIFT, the stagger of the center chain along $\underline{c}$ with respect to the chain at the origin;
2) ϕ, the rotation of the origin chain about its helix axis;
3) ϕ', the rotation of the center chain about its helix axis;
4) χ, the orientation of the $-CH_2OH$ groups of the origin chain;
5) χ', the orientation of the $-CH_2OH$ groups of the center chain;
6) K, a scale factor for comparison of the observed and calculated structure amplitudes, and
7) B, the isotropic temperature factor.

The refinement will be discussed in terms of the residuals, which give a measure of the agreement between the observed and calculated structure amplitudes. These are defined:

$$R = \frac{\Sigma||F_o|-|F_c||}{\Sigma|F_o|} \qquad R' = \frac{\Sigma w(|F_o|-|F_c|)^2}{\Sigma w|F_o|} \qquad R'' = \frac{\Sigma w(|F_o|-|F_c|)^2}{\Sigma w F_o^2}$$

where F_o and F_c are the observed and calculated structure amplitudes and w is a weigh assigned to each observed structure amplitude.

Cellulose I

The initial refinement was done for models where both chains had the same orientation for the CH_2OH groups i.e. $\chi = \chi'$. In later work it was found that models with all but very small differences between χ and χ' are not compatible with the x-ray data for stereochemical requirements. These small differences do not give significant improvement in the fit between the observed and calculated structure amplitudes, and hence in the structures described below, $\chi = \chi'$, and the refinement is for 6 variables. When the models were refined against the 36 observed reflections only, the resulting R values were R_{a1}=0.207, R_{a2}=0.249, R_{p1}=0.179, and R_{p2}=0.202. In all four refined models, the 'planes' of the pyranose rings are approximately in the a c plane and the value of SHIFT staggers the glycosidic oxygens by ∿c/4. The χ value for models a_1, a_2, and P_1 places the $-CH_2OH$ groups near the tg position, such that O(2')-H···O(6) intramolecular and reasonable intramolecular hydrogen bonds can be formed. For model p_2, and χ value places the CH_2OH group intermediate between the tg and gt positions, which does not allow for hydrogen bonding of the O(2)-H groups. Statistical tests (15) indicate the model p_1 gives significantly better agreement with the x-ray data than model p_2. Model a_2 can be rejected in favor of model a_1 on the same basis. Thus models a_1 and p_1 were the most likely antiparallel and parallel chain models and were considered for further refinement. At this stage the unobserved reflections were included in the refinement where the calculated structure amplitude was larger than the threshold value, under these circumstances. A weighting scheme of w=1 for observed and w=1/2 for unobserved reflections was used at this point. The final residuals were R'_{p1}=0.233, R'_{a1}=0.299, R''_{p1}=0.215 and R''_{a1}=0.270. Application of the Hamilton statustical test (16) to these data indicate a preference for the parallel chain model (p_1) at a significance level of 0.5%, i.e., the parallel model is prefered by a factor of more than 200 to 1. The ab and ac projections of the structure are shown in Figure 4. The structure has no bad contacts on the basis of accepted stereochemical criteria. The refined value of ϕ' is 0.4° from that of ϕ, and the Hamilton test indicates that the constrained model with $\phi = \phi'$ is in as good agreement with the data as the model with ϕ' as a separate variable. The final value of ϕ=19.4° (arbitrary origin) places the chains so that the "planes" of the rings are approximately in the ac plane (see Figure 4). The refined value of SHIFT=0.266c. This deviation from a perfect quarter stagger is not unexpected: the weak 002 meridional would be absent for a stagger of 0.25c. The refined value of χ=80.3° places the CH_2OH groups within ∿20° of the tg position (χ=60°). This orientation did not shift significantly from that refined for the observed reflections only. The isotropic temperature factor is B=2.50.

Hydrogen Bonding in Cellulose I

The hydrogen bonding network in cellulose I is shown in Figure 4c. All the hydroxyl groups form hydrogen bonds with acceptable bond lengths and angles. In addition to the

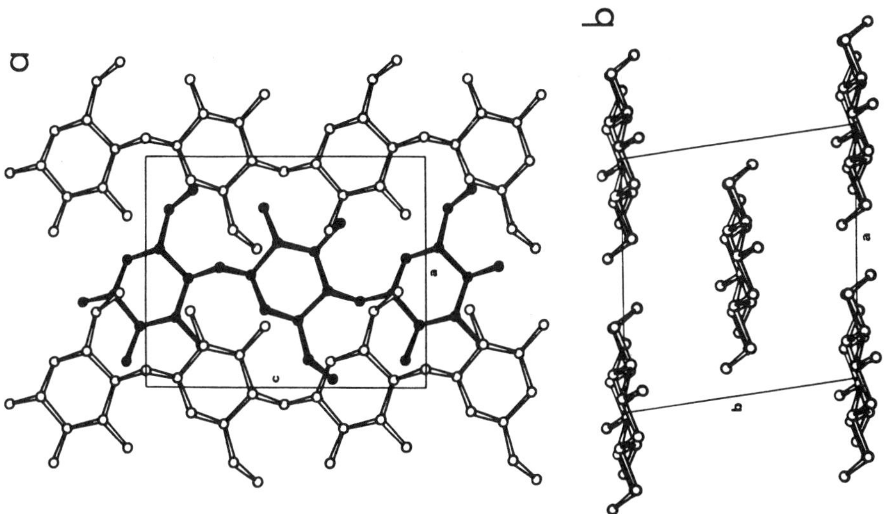

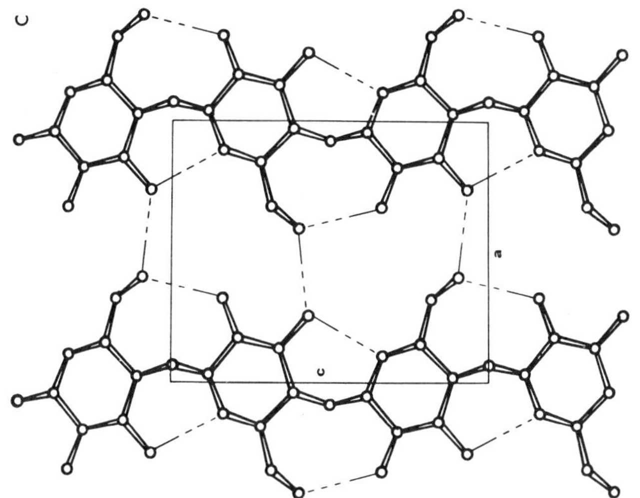

Figure 4. Projections of the parallel chain model for cellulose I. (a) Projection perpendicular to the ab *plane. The center chain (black) is staggered with respect to that at the origin. (b) Projection perpendicular to the* ab *plane, looking along the fiber axis. (c) Hydrogen bonding network in the 020 plane*

O(3)-H•••O(5') hydrogen bond of length 2.75Å defined in the model, there is a second intramolecular bond: O(2')-H•••O6 of length 2.87Å. These intramolecular bonds run on both sides of the cellulose chain. In addition, there is an interchain hydrogen bond between O(6)-H and O(3) of the next chain along the *a* axis; this bond is 2.79Å in length. No hydrogen bonding occurs along the unit cell diagonals, rather the hydrogen bonding is all in the 020 planes, and the structure is seen as a series of hydrogen bonded sheets of chains, where successive sheets are staggered and all the chains have the same sense.

Cellulose II

For cellulose II, study of molecular models indicated that the two chains probably have different conformations for the $-CH_2OH$ groups, and hence all seven variables were considered in the refinement. Refinement against the 44 observed reflections gave residuals of R_{p1}=0.254, R_{p2}=0.188 and R_{a1}=0.195, R_{a2}=0.171. Of these four models, only a_2 is stereochemically acceptable, and this gives the lowest residual. Model P_1 contains four bad contacts and models p_2 and a_1 contain two each (The worst of these contact distances are nonbonded oxygen-oxygen distances of 2.17Å, 2.05Å and 2.11Å in p_1, p_2 and a_2 respectively, which are totally unacceptable). Efforts to remove these contacts by incorporation of non-bonded constraints were not sucessful: the R values increased to R_{p1}=0.272, R_{p2}=0.219 and R=0.230, but although the contact distances were lengthened, the stereochemical criteria were still not satisfied.

All four models were then refined against the total observed and unobserved data, as was done for cellulose I. The bad contacts for models p_1, p_2 and a_1 were not removed and these structures remain unacceptable. For model a_2, a short oxygen-oxygen contact of 2.49Å was introduced, but this was erradicated with an appropriate nonbonded constraint. The final R values were R'=0.219 and R"=0.167.

Thus an antiparallel chain model is proposed for cellulose II. The *ab* and *ac* projections of the structure are shown in Figure 5a and 5b. The refined values of ϕ and ϕ' orient the chains so that the rings are almost parallel to the *ac* planes, although not quite so close as for cellulose I. The relative stagger of the chains is 0.216*c*. The side chains have different conformations for the corner "up" chains (through 0,0), χ=186.3°, placing the $-CH_2OH$ group close to the *gt* position (χ=180°), for the center "down" chains (through 1/2,1/2), χ'=70.2°, placing these $-CH_2OH$ groups close to the *tg* position (χ=60°). The refined isotropic temperature factor is B=19.96.

Hydrogen Bonding in Cellulose II

The hydrogen bonding network in cellulose II is more complex than in cellulose I, and is shown in Figure 5b-e. All of the hydroxyl groups form hydrogen bonds with acceptable bond lengths and angles. Each chain has the 0(3)-H•••0(5') intramolecular bond of length 2.69Å, as defined in the model. The $-CH_2OH$ groups of the center "down" chains are close to the tg position and these chains have a second intramolecular 0(2)-H•••0(6) bond of length 2.73Å. The 0(6)-H group of this chain also forms an intermolecular 0(6)-H•••0(3) bond of length 2.67Å to the next ("down") chain along the a axis, with a result that the "down" chains form hydrogen bonded sheets in the 020 planes similar to those in cellulose I. The sheet of down chains is shown in Figure 5c.

For the "up" corner chains the $-CH_2OH$ groups are close to the gt position, and form 0(6)-H•••0(2) intermolecular bond of length 2.73Å to the next chain along the a axis, again in the 020 plane. The sheet of "up" chains is shown in Figure 5d. For the gt conformation the 0(2)-H group cannot form an intramolecular bond, but is involved in an intermolecular 0(2)-H•••0(2') bond of length 2.77Å to the next "down" chain on the diagonal along the 110 plane, as shown in Figure 5e. Hence the cellulose II structure is an array of staggered hydrogen bonded sheets. The chain sense is the same within the sheets, but the sheets have alternating polarities and are hydrogen bonded together along the long diagonal of the unit cell.

Discussion

A parallel chain structure for cellulose I effectively rules out chain folding during synthesis of cellulose microfibrils. Native microfibrils are therefore shown to be extended-chain polymer single crystals, which are highly desirable structures in terms of mechanical properties. For cellulose II, the chains are antiparallel, which is certainly compatible with folded chains, although there is no definite evidence for such a crystallization process.

Cellulose II is the stable polymorphic form, in that it is possible to convert form I to form II, but not vica versa. The cellulose II structure contains the attractive feature of a hydrogen bond between adjacent sheets of chains, which may account for this stability. In addition the hydrogen bonds have an average length of 2.72Å in cellulose II, as compared to 2.80Å in cellulose I. The resolution attainable in the x-ray refinements is not sufficient to determine individual bond lengths, but this difference in the average bond lengths is probably significant, and reflects a tighter chain packing in cellulose II, consistant with the higher stability of this form.

Of the various possibilities for the chain polarities in the two forms, the parallel I-antiparallel II solution seems to be the most reasonable. Results from x-ray and packing analyses by Sarko and Muggli (17) also favor these polarities. If

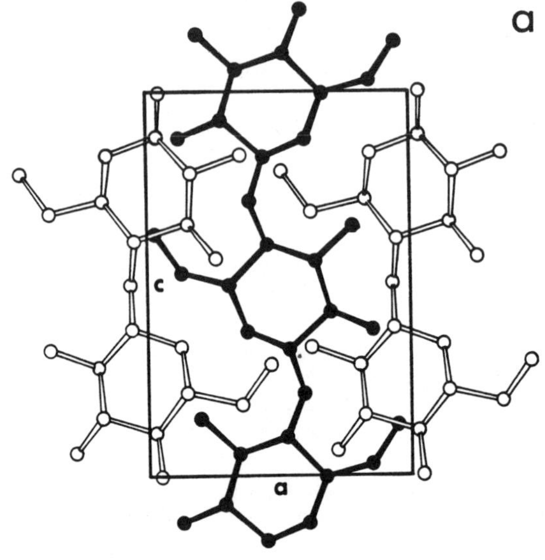

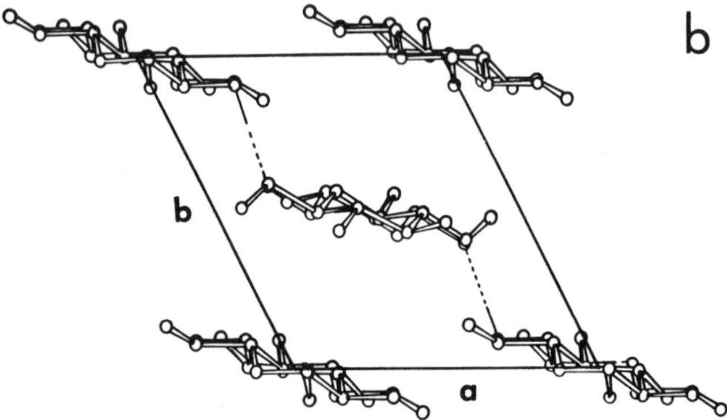

Figure 5. Projections of the antiparallel chain model for cellulose II. (a) Projection perpendicular to the ac *plane. The center "down" chains (dark) are staggered with respect to the corner "up" chains. (b) Projection perpendicular to the* ab *plane along the fiber axis. The O(2)–H···O(2′) hydrogen bond along the 110 planes is indicated.*

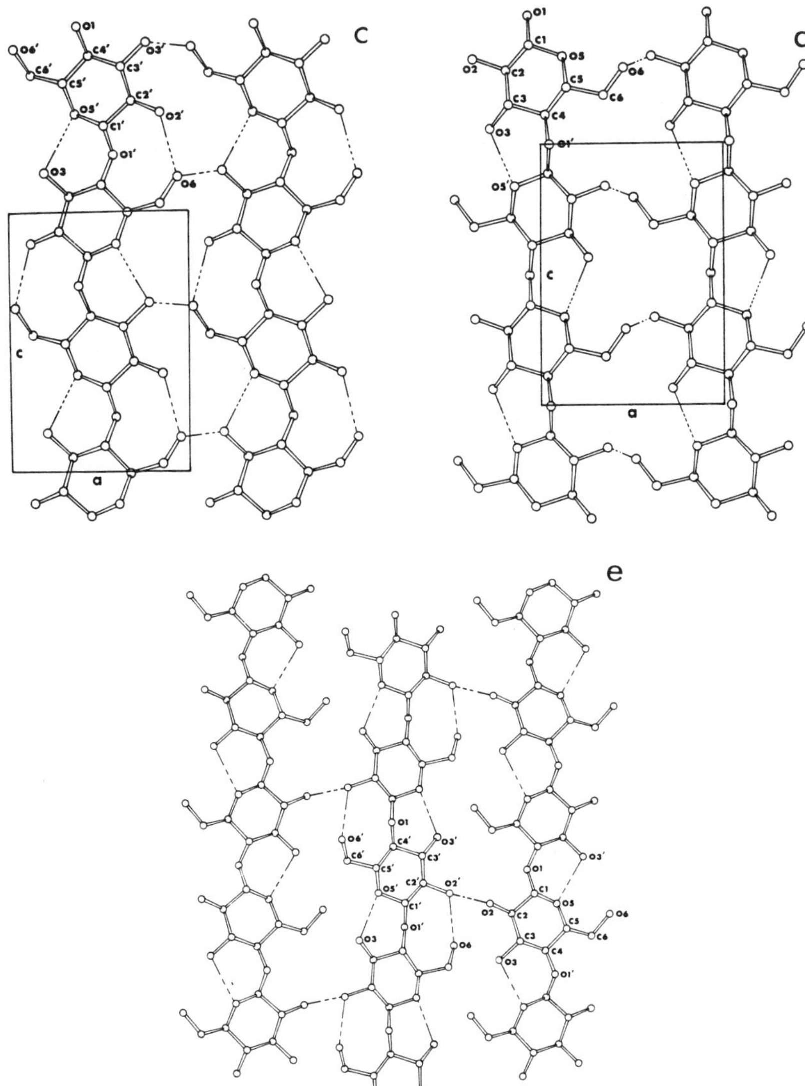

Figure 5. (c) Hydrogen bonding network in the 020 plane for the center "down" chains. These sheets are very similar to those for cellulose I. (d) Hydrogen bonding network in the 020 plane for the corner "up" chains. (e) Hydrogen bonding between antiparallel chains in the 110 plane.

cellulose I was an array of antiparallel chains, it is difficult to see why these would not adopt the cellulose II lattice. A consequence of the parallel chain structure, however, is that it requires a relatively complex biosynthesis mechanism with polymerization followed very closely by crystallization. If the two steps were to be well separated then a rayon-like structure would be produced.

The results for cellulose II were obtained for rayon. There is no reason to believe they do not apply to mercerized cellulose, although we are currently reinvestigating the latter structure. The mercerization process involves swelling in caustic soda solution and is accompanied by only a small change in length. Chanzy *et al.* (18) have recently shown that shish-kabob structures of low molecular weight cellulose with the form II lattice will epitaxially crystallize on cellulose I fibers. Such epitaxial crystallization is to be expected since half of the sheets in cellulose II are the same as those in cellulose I. Mercerization proceeds showly and never goes to completion. The unconverted cellulose I could maintain the fiber dimensions and serve as a template for crystallization of cellulose II.

Acknowledgements

This work was supported by N.S.F. Grant No. DMS 75-01028 and N.I.H. Career Development Award No. AM 70642 (to J.B.).

Abstract

The crystal structures of native and regenerated celluloses have been determined using x-ray diffraction and least squares refinement techniques. Both structures have monoclinic unit cells containing sections of two chains with 2_1 screw axes. Models containing both parallel and antiparallel chains were refined in each case by comparison with the x-ray intensities for *Valonia* cellulose I and rayon cellulose II. For native cellulose, the results show a preference for a system of parallel chains (i.e. all the chains have the same sense). The refinement orients the $-CH_2OH$ groups close to the *tg* conformation such that an 0(6)···H-0(2') intramolecular hydrogen bond is formed. The structure also contains an 0(3)-H···0(3) intermolecular bond along the *a* axis. All these bonds lie in the 020 plane, and the native structure is an array of staggered hydrogen bonded sheets. In contrast, for regenerated cellulose the only acceptable structure contains antiparallel chains (i.e. the chains have alternating sense). The CH_2OH groups of the corner chain are oriented close to the *gt* position; those of the center chain are close to the *tg* position. Both center and corner chains have the 0(3)-H···0(5') intramolecular bond and the center chain also has an 0(2')-H···0(6) intramolecular bond. Intermolecular hydrogen bonding occurs along the 020 planes: 0-(6)-H···0(2) bonds for the corner chains and 0(6)-H•••0(3) bonds for the center chains, and also along the 110 planes, with a hydrogen bond between 0(2)-H of the corner chain and 0(2') of the center chain. The major consequence of these structures is that native cellulose is seen as extended chain polymer single crytals. The cellulose II structure is compatible with regular chain folding, although there is no direct evidence for such folding.

Literature Cited

1. Meyer, K.H., and Misch, L. Helv. Chim. Acta. (1937) 20, 232-244.
2. Wellard, N.J., J. Polymer Sci. (1954) 13, 471-476.
3. Jones, D.W., J. Polymer Sci. (1958) 32, 371-394.
4. Jones, D.W., J. Polymer Sci. (1960) 42, 173-188.
5. Liang, C.Y. and Marchessault, R.H., J. Polymer Sci. (1959) 37, 385-395.
6. Frey-Wyssling, A., Biochim, Biophys. Acta. (1955) 18, 166-168.
7. Hermann, P.H., DeBooys, J., and Maan, C., Kolloid-Z. (1943) 102, 169-180.
8. Rao, V.S.R., Sundararajan, P.R., Ramakrisnan, C., and Ramachandan, G.N., in Conformation of Biopolymers, (G.N. Ramachandran, ed.), (1967), Vol. 2, p. 271. Academic Press, New York.
9. Arnott, S., and Wonacott, A.J., Polymer (1966) 7, 157-166.
10. Gardner, K.H., and Blackwell, J., Biopolymers (1974) 14, 1975-2001.
11. Kolpak, F.J., and Blackwell, J. (1976) 9, 273-278.
12. Cella, R.J., Lee, B., and Hughes, R.E. Acta Cryst. (1970) A26, 118-124.
13. Honjo, G., and Watanabe, M., Nature (1958) 181, 326-328.
14. Arnott, S., and Scott, W.E., J. Chem. Soc. Perkin. Trans. II (1972) 324-335.
15. Sundaratingham, M., Biopolymers (1966) 6, 189-213.
16. Hamilton, W.C., Acta Cryst. (1965) 18, 502-510.
17. Sarko, A., and Muggli, R. Macromolecules (1974) 7, 486-494.
18. Chanzy, H., Roche, E., and Vuong, R. Appl. Polymer Sci. Symposia (in press).

5

X-Ray Diffraction by Bacterial Capsular Polysaccharides: Trial Conformations for *Klebsiella* Polyuronides K5, K57, and K8

E. D. T. ATKINS, K. H. GARDNER, and D. H. ISAAC

H. H. Wills Physics Laboratory, University of Bristol, Tyndall Avenue, Bristol, BS8 1TL, UK

Interest in the polyuronides, in this laboratory, stems back to the early 1970's when we applied x-ray diffraction methods to elucidate the molecular structures of the plant polysaccharide alginate components, polymannuronic acid and polyguluronic acid (1,2). In 1971 we turned our attention to the more complicated connective tissue linear polydisaccharides. Starting with hyaluronic acid (3) we systematically explored the conformations of the connective tissue polyuronides (4), also including the capsular bacterial polyuronide pneumococcus type III for comparison and encompassing the highly sulphated blood anticoagulant heparin (5,6). During this period we developed and extended our computerised model building procedures and with these improved aids have naturally become interested in the even more complex microbial polysaccharides. In this contribution we wish to report on some selected complex polyuronides from the Klebsiella serotypes.

Becteria of the genus Klebsiella produce a capsular polysaccharide which is antigenic. Approximately eighty different serotypes are recognized and Nimmich (7,8) has proved qualitative analyses of these capsular materials. The chemical covalent repeating sequences of a number of these serotypes is already established; others are in the process of being elucidated, while the remainder are only partially known.

We have induced a number of these serotypes to form oriented, crystalline fibres suitable for x-ray diffraction analysis. These x-ray data enable helical molecular models to be computer generated while maintaining standard bond lengths, bond angles, ring geometries and side chain conformations and also avoiding any undesirable stereo-chemical clashes, yet preserving the known helix symmetry and repeating axial dimensions. In addition any stabilizing influences, particularly intra-chain hydrogen bonds across glycosidic linkages, are monitored and incorporated if a satisfactory solution is indicated.

We have chosen the three serotypes K5, K57 and K8 from our selection because they are all polyuronides and we have arranged them in order of increasing complexity. Our intention in this short paper is two-fold. First we wish to establish a continuum between our previous studies on the connective tissue polyuronides and secondly we wish to gain an understanding of the rules which govern the molecular geometry of the polysaccharides as we advance up the hierarchy of complexity.

Klebsiella K5

Of the three Klebsiella serotypes to be discussed in this particular contribution the K5 polysaccharide has the least complicated chemical covalent repeat. It is a linear polytrisaccharide of the form $(\text{-A-B-C-})_n$, and the detailed chemical constitution has been reported by Dutton and Mo-Tai Yang (9), as shown schematically in Figure 1. The essential backbone structure consists of two neutral sugars, a 1,3-linked β-D-mannose and a 1,4-linked β-D-glucose, sandwiching a 1,4 - linked β-D-glucuronic acid residue. There are two appurtenances worth mentioning: pyruvic acid is linked to the D-mannopyranose as a 4,6 acetal, and an O-acetate group is attached to the 2-position of the glucopyranose ring. Thus the repeating sequence contains two charged carboxylate groups and the glycosidic linkage geometry is illustrated in Figure 1. We would expect all three monosaccharides to exist in the normal 4C1 chair conformation resulting in a pair of 1→4 - diequatorial glycosidic linkages together with a single 1→3 - diequatorial linkage. Both these linkage goemetries are common to the simpler plant and animal polyuronides. If the vectors between each glycosidic oxygen atom were to align precisely the maximum theoretical extension, using standardized coordinates (10), per covalent repeat would be 1.56nm. Of course it is extremely unlikely that a stereochemical acceptable model could exist with such a repeat but at least it gives us an idea of the upper limit that we should expect. For the simpler polyuronides such as the alginates and connective tissue polyuronides our observed axially projected repeats in the solid state were within 18% of the theoretical limit and typically extended conformations gave repeats centred around values 10% less than the maximum.

The x-ray diffraction pattern obtained from the sodium salt of K5 is shown in Figure 2. The layer line spacing is 2.70 nm and meridional relfections occur only on even layer lines suggesting a two-fold helical conformation of the molecule. The axially projected repeat of 2.70/2 = 1.35 nm correlates well with the covalent trisaccharide repeat and is only 14% below the theoretical limit suggesting an extended chain conformation.

At this stage it is worthwhile to systematically examine the three types of glycosidic linkage geometry using Ramachandran type hard sphere plots (11). These maps (Figure 3) show the combination of glycosidic torsion angles that could produce a stereochemically allowable linkage geometry. In addition, the

CH_3-C-COOH (4, 6)

→3)-β-D-Man-(1→4)-β-D-GlcUA-(1→4)-β-D-Glc-(1→

2-OAc

(a)

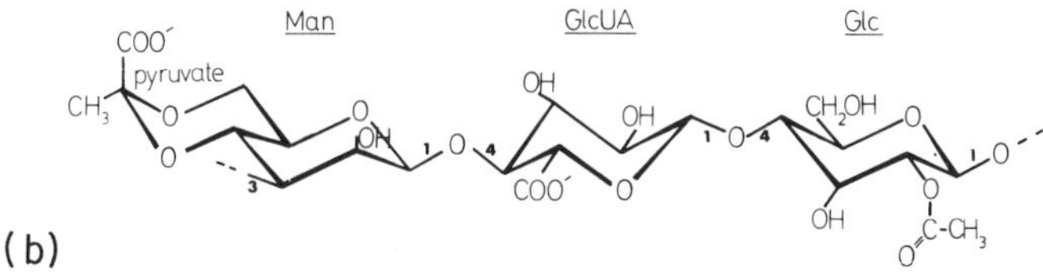

(b)

Figure 1. Klebsiella *serotype K5: (a) chemical repeat, (b) schematic. Note the chair shape of the pyruvic acid.*

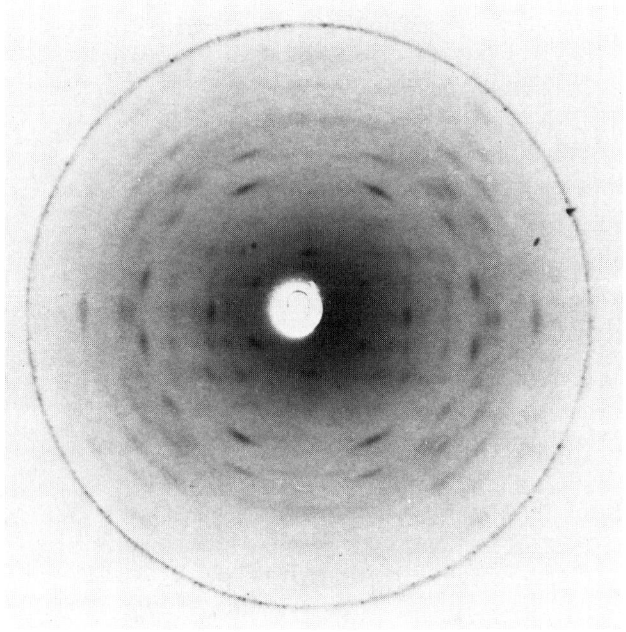

Figure 2. X-Ray fibre diffraction pattern from the sodium salt K5 polysaccharide. This shows a layer line spacing of 2.70 nm with meridional reflections on even layer lines only indicating a two-fold conformation for the molecule.

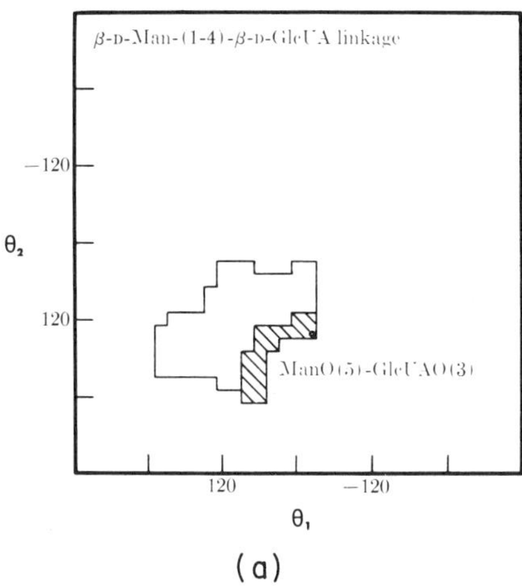

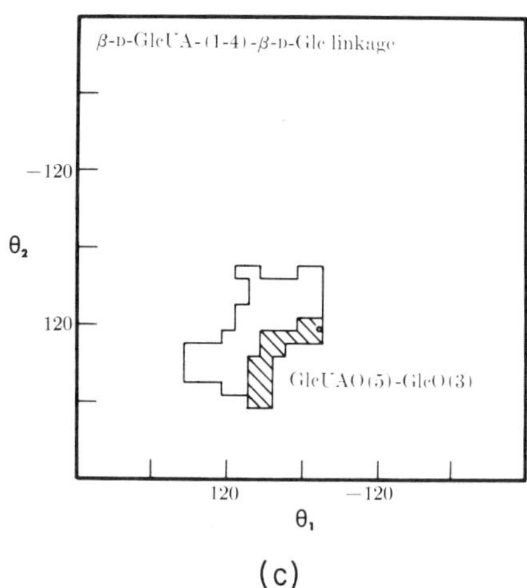

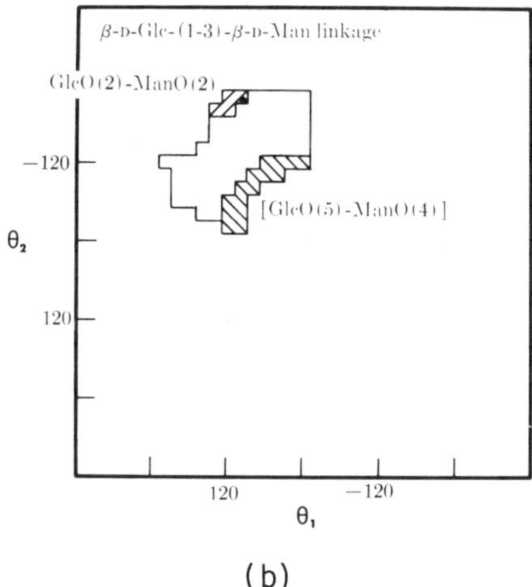

Figure 3. Conformational maps of the glycosidic linkages of K5 showing the stereochemically allowed regions and possible hydrogen bonds. In each case the final position of our trial model is marked (o). The bracketed hydrogen bond cannot be formed in the specific case of K5 since Man–O(4) forms a ketal pyruvate with O(6).

(a) $\theta_1 = C(2) - C(1) - O - C(4)$; $\theta_2 = C(1) - O - C(4) - C(3)$.
(b) $\theta_1 = C(2) - C(1) - O - C(4)$; $\theta_2 = C(1) - O - C(4) - C(3)$.
(c) $\theta_1 = C(2) - C(1) - O - C(3)$; $\theta_2 = C(1) - O - C(3) - C(2)$.

cross-hatched regions indicate areas of the allowed conformation regions where oxygen-oxygen distances (≤0.3 nm) are such that specific hydrogen bonds could form. In the calculation of the steric maps pendent groups whose orientation could not be explicitly defined were not considered in the preliminary model building, but were positioned after the backbone conformation had been determined.

Stereochemical acceptable structures with the required helical parameters have been computer generated and it has been found possible to form intra-chain hydrogen bonds across all three glycosidic linkages and fit the experimentally recorded helical parameters.

The torsion angles incorporated in the helix model are indicated on the steric maps (Figure 3). Computer drawn projections of the helix are shown in Figure 4. The structure incorporates a O(5) - - - H - O (3) hydrogen bond at both the β-D-Man-(1→4) -β -D - GlcUA and β - D - GlcUA - (1→4) - β - D - Glc glycosidic linkages. This hydrogen bond is present in many homopolysaccharides such as cellulose (12) and the connective tissue polydisaccharide structures (13). It is gratifying to find it can also appear in this polytrisaccharide. At the third glycosidic linkage (β - D - Glc - (1→3) - β - D - Man)there is an O(2) - - - O(2) hydrogen bond. This linkage is not found in any of the structures with mono or disaccharide repeats and it is of interest to consider its role in this particular structure. First, the incorporation of the O(3) oxygen in a glycosidic linkage has freed the O(4) which is needed for the formation of the 4,6 acetal pyruvate group. Secondly the geometry of the 1→3 - diequatorial glycosidic linkage is such that the carboxyl group of the pyruvate is placed at the maximum distance from the helix axis and the axial O(2) on the mannopyranose residue allows formation of an O(2) - - - O(2) stabilizing intra-chain hydrogen bond. If the pyruvated residue had been other than manupyranose this linkage conformation would have been sterically disallowed.

The other interesting feature of the K5 polysaccharide is the presence of an O-acetyl substituent on the C(2) of the glucopyranose residue. The O-acetyl group was positioned following the arrangement determined from model compounds (14). It has long been known that noncarbohydrate constituents such as O-acetyl groups and pyruvate may function as antigenic determinants. We find that in the three dimensional structure for K5 the O-acetyl and pyruvate are in close proximity (see Figure 4) and could well represent the determinant site.

Klebsiella K57

K57 is a polytetrasaccharide consisting of three sugar residues in the backbone and with one single residue in the side chain. The repeat is therefore of the form:

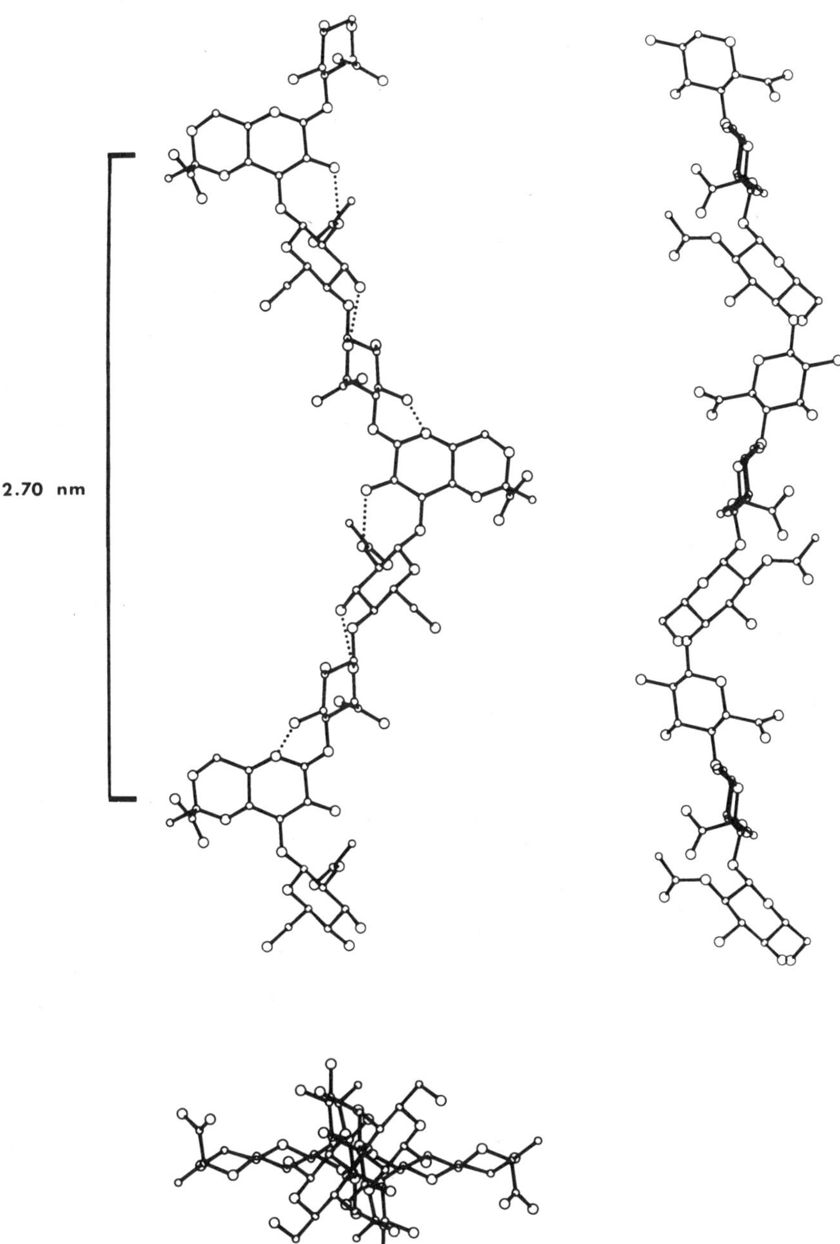

Figure 4. Projections of proposed molecular conformation of the K5 polysaccharide. Hydrogen bonds are indicated by dotted lines.

$$\begin{array}{c} S \\ | \\ (-A-B-C-)_n \end{array}$$

, where S is an α - D - mannose residue. The detailed chemical covalent repeat has been established by Kamerling et al. (15) and is given in Figure 5. As in the case of the K5 serotype the backbone consists of two neutral sugars and a uronic acid residue. This 1,3 - linked - α - D - galacturonic acid residue is attached to a 1,2 - linked - α - D - mannose residue followed by a 1,3 -linked β - D - galactose residue. The side group (S) is attached to the uronic acid residue. Again we would anticipate that all the sugar residues exist in the normal 4C1 chair conformation resulting in one 1→3 - diequatorial glycosidic linkage a 1→2 - diaxial linkage and one 1 ax →3eq linkage (see, Figure 5). In addition the mannopyranose appurtenance is 1→4 diaxially attached. *A priori* this structure presents us with some novel glycosidic linkage geometries to examine and with the added complication of a small side chain. The maximum theoretical extension for the chemical repeat, following the method described for K5, is 1.27 nm.

The x-ray diffraction pattern from the K57 polysaccharide is shown in Figure 6. The layer line spacing was measured to be 3.429 nm with meridional reflections present only on layer lines l = 3n. The simplest interpretation of this pattern is that the polysaccharide backbone forms a three-fold helix with a projected axial repeat of 1.143 nm. This value, 10% less than the maximum permissible, correlates with a single covalent repeat and suggest a fairly extended structure.

Trial models have been computer generated conforming to the helical parameters. Both left and right-handed models have been generated using the techniques and criteria outlined earlier.

Attempts were made to form the maximum number of intra-chain hydrogen bonds. It was found that no model could be constructed that included hydrogen bonds across all three backbone glycosidic linkages. Only a left-handed helix allowed the formation of two intra-chain hydrogen bonds in the backbone: α - D - GalUA - O(2) - - - O(3) - α - D - Man and α - D - Man O(5) - - - H - O(2) - β - D - Gal. The torsion angles at the glycosidic linkages of this model are marked on the steric maps in Figure 7. Computer drawn projections of the three-fold structure are shown in Figure 8.

It is interesting to note the 1→3 - diequatorial glycosidic linkage adjacent to the attachment of the mannopyranose side chain appears unable to form a suitable intra-chain hydrogen bond. The introduction of the mannopyranose side chain prevents the formation of a GalUA - O (4) - - - Gal O(5) intra-chain hydrogen bond that is possible in the K8 polysaccharide (see below). Thus the side chain may have a controlling influence on the backbone geometry. At the linkage to the side group, namely α - D - Man - (1→4) - α - D - GalUA, no hydrogen bonds are possible, irrespective of the backbone conformation, and since the torsion angles at this linkage do not affect the helical parameters a conformation was

(a)

α-D-Man
$\downarrow$ (1→4)
→3)-α-D-GalUA-(1→2)-α-D-Man-(1→3)-β-D-Gal-(1→

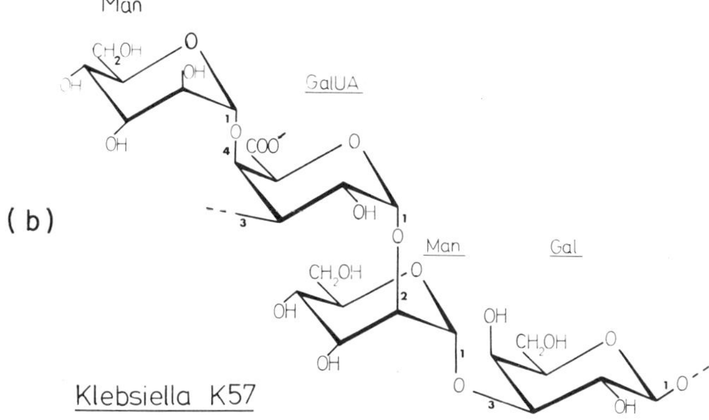

Figure 5. Klebsiella *serotype K57: (a) chemical repeat, (b) schematic.*

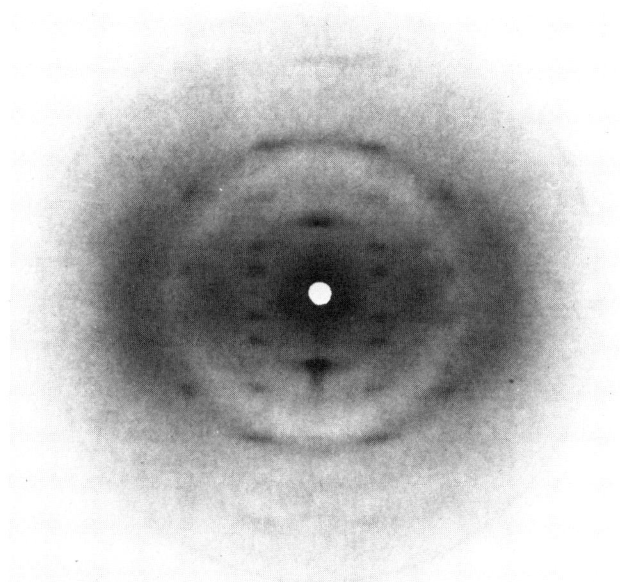

Figure 6. X-ray fibre diffraction pattern of K57 polysaccharide. This shows a layer line spacing of 3.429 nm with meridional reflections on layer lines (b) governed by $l = 3n$. The simplest interpretation is a three-fold helix.

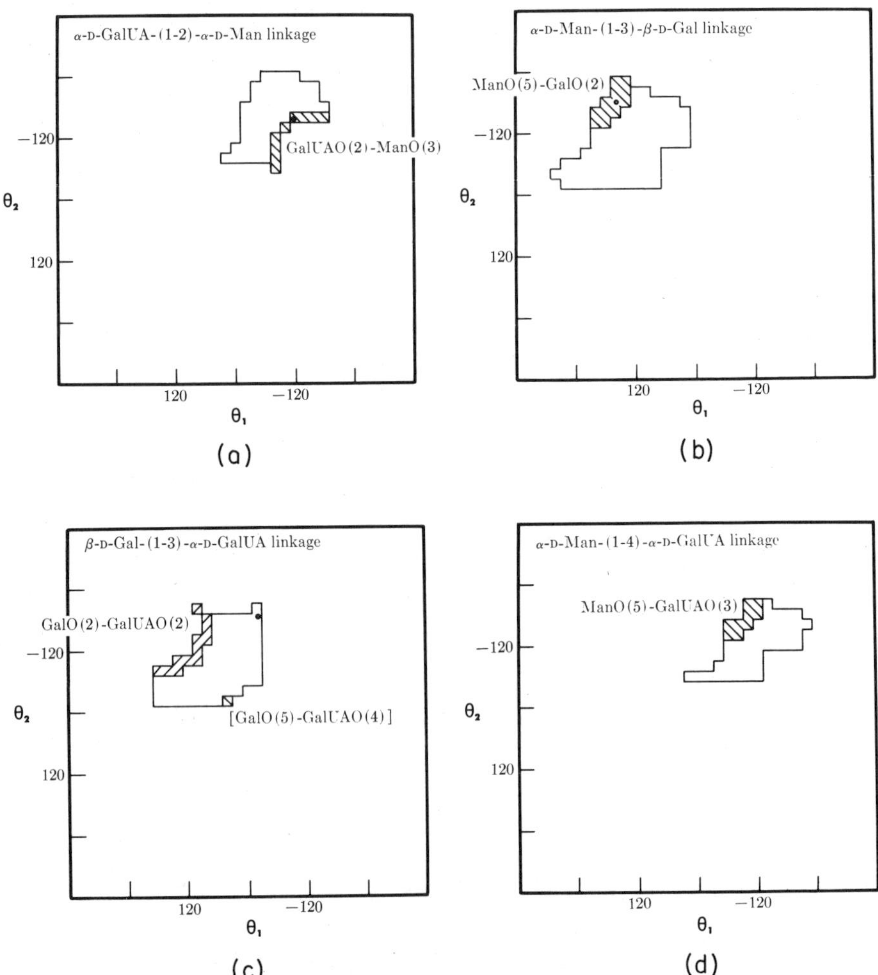

Figure 7. Steric maps and possible hydrogen bonds for the K57 polysaccharide. In each case the final position of our trial model is marked (o). The bracketed hydrogen bond cannot be formed in the specific case of K57 since the GalUA-O(4) atom is glycosidically linked to the Man side chain.

(a) $\theta_1 = C(2) - C(1) - O - C(2)$; $\theta_2 = C(1) - O - C(2) - C(1)$.
(b) $\theta_1 = O(5) - C(1) - O - C(3)$; $\theta_2 = C(1) - O - C(3) - C(2)$.
(c) $\theta_1 = C(2) - C(1) - O - C(3)$; $\theta_2 = C(1) - O - C(3) - C(2)$.
(d) $\theta_1 = C(2) - C(1) - O - C(4)$; $\theta_2 = C(1) - O - C(4) - C(3)$.

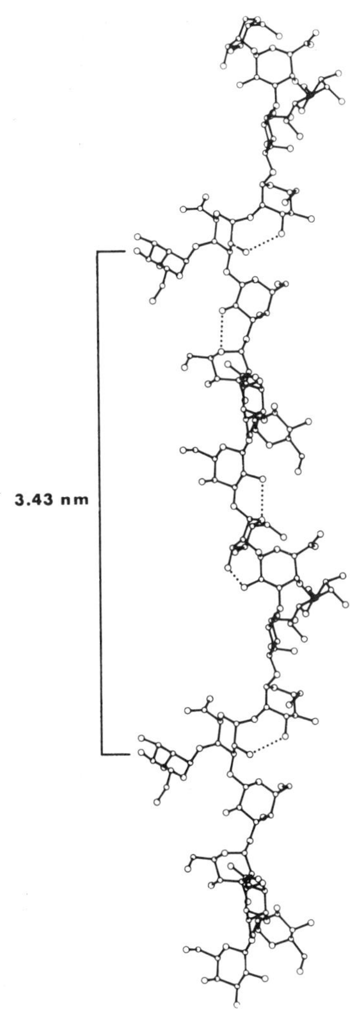

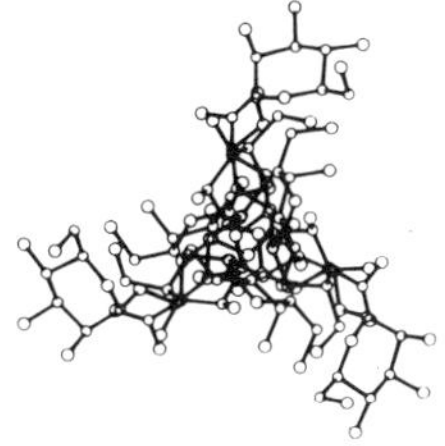

Figure 8. Projections of proposed molecular conformation of the K57 polysaccharide. Hydrogen bonds are indicated by dotted lines.

chosen to be near the centre of the allowed region as marked (Figure 7(d)).

Klebsiella K8

The covalent chemical repeating sequence has been established by Sutherland (16) and is given in Figure 9. It is similar to the K57 polysaccharide with a polytrisaccharide backbone and a single monosaccharide chain. Both structures have a uronic acid in the repeat, however, in K57 the uronic acid is incorporated in the backbone and the side chain is a neutral sugar. The converse is true in K8 which has an uncharged backbone and an α - D - glucuronic acid and side chain. The backbone consists of three commonly occurring neutral sugars: a 1,3 - linked β - D - galactose followed by a 1,3 - linked β - D - galactose and finally a 1,3 - linked α - D - glucose. All four monosaccharides are expected to exist in the normal 4C1 chair conformation and a schematic diagram representing the glycosidic linkage geometry is shown in Figure 9. Thus two 1→3 - diequatorial glycosidic linkages straddle a 1 ax→3 eq-glycosidic linkage in the backbone.

The maximum extension for the chemical repeat, following the method described previously, is 1.38 nm falling partway between the values for K5 and K57. The x-ray diffraction pattern for the sodium salt of the K8 polysaccharide is shown in Figure 10. The material is highly oriented and crystalline and from the systematic absences of odd order meridional reflections it can be seen that the molecule forms a 2_1 helix. However the layer line spacing of 5.078 nm is far too large for a repeat of two asymmetric units. In fact this value is very close to the theoretical maximum extension for four complete covalent repeats, i.e. 4 x 1.38 = 5.52 nm. Thus the observed repeat is only 10% less than maximum extension.

For preliminary model building it appeared reasonable to assume that the structure of the isolated molecule is a perfect four fold helix with an axial advance per covalent repeat 1.27 nm. Perturbations from an idelized four-fold helix would be expected to result in a lower symmetry and consequently the packing of the molecular chains in an orthorhombic rather than a tetragonal unit cell. This phenomenon has also been observed in hyaluronic acid (13).

Models were constructed imposing the experimentally determined fibre repeat and the assumption that the isolated molecule is a four-fold helix. Attempts were made to construct a stereochemically acceptable structure incorporating the maximum number of intramolcular hydrogen bonds. Both right-handed and left-handed structures were considered. It was impossible to construct a model which incorporated hydrogen bonds at all three backbone linkages. Only a left-handed helix allowed the formation of two hydrogen bonds in the backbone. The glycosidic torsion angles that generate this model are indicated on the steric maps in

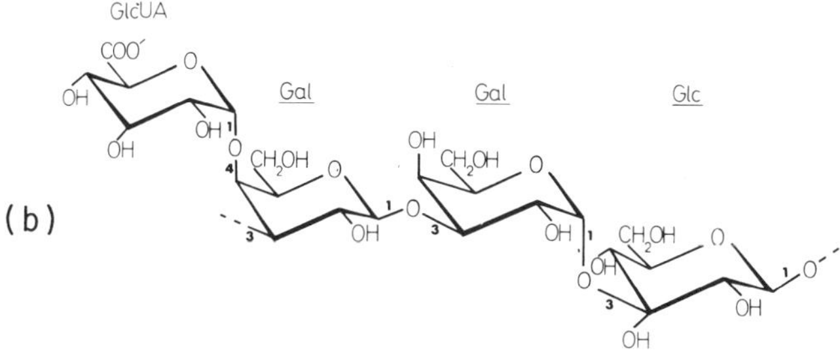

Figure 9. Klebsiella *serotype K8: (a) chemical repeat, (b) schematic*

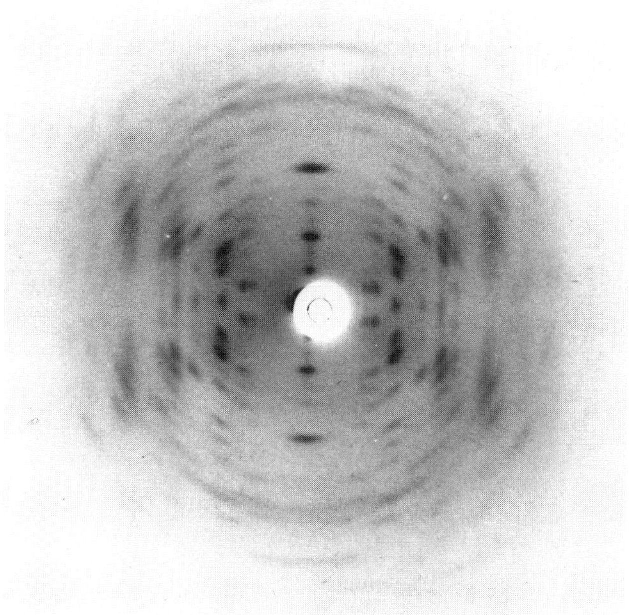

Figure 10. X-Ray fibre diffraction pattern of K8 polysaccharide. This shows a layer line spacing of 5.078 nm with meridional reflections on even layer lines only. Such a spacing is compatible with four covalent repeats, and we interpret this as a slightly perturbed four-fold helix.

Figure 11. Projections of this left-handed helical model are shown in Figure 12. The structure contains hydrogen bonds O(5) - - - O(4) and O(5) - - - O(2) at the β-D - Gal - (1→3) - α - D - Gal and α - D - Gal - (1→3) - β - D - Glc linkages respectively. It was found to be impossible to form a four-fold helix (with the observed extension) which contained a hydrogen bond at the β - D - Glc - (1→3) - β - D - Gal glycosidic linkage. This linkage adjacent tc the attachment site of the uronic acid side chain. This feature is analogous to the situation in K57.

The steric map for the glycosidic linkage that defines the orientation of the side chain is shown in Figure 11(d). Lacking any information to define the position of the side chain, the glycosidic torsion angles were set at values corresponding to the centre of the allowed region on the steric map. One possible explanation for the two-fold rather than four-fold nature of the molecule in the crystalline state is that adjacent covalent repeats have the side chain in different orientations.

Discussion

We have presented examples of trial model building on a small selection of the large variety of microbial polysaccharides. This has recently become possible with the crystallization of these molecules in a form suitable for x-ray diffraction studies. It is only through these x-ray studies that the helical parameters necessary for meaningful model building can be obtained. Even though the examples we have chosen have substantially different primary structures, certain salient points are apparent. For example, all the structures presented exist in extended conformations. This feature appears to be independent of the helical symmetry exhibited by the molecule.

We have presented preliminary structural models computed on the basis of a search for stabilising intrachain hydrogen bonds. This approach is a useful start in determining helical structures for further refinement. The investigation of conformational maps and potential hydrogen bonds gives us insight into the molecules concerned and concurrently provides us with various starting models for further refinement using terms approximating more closely to potential energy functions, (see e.g.Guss *et al.* (13)). By selecting the models containing the maximum number of hydrogen bonds consistent with the stereochemistry we are choosing the most likely structure at this first level hard sphere approximation. As we include more complex interaction terms we would hope that the model will not change significantly. If this indeed turns out to be the case, as our continuing calculations have so for indicated, then such a simple approach could be extremely useful in model building.

Clearly, the quality of diffraction photographs is going to determine the eventual detail of the refined structures. In the case of K8 for which a highly crystalline pattern has been

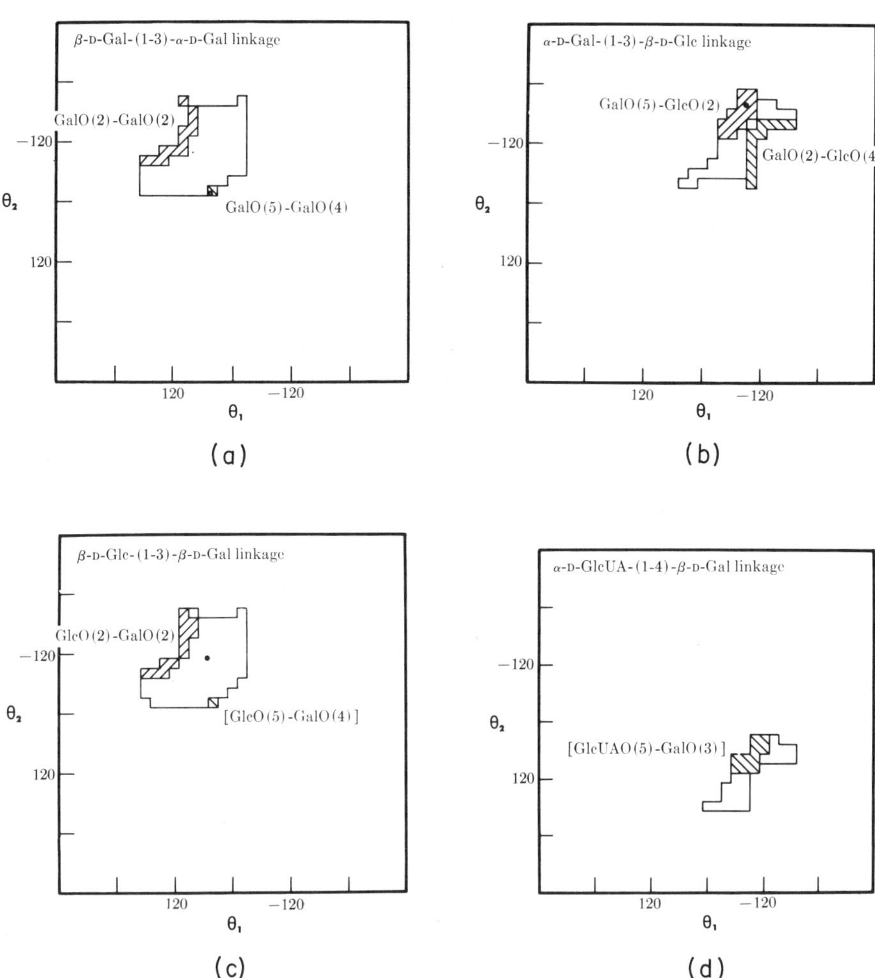

Figure 11. Steric maps and possible hydrogen bonds for the K8 polysaccharide. In each case the final position of our trial model is marked (o).

(a) $\theta_1 = C(2) - C(1) - O - C(3)$; $\theta_2 = C(1) - O - C(3) - C(2)$.
(b) $\theta_1 = C(2) - C(1) - O - C(3)$; $\theta_2 = C(1) - O - C(3) - C(2)$.
(c) $\theta_1 = C(2) - C(1) - O - C(3)$; $\theta_2 = C(1) - O - C(3) - C(2)$.
(d) $\theta_1 = C(2) - C(1) - O - C(4)$; $\theta_2 = C(1) - O - C(4) - C(5)$.

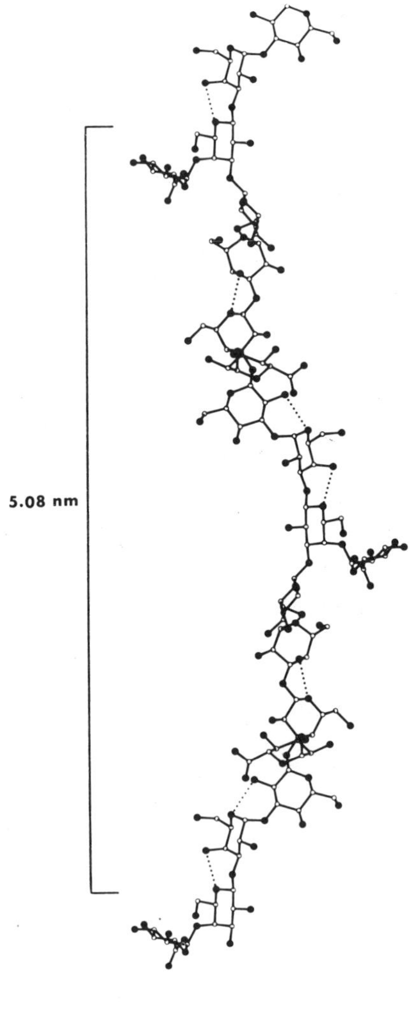

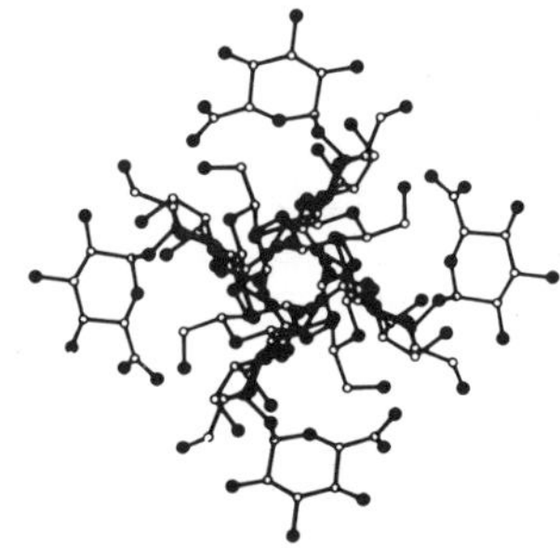

Figure 12. Projections of proposed molecular conformation of the K8 polysaccharide. Hydrogen bonds are indicated by dotted lines.

obtained a complete structure determination is currently in progress. For K5 a somewhat poorer quality photograph was obtained and the final determination is unlikely to be so detailed unless better experimental information is obtained. In the case of K57 where some discrete Bragg reflections are observed, but in addition there is substantial diffuse scattering, our structural refinements will be limited to intrachain interactions, possibly supported by a comparison of the observed data with cylindrically averaged Fourier transform calculations. We anticipate that the molecular model building presented here will be a useful guide to further, more elaborate, refinements.

Acknowledgements

We wish to express our thanks to Professors W. Burchard, G. Dutton, S. Stirm and Miss C. Wolf for the collaboration in this investigation. The work was supported by the Science Research Council.

Abstract

Structural information form x-ray diffraction by oriented samples of some Klebsiella polysaccharides has enabled trial conformations to be deduced for the molecules in the inspissated phase. Three serotypes K5, K57 and K8 have been selected (all are polyuronides) and examined in increasing order of complexity. Type K5 is a linear trisaccharide of the form $(-A - B - C -)_n$, whilst K57 and K8 are both polytetrasaccharides with three sugar units in the backbone and one in the side chain. Our analyses show that all three polysaccharides exhibit extended chain conformations.

The three structures also collectively offer a variety of helical symmetries giving in succession two-fold (K5), three-fold (K57) and four-fold (K8) structures. Trial models have been computer generated conforming to the helical parameters and including the maximum number of intrachain hydrogen bonds. Various detailed aspects of the three conformations are compared and contrasted.

Literature Cited

(1) Atkins, E.D.T., Mackie, W., Parker, K.D. and Smolko, E.E. J. Polym. Sci (B) (1971) 9, 311.

(2) Atkins, E.D.T., Nieduszynski, I.A., Mackie, W., Parker,K.D. and Smolko, E.E Biopolymers (1973) 12, 1865.

(3) Atkins, E.D.T, and Sheehan, J.K. Biochem. J. (1971) 125, 92; Nature (New Biol) (1972), 235,253.

(4) Atkins, E.D.T., Isaac, D.H., Nieduszynski, I.A., Phelps, C.F. and Sheehan, J.K. Polymer (1974) 15, 263.

(5) Atkins, E.D.T and Nieduszynski, I.A. in "Heparin, Structure, Function and Clinical Implications", edits. R.A. Bradshaw and S. Wessler, Plenum Press, New York and London (1975) 19; Atkins, E.D.T. and Nieduszynski, I.A. Fed. Proc. (in press).
(6) Nieduszynski, I.A., Gardner, K.H. and Atkins, E.D.T. (this issue).
(7) Nimmich, W.Z. Med Mikrobiol. Immunol. (1968) 154, 177.
(8) Nimmich, W. Acta. Biol. Med. Germ. (1971) 26, 397.
(9) Dutton, G.G.S. and Mo-Tai Yang. Can.J.Chem. (1973) 51, 1826.
(10) Arnott, S. and Scott, W.C. J.Chem.Soc. (Perkins Transactions II) (1972) 324.
(11) Ramachandran, G.N. Ramakrishnan, C. and Sasisekharan, V. in Aspects of Protein Structure edit. G.N. Ramachandran, Academic Press, New York (1963) 121.
(12) Gardner, K.H. and Blackwell, J. Biopolymers, (1974) 13, 1975.
(13) Guss, J.M. Hukins, D.W.L., Smith, P.J.C., Winter, W.T. Arnott S., Moorhouse, R. and Rees, D.A. J. Mol. Biol. (1975) 95,359.
(14) Leung, F. and Marchessault, R.H., Can.J.Chem. (1973) 51, 1215.
(15) Kamerling, J.P., Lindberg, B., Lonngren, J. and Nimmich, W. Acta.Chem. Scand. (1975) B29, 593.
(16) Sutherland, I.W. Biochemistry (1970) 9, 2181.

6

X-Ray Diffraction Studies of Heparin Conformation

I. A. NIEDUSZYNSKI,[1] K. H. GARDNER, and E. D. T. ATKINS

H. H. Wills Physics Laboratory, University of Bristol, Bristol BS8 1TL, U.K.

Introduction. Heparin is a sulphated polysaccharide which occurs as intra-cellular packets of material in mast cells and has a function in fat-clearing and as a blood anti-coagulant. In this latter role, heparin finds great application in medicine and is isolated on a commercial scale.

The detailed chemical structure of heparin is still under investigation, but it is believed to have a glycosamino-glycuronan backbone $(\text{-H-U-})_n$ where H is generally 2-deoxy-2-sulphamino-α-D-glucose-6-sulphate and U is mainly 2-sulphated α-L-iduronic acid but also contains some β-D-glucuronic acid. Heparins generally have 2-3 sulphate ester groups per disaccharide and a small proportion of N-acetyl groups.

Since the known functions of heparin are mediated by specific binding to proteins such as lipoprotein lipase (1) and the various blood coagulation factors (2) a knowledge of the detailed shape of the molecule is of great importance. This point may be emphasized by comparison (3,4) of the anticoagulant activities of heparin (about 160 IU/mg.) and certain heparan sulphates (e.g. about 16 IU/mg.); the latter molecules being chemically similar to heparin and differing principally in having a higher proportion of -D-glucuronic acid and one sulphate ester group fewer per disaccharide.

The principal problem in determining the shape of the heparin molecule residues in the conformation of the α-L-iduronate residue, since the hexosamine residue is expected to exist in the normal $^4C_1$ chair in which all of the bulky substituents would be equatorially disposed. In the sulphated α-L-iduronate residue

[1] Current address: Dept. Biol. Sciences, University of Lancaster, Lancaster LA1 4YQ, Lancashire, England.

(Fig.2), however, the normal $^4C_1$ chair would require an axial carboxyl group and the alternative $^1C_4$ chair requires axial sulphate and linkage positions. It is consequently difficult to choose between these two residue conformers and other forms also need to be considered.

Recently, X-ray fibre diffraction patterns have been obtained from pig mucosal heparin (5) and macromolecular heparin from rat skins (6) and these give the first information of the molecular repeat distances in heparin. On the basis of the observed tetrasaccharide repeating distance these patterns have been interpreted as favouring the $^1C_4$ residue conformation whose axial projection is shorter rather than the $^4C_1$ conformation for α-L-iduronate.

In this paper all of the 0.3 nm. X-ray data is used in a more comprehensive study of the α-L-iduronate residue conformation in heparin.

The α-L-Iduronate Residue. The α-L-iduronate residue occurs in the glycosaminoglycans, dermatan sulphate, heparan sulphate and heparin. In dermatan sulphate interpretations of the X-ray (7-9) have favoured the $^4C_1$ chair but the nuclear magnetic resonance (NMR) data (11) have favoured the $^1C_4$ model. In heparan sulphate the proportions of α-L-iduronate are generally low and conformational assignments should therefore be withheld until dermatan sulphate and heparin are better understood. In heparin the α-L-iduronate residue is generally sulphated in the 2-position and the interpretation of both the X-ray (5) and NMR data (11) has favoured the $^1C_4$ over the $^4C_1$ conformer, but no definitive statement has been made eliminating other possible conformers.

In principle, pyranoses may adopt any of 26 standard residue conformations (see (12) for a comprehensive nomenclature system). These include 2 chairs, 6 boats, 6 skew-boats and 12 half-chairs. A molecular model of 2-sulphated α-L-iduronate in each of these conformations was constructed and examined on the basis of three criteria: (i) the molecular extension observed for heparin; (ii) the steric requirements of the bulky substituents on the ring; (iii) the NMR data.

(The NMR data which are most sensitive to residue conformation are the proton-proton coupling constants which may be measured with a high-field proto spectrometer and then related to residue conformation via the Karplus relationship for dihedral angles).

These criteria indicated that the $^{1}S_{3}$ skew-boat should be considered together with the commonly occurring chair forms $^{4}C_{1}$ and $^{1}C_{4}$ (see Fig. 1).

METHODS. Materials - X-ray Diffraction. The sodium salt of heparin of mucosal origin was kindly provided by Dr. Johnson of the National Institute for Medical Research, Mill Hill, London.

Films of this material were cast from aqueous solution and stretched at 76% relative humidity in a manner previously described (13). The X-ray fibre diagram shown in Fig. 2 was taken in a hydrogen-filled fibre camera with humidity controlled at 76% and the reflection spacings were calibrated with calcite.

Intensities were measured with a Joyce-Loebl microdensitometer and corrected for Lorentz polarisation and multiplicity factors and for oblique incidence on the film. A temperature factor of $B = 0.05\ nm^2$ was applied to the calculated structure factors.

Model-Building. The coordinates of the trial models were set up on the assumptions that the chemical structure of heparin is described by a repeating disaccharide consisting of 1,4-linked 2-deoxy-2-sulphamino-α-D-glucose-6-suophate and 1,4-linked 2 sulphated α-L-idopyranosyluronic acid, and that the helix formed has n (the number of disaccharides per turn) = 2 and h (the axial rise per disaccharide) = 0.84 nm.

Trial molecular conformations were built according to the "linked-atom" system (14) in which covalent bond lengths and angles have fixed values. Internal coordinates for the pyranose rings and sulphate ester groups were derived from standard coordinates (15), leaving the torsional angles defining the glycosidic bridge conformations and side chain orientations as explicit variables. Helix pitch, symmetry and continuity were maintained by means of Lagrangian constraints.

The steric permissibility of the starting conformations was imposed by ensuring that all distances

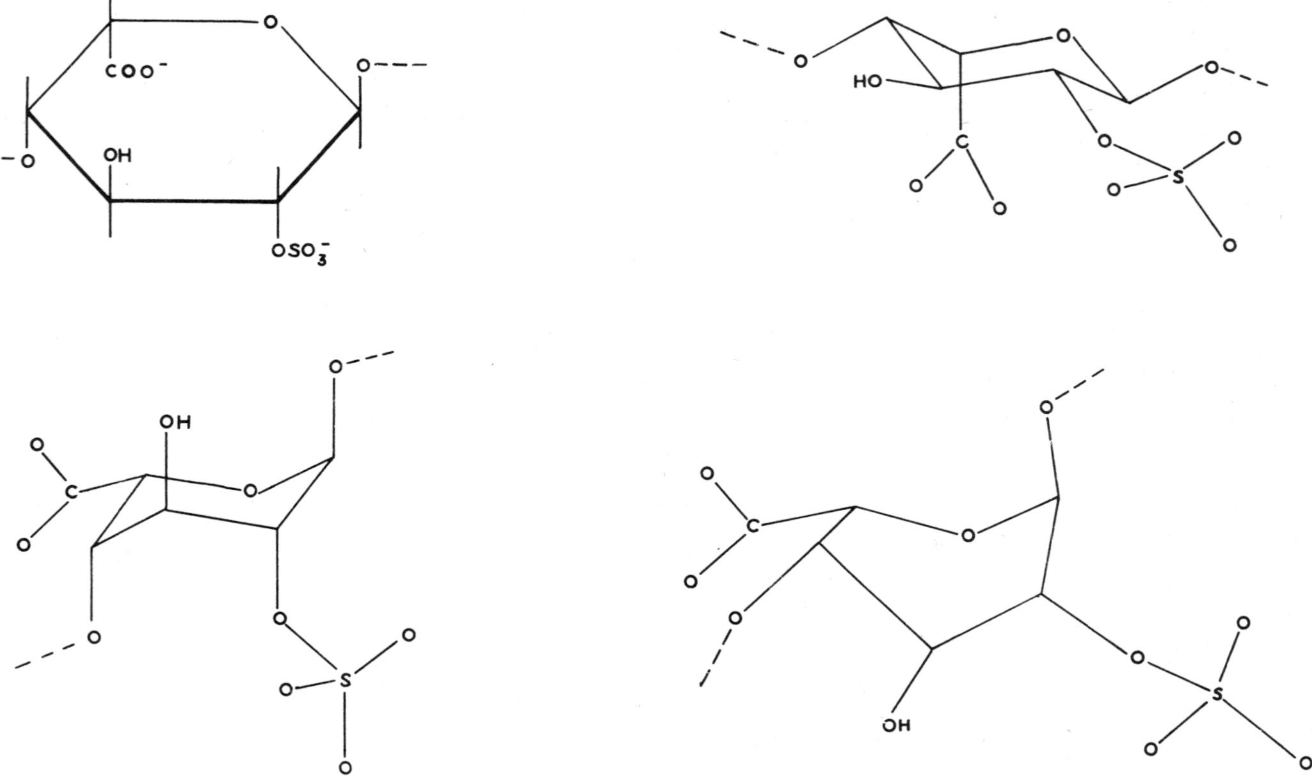

Figure 1. *The 2-sulfated* α-L-*iduronate residue in (i) Haworth formula (top left); (ii)* $^{4}C_{1}$ *chair conformation (top right); (iii)* $^{1}C_{4}$ *chair conformation (lower left); (iv)* $^{1}S_{3}$ *skew-boat conformation (lower right).*

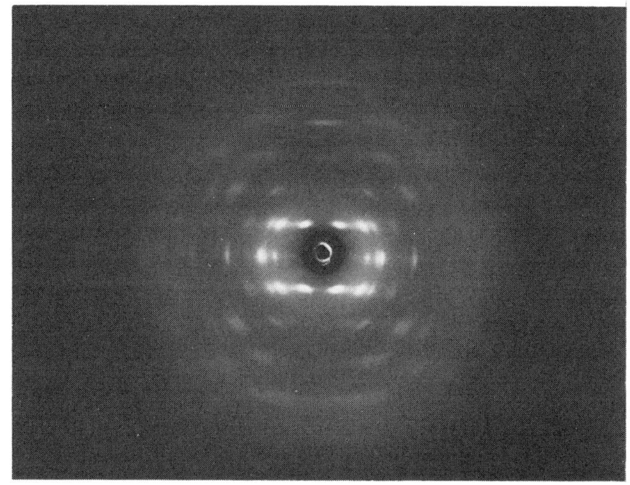

Figure 2. X-Ray fibre diffraction pattern obtained from the sodium salt of heparin. Note the layer lines with spacing of 1.68 nm. The fibre axis is vertical.

between non-bonded atoms were kept greater than extreme limit contact distances (16). Sulphur atoms were added to the extreme limit contact table by assuming a van der Waals contribution to their contact distances of 0.165 nm. The same contact table was used for intermolecular "packing" interactions.

The initial conformations for the N-sulphate and iduronate 2-sulphate were defined by having the N-S and O-S bonds respectively *cis* to the C(2)-H(2) bonds on hexosamine and iduronate respectively, and one of the S-O bonds of the SO_3 group *trans* to C(2) - and C(2) - O respectively. These settings put the groups in the middle of their single allowed regions which permitted them approximately 30° free rotation on either side of this median position. The 6-sulphate group was placed initially in the *gt* conformation, that is, with C(6) - O (6) *trans* to C(4) -C(5) and the remaining atoms of the sulphate ester group also in a *trans* configuration. However, the 6-sulphate group may lie in three separate allowed areas as defined by rotation about the first C(5) - C(6) bond which are labelled: (i) *gt* gauche to C(5) - O(5) and trans to C(4) - C(5); (ii) *tg* trans to C(5) - O(5) and gauche to C(4) - C(5); (iii) *gg* gauche to C(5) - O(5) and gauche to C(4) - C(5).
The initial conformation of the carboxyl group was established with the torsional angles C(4) - C(5) - C(6) - O as +90° and -90_1 for the two carboxyl oxygens.

The $^1S_3$ skew-boat conformation for α-L-iduronate was constructed using bond lengths from Arnott and Scott (15) coordinates. Bond angles and torsional angles were initially set at calculated values for a skew-boat conformation given by Hendrickson (17) but were used as variables to which the constraints of ring closure and the planarity of atoms C(2), C(4), C(5), and O(5) were imposed.

All intra-molecular hydrogen-bonds in these models were set at 0.280 nm between the non-hydrogen atoms involved.

Refinement Method. The possible structural models for heparin were refined by using the "linked-atom" constrained least-squares method of Arnott and Wonacott (14). By refining parameters in a least-squares manner the function Θ was minimised, where

$$\Theta = \sum_{M} W_m (\Delta F_m^2) + \sum_{N} \lambda_n G_n$$

(ΔF_m^2) is the difference between the squares of the observed F_m(obs) and calculated F_m(calc) structure amplitudes for the mth of M reflections. A refinable scale factor and a fixed isotropic temperature factor were incorporated in values of F_m(calc). The constraint equations Gn were applied by Lagrange multipliers λ_n. A unit weighting scheme was used, and where more than one hkl plane contributed to an observed reflection, the calculated structure amplitude was given by

$$F_m(\text{calc}) = \left[\sum_{R} F_r(\text{calc})^2 \right]^{1/2}$$

where F_r(calc) is the calculated structure amplitude for the rth of the R planes involved.

In order to assess alternative structural models, two kinds of residual index were used, the common crystallographic residual factor.

$$R = \frac{\sum_{M} \left| |F_m(\text{obs})| - |F_m(\text{calc})| \right|}{\sum_{M} |F_m(\text{obs})|} \times 100$$

and

$$R'' = \left\{ \frac{\sum_{M} W_m (|F_m(\text{obs})| - |F_m(\text{calc})|)^2}{\sum_{M} W_m |F_m(\text{obs})|^2} \right\}^{1/2} \times 100$$

which can be used for Hamilton's statistical tests(18).

DESCRIPTION OF STARTING CONFORMATIONS. The heparin trial conformations resulting from the incorporation of the different α-L-iduronate conformers mentioned above will now be briefly described (see Fig. 3 and Table I).

Heparin models with the sulphated α-L-iduronate residue adopting a $^4C_1$ chair conformation are only just sterically acceptable for a two-fold helix with rise per disaccharide as low as 0.84 nm. No intramolecular hydrogen-bonds can be formed and detailed structure factor calculations (19) have shown that this model gives a significantly poorer agreement with the observed intensity data than the model described below. The model is shown (Fig. 3(i)) for the sake of completeness but will not be considered further.

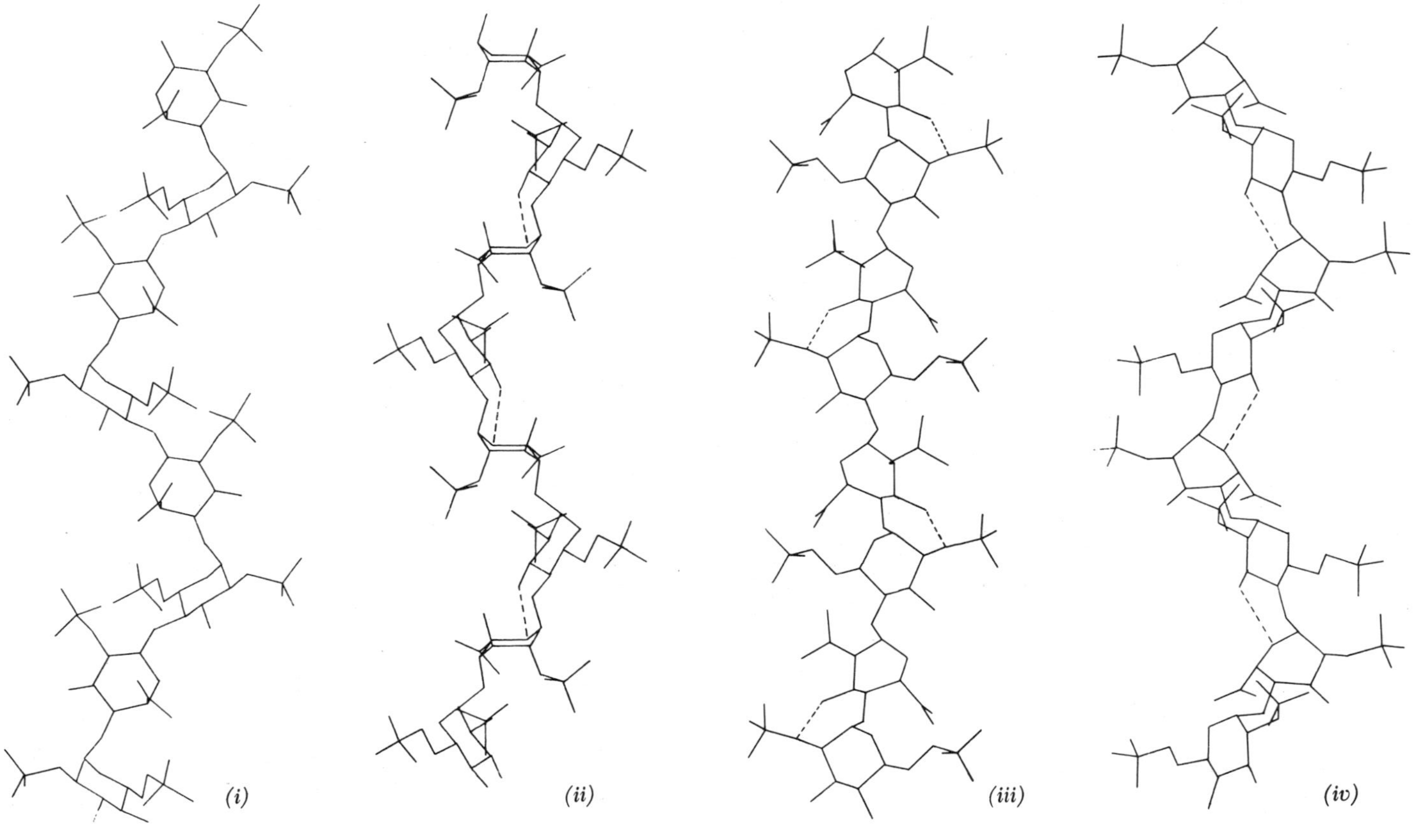

Figure 3. Trial conformation of heparin molecule with n *(no. of disaccharides per turn)* = 2 *and* h *(axial rise per disaccharide)* = 0.84 *nm, corresponding to the sulfated* α-L*-iduronate residue in (i)* $^4C_1$ *chair conformation (left); (ii)* $^1C_4$ *chair conformation with hexosamine O(3)–H . . . O(5) uronic acid hydrogen-bond; (iii)* $^1S_3$ *skew-boat conformation with hexosamine N –H . . . O(3) uronic acid hydrogen bond; (iv)* $^1S_3$ *skew-boat conformation with hexosamine O(3) –H . . . O(5) uronic acid hydrogen bond (bottom).*

Heparin models with the sulphated α-L-iduronate residue in the $^1C_4$ chair conformation (Fig. 3 (ii)) can readily form a two-fold helix with axial rise per disaccharide of 0.84 nm and may be stabilised by an O(3) - H.....O(5) intramolecular hydrogen-bond from hexosamine to uronic acid residue.

Two heparin models with the sulphated L-iduronate residue in the $^1S_3$ skew-boat have been established (see Fig. 3 (iii) and (iv)). The first contains a hydrogen-bond from the hydrogen on the hexosamine N-sulphate to O(3) on the iduronate residue and cannot be formed in the previously considered models. The second contains an O(3) - H.....O(5) hydrogen-bond from hexosamine to uronic acid residue. A possible advantage of the skew-boat conformation is that the two charged groups, the carboxyl and O-sulphate groups are sited further apart than in the other residue conformations.

RESULTS. The sodium salt of heparin crystallises in a triclinic unit cell with dimensions, $\underline{a}$ = 1.102 nm, $\underline{b}$ = 1.201 nm, $\underline{c}$(fibre axis) = 1.680 nm, α = 90.0°, β = 116.2°, and γ = 107.3°. The density of the sample was determined by floatation to be 1.74 g. cm^{-3} which is consistent with one hydrated tetrasaccharide chain segment passing through the unit cell. Since the sodium salt sample used in this experiment is also capable of yielding another X-ray pattern (identical to that found for macromolecular heparin (6)) which clearly shows a two-fold helix with two disaccharides per turn, it is assumed that the backbone of the triclinic heparin conformation is also a two-fold helix.

Thus, the three initial conformational models for heparin were defined by n (number of disaccharides per turn) = 2 and h (rise per disaccharide) = 0.84 nm. Since only one chain segment passes through the unit cell, chain rotation, but not translation, was considered. Each initial conformation is represented as two models because the polarity of the chain with respect to the $\underline{c}$ axis of the unit cell must be defined, so the models are labelled with arrows and the symbols "U" and "D" for "up" and "down" pointing chains.

Initially, the principal variable of chain rotation was explored. The starting models were rotated

Table I. Initial Trial Conformations of Heparin*

Model	Iduronate Residue Conformation	Stabilising Hydrogen-Bond
0	$^{4}C_{1}$	None
1	$^{1}C_{4}$	O(3)-H.....O(5) hexosamine uronic acid
2	$^{1}S_{3}$	N-H......O(3) hexosamine uronic acid
3	$^{1}S_{3}$	O(3)-H.......O(5) hexosamine uronic acid

* *The initial conformations have side groups in standard positions (see text);* n = 2 *and* h = 0.84 *nm.*

about the c axis and their ability to fit the X-ray intensity data and to pack with neighbouring chains was monitored (see Table II).

Next, two more variables were introduced. Independent rotations were made about the two hexosamine C(5) - C(6) bonds at 120° intervals whilst constraining the chain rotation to stay within sterically allowed zones. Thus, the permitted gt, tg, and gg conformations of the 6-sulphate groups were tested giving the first opportunity for the heparin conformations to depart from the two-fold helical symmetry initially imposed upon them. Table III shows the best agreement obtained with the various models.

At this stage, each model was subjected to a least-squares refinement in which 11 structural variables were tested against the 39 data points. These 11 variables included the main chain rotation and rotations about the following bonds for each disaccharide: (i) Uronic acid C(5) - C(6) permitting rotation of the carboxyl group; (ii) Hexosamine C(2) - N permitting rotation of the N-sulphate group; (iii) Uronic acid C(2) - O(2) permitting rotation of the 2-sulphate group; (iv) Hexosamine C(5) -C(6) permitting rotation of the 6-sulphate group; (v) Hexosamine C(6) - O(6) also permitting rotation of the 6-sulphate group.

For this refinement, several starting positions depending on 6-sulphate conformation were used for each model and no steric permissibility constraints were incorporated at this stage as it was considered that this would reduce the radius of convergence of the refinement. The results re-inforced the conclusion in evidence in Table II that 3 of the models, 2U, 2D, and 3D should now be discarded. None of these models yielded a R'' lower than 54.

For the final stage of assessment of the models, the best trial structures from the previous stage were again subjected to least-squares refinement against the X-ray data using the 11 variables described above, but this time constraints on the steric permissibility, both with respect to intra- and inter-molecular contacts, were re-imposed. Model 1U, with iduronate in the $^1C_4$ chair conformation, yielded values of R = 33.1 and R'' = 42.6 for a final structure with both 6-sulphate groups in the gg conformation. Model 1D yielded

Table II*

Trial Model	Test	0	10	20	30	40	50	60	70	80	90	100	110	120	130	140	150	160	170	180
		Chain Rotation (in degrees)																		
1U $^{1}C_{4}$ ↑	R''	68	81	79	80	76	77	71	64	75	69	67	49	52	60	66	69	71	64	68
	Packing	X	X	X	X	X	X	X	X	X	X	X		X	X	X	X	X	X	X
1D $^{1}C_{4}$ ↓	R''	60	62	70	76	72	72	65	66	79	75	77	80	70	77	69	55	49	63	60
	Packing	X	X	X	X	X	X	X	X	X	X	X	X	X	X	X		X	X	X
2U $^{1}S_{3}$ ↑	R''	80	78	82	79	75	71	66	56	62	69	69	64	56	56	62	66	67	75	80
	Packing	X	X	X	X	X	X	X	X	X	X	X			X	X	X	X	X	X
2D $^{1}S_{3}$ ↓	R''	64	66	74	73	67	65	62	61	66	68	70	76	83	73	71	69	67	62	64
	Packing	X	X	X	X	X		X	X	X	X	X	X	X	X	X	X	X	X	X
3U $^{1}S_{3}$ ↑	R''	69	69	67	67	66	80	71	64	59	56	55	54	48	60	71	73	79	70	69
	Packing	X	X	X	X	X	X	X	X		X	X	X	X	X	X	X	X	X	X
3D $^{1}S_{3}$ ↓	R''	65	66	65	62	64	65	74	63	64	53	57	69	73	76	75	62	62	69	65
	Packing	X	X	X	X	X	X	X	X		X	X	X	X	X	X	X	X	X	X

* *Residue conformation designation follows the convention given in Ref. 12. Arrows represent chain sense, up or down, relative to the unit cell axes. Packing refers to extreme limit contacts. X represents positions sterically disallowed. Blank positions are allowed. Starting models are two-fold helices; hence rotations are equivalent after 180°.*

Table III. Best Fit Obtained against the X-Ray Data for Each Heparin Model Tested with the Nine Possible 6-Sulfate Positions

Model	Chain Rotation (Degrees)	First 6-Sulphate Conformation	Second 6-Sulphate Conformation	R	R''
1U	110	gg	gg	35.2	43.3
1D	150	tg	gg	41.3	50.2
2U	120	gt	tg	44.0	56.9
2D	50	gt	gt	57.3	65.1
3U	80	tg	gg	45.8	53.2
3D	80	gg	tg	49.8	56.7

values of R = 32.4 and R¨ = 36.7 for a final structure with both 6-sulphate groups in the gt conformation. Model 3U, with iduronate in a $^{1}S_{3}$ skew-boat conformation gave values of R = 33.6 and R¨ = 39.1 for a final structure with one 6-sulphate in the gt and one in the gg conformation. The performance of these models is summarised in Table IV and their b c projections are shown in Fig. 4.

DISCUSSION. It is clear from examination of Table IV that none of the three models represented as 1U, 1D, and 3U can be uniquely selected as best fitting the observed X-ray data, since the residual indices are all comparable. On the basis of the triclinic shape of the unit cell, it would be expected that the heparin conformation would depart somewhat from a two-fold helix but only the skew-boat model displays its 6-sulphate groups in markedly different side group conformations.

It seems improbable, therefore, that this X-ray data is adequate to resolve the detailed residue conformation of the sulphated -L-iduronate residue in heparin, and this situation is compounded by several problems which are especially marked in heparin. First, water molecules and sodium counter-ions comprise about 40% of the scattering matter in this heparin unit cell and these cannot satisfactorily be positioned or accounted for. Then, there are several partial occupancies in the chemical structure, and it is not clear how the gross chemical description of heparin relates to the matter which is crystallising. Thus, the overall composition of sulphate ester per disaccharide in the sample is 2.4 and not 3 as assumed in the calculations and a certain partial occupancy of both N-acetyl and glucuronic acid residues is likely.

Finally, the fact that models with both "up" and "down" pointing chains can give reasonable agreement with the X-ray data in positions which are sterically viable suggests that a statistical distribution of chain senses is a possibility which must be further explored when better X-ray data for heparin becomes available.

CONCLUSION. Structure factor calculations using 0.3

Table IV. Residual Indices for the Three Heparin Conformational Models Which Best Match Observed X-Ray Data

Model		First 6-Sulphate Conformation	Second 6-Sulphate Conformation	R	R''
1U	$^{1}C_{4}$ ↑	gg	gg	33.1	42.6
1D	$^{1}C_{4}$ ↓	gt	gt	32.4	36.7
3U	$^{1}S_{3}$ ↑	gt	gg	33.6	39.1

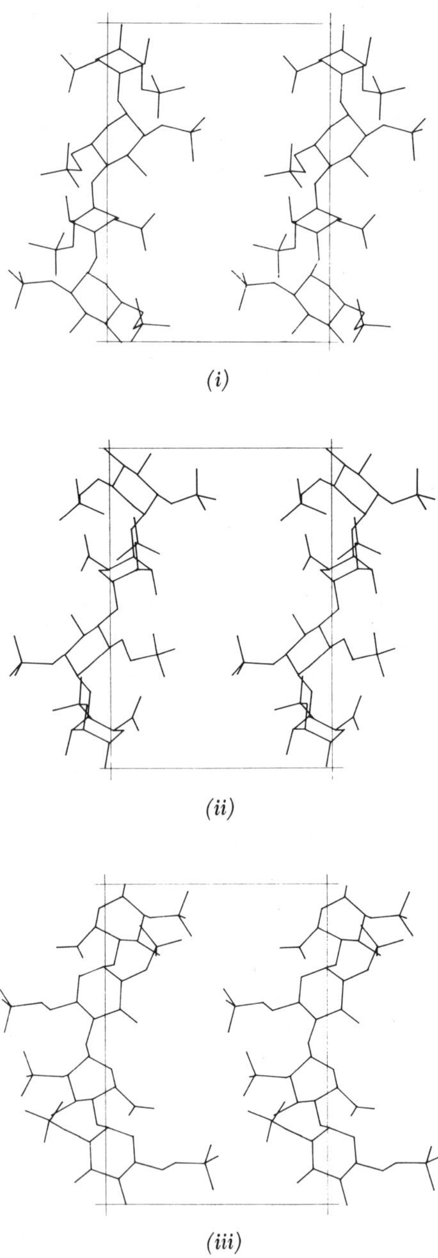

Figure 4. Projections (b c) of the possible heparin structures which give the best agreement with the x-ray data. (i) Model 1U (top); (ii) Model 1D (middle); (iii) Model 3U (bottom).

nm X-ray intensity data were unable to distinguish convincingly between the $^{1}C_{4}$ and $^{1}S_{3}$ residue conformation for the sulphated -*L*-iduronate in heparin.

It is hoped that higher resolution X-ray and NMR data may be able to resolve what must currently remain an open question.

Acknowledgements. The suthors would like to thank the Science Research Council, the Arthritis and Rheumatism Council and the Wellcome Trust for research support.

Abstract. A wide variety of heparins in the sodium salt form yield a characteristic X-ray pattern with a triclinic unit cell having dimensions, *a* = 1.102 nm, *b* = 1.201 nm, *c* (fibre axis) = 1.680 nm, = 90.0°, = 116.2°, and = 107.3°. The molecular conformation of such heparins has been studied by packing analysis and X-ray structure factor methods, including linked-atom least-squares refinement against the 0.3 nm data.

The heparin conformations resulting from the incorporation of the sulphated -*L*-iduronate residue in either of the two chairs $^{4}C_{1}$ and $^{1}C_{4}$ or the skew-boat $^{1}S_{3}$ conformation were examined. Those conformations with the sulphated -*L*-iduronate residue in the $^{4}C_{1}$ chair form could be excluded, but models containing $^{1}C_{4}$ and $^{1}S_{3}$ conformers were able to fit the packing and X-ray structure data satisfactorily and it did not prove possible to discriminate between them.

Literature Cited

1. Olivecrona, T., Egelrud, T., Iverius, P. H., Lindahl, U., Biochem. Biophys. Res. Commun. (1971) 43, 524.
2. Rosenberg, R. D., Damus, P. S., J. Biol. Chem. (1973) 248, 6490.
3. Taylor, R. L. Shively, J. E., Conrad, H. E., Cifonelli, J. A., Biochemistry (1973) 12, 3633.
4. Hook, M., Lindahl, U., Iverius, P. H., Biochem. J. (1974) 137, 33.
5. Nieduszynski, I. A., Atkins, E. D. T., Biochem. J. (1973) 135, 729.
6. Atkins, E. D. T., Nieduszynski, I. A., Horner, A.A. Biochem. J. (1974) 143, 251.

7. Atkins, E. D. T., Laurent, T. C., Biochem. J. (1973) 133, 605.
8. Atkins, E. D. T., Isaac, D. H., J. Mol. Biol. (1973) 80, 773.
9. Arnott, S., Guss, J. M., Hukins, D. W. L., Mathews, M. B., Biochem. Biophys. Res. Commun. (1973) 54, 1377.
10. Nieduszynski, I. A., Atkins, E. D. T., in "Structure of Fibrous Biopolymers" E. D. T. Atkins and A. Keller, Eds., Colston Papers No. 26, Butterworths, London, 1975, p. 323.
11. Perlin, A. S., Casu, B., Sanderson, G. R., Johnson, L. F., Canad. J. Chem. (1970) 48, 2260.
12. Carbohydrate nomenclature committee, J. Chem. Soc., Chem. Commun. (1973) p. 505.
13. Atkins, E. D. T., Phelps, C. F., Sheehan, J. K., Biochem. J. (1972) 128, 1255.
14. Arnott, S., Wonacott, A. J., Polymer (1966) 7, 157.
15. Arnott, S., Scott, W. E., J. Chem. Soc. Perkins Trans. II (1972), 324.
16. Ramachandran, G. N., Ramakrishnan, C., Sassisekharan, V., in "Aspects of Protein Structure", G. N. Ramachandran, Academic Press, 1963, p. 121.
17. Hendrickson, J. B., J. Am. Chem. Soc. (1961) 83, 4537.
18. Hamilton, W. C., Acta Cryst. (1965) 18, 502.
19. Atkins, E. D. T., Nieduszynski, I. A., Fed Proc. (1976), In press.

7

Hyaluronic Acid Conformations and Interactions

W. T. WINTER, J. J. CAEL, P. J. C. SMITH, and STRUTHER ARNOTT
Department of Biological Sciences, Purdue University, West Lafayette, IN 47907

For some time now we have been interested in glycosaminoglycan conformations, the factors controlling selection of particular conformational modes, and the possible relationships of structure and structural changes to biological function. In this paper we would like to present our results with the least complex of the glycosaminoglycans, hyaluronic acid.

Hyaluronic acid is found in most connective tissues with the largest amounts in the softer tissues e.g. Wharton's jelly, vitreous and synovial fluids. The principle activities in which hyaluronic acid is suspected to play a significant role are morphogenetic regulation (1) and shock absorbancy (2). Chemically, hyaluronic acid is a polydisaccharide -(A-B-)$_n$ where A is D-glucuronic acid and B is 2-acetamido-2-deoxy-D glucose. The glycosidic linkages A-B and B-A are β(1→3) and β(1→4) respectively as shown in Figure 1. Linkages in dermatan, chondroitin and keratan sulfates are geometrically equivalent to these and we therefore expect many of our observations with hyaluronic acid will obtain to these sulfated polysaccharides. Light scattering results of Mathews as reported by Swann (3), viscosity (3) and magnetic resonance (4) all suggest the existence, in dilute solution, of stiff segments, involving more than 50% of the residues, stabilized by cooperative long range interactions extending over several residues. The existence of such segments implies that considerable conformational regularity persists in solution. Since hyaluronic acid concentrations in soft connective tissue are sufficiently large that the calculated excluded volume exceeds the total volume of solution (5), it seems reasonable that more compact and orderly structures also pertain *in vivo*. We have used X-ray diffraction from oriented films to investigate the nature of such ordered conformations,

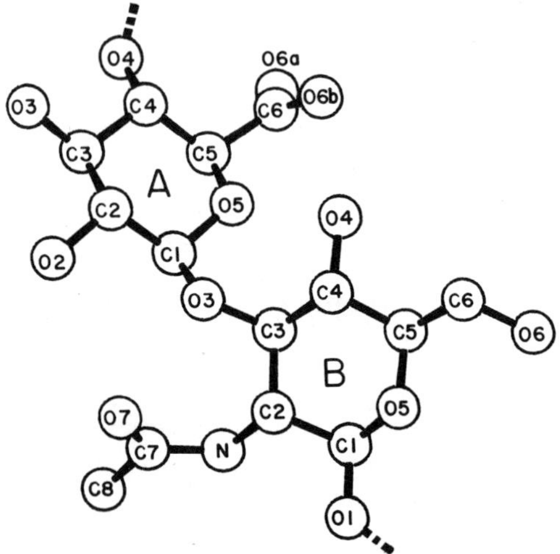

Figure 1. Chemical structure of the repeating unit in hyaluronates. A *is* D *glucuronate and* B *is 2-acetamido-2-deoxy-*D *glucose.*

the forces which stabilize them and the various factors influencing conformation and spatial organization.

Hyaluronate Geometries

Studies in our laboratory and in Bristol by E.D.T. Atkins and co-workers suggest that hyaluronates exhibit considerable conformational versatility (Figure 2). The different conformers are most simply classified in terms of their helix symmetry (n) and axial periodicity per disaccharide unit (h). These parameters are readily obtainable from the observed intensity distribution and layer line separation (Figure 2). Table I summarizes these parameters for the known hyaluronate conformations.

TABLE I. Summary of Observed Hyaluronic Acid Conformations

Helix Symmetry[a]	Disaccharide pitch (nm)	Conformational Effector	Reference
2_1	0.98	free acid	13
3_2	0.95	divalent cation	7,8,13,18
4_3	0.85	monovalent cation	6,20,21
	0.93	chondroitin-6-sulfate (≃10%)	21,22
6_4	0.91	hemiprotonated	*b*

[a] Helix symmetries are assigned assuming a regular helix of disaccharides. In some instances such as calcium hyaluronate or sodium hyaluronate, at high humidities the crystal symmetry indicates that successive disaccharides are not exactly equivalent.

[b] This paper.

Conformational Features

We have examined several packing variants of each of the two principal hyaluronate conformational motifs, the sinuous 4-fold forms with h = 0.85nm (6) and the extended 3-fold conformers with h = 0.95nm (7,8). In each case we have succeeded in defining the structural

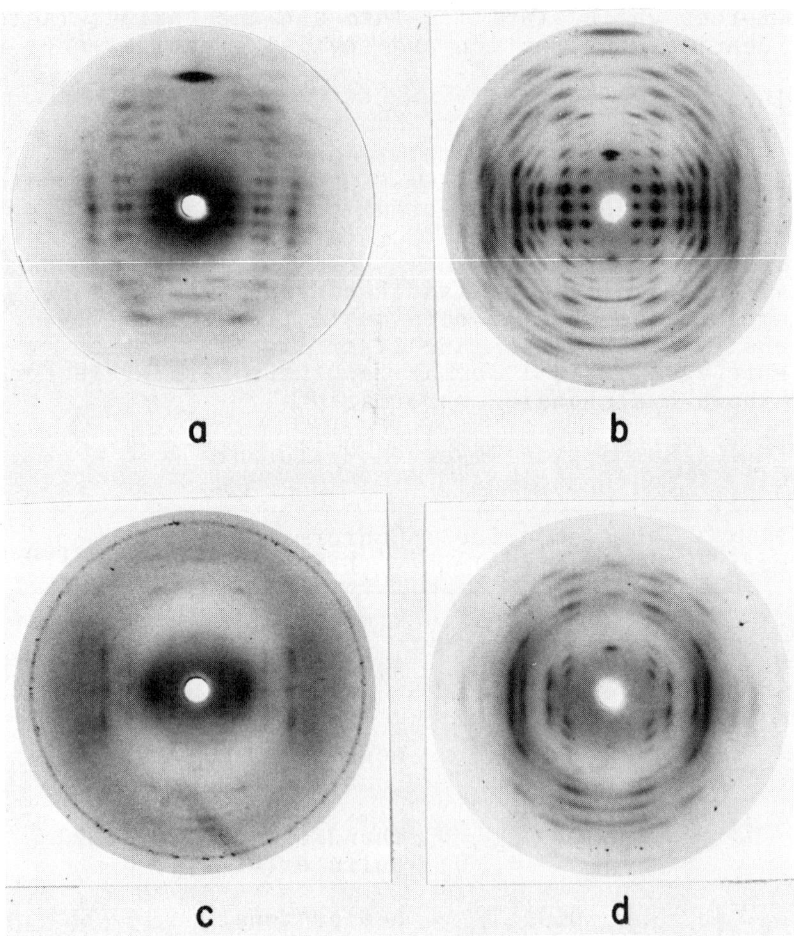

Figure 2. Representative x-ray diffraction patterns from hyaluronates: (a) three-fold helical form obtained from monovalent salts containing traces of divalent cation. Two chains pass through a unit cell with dimensions a = b = *1.70 nm,* c = *2.850 nm, and* γ = *120°; (b) three-fold helical calcium hyaluronate. Six chains pass through a unit cell with dimensions* a = b = *2.09 nm,* c = *2.83 nm, and* γ = *120°; (c) "six-fold" hemiprotonated potassium salt. Six chains pass through a unit cell with dimensions* a = b = *2.12 nm,* c = *5.47 nm, and* γ = *120°; and (d) four-fold sodium salt. Two chains pass through a cell with dimensions* a = b = *0.99 nm,* c = *3.39 nm, and* γ = *90°.*

details of the polyanion geometry and packing, the specific sites occupied by cations, and the locations of at least those water molecules which appear to be most highly ordered. Within each structural family we observe that packing variations only perturb the detailed molecular conformation but do not alter the general structural conclusions to any serious degree.

In all cases the polyanion chains exhibit left-handed chirality and are packed so that nearest neighbor chains are antiparallel. As shown in Figure 3, both conformations permit the formation of O(3)^{A}-H---O(5)^{B} and O(4)^{B}-H---O(5)^{A} intramolecular hydrogen bonds across the β(1→4) and β(1→3) linkages. Side chain orientations are similar to those observed in related carbohydrate materials with hydroxymethyl groups oriented either *gt* or *gg* and the acetamido side chain rotated 20 to 40° from eclipsing H(2)^{B}. This arrangement places the plane of the acetamido group nearly perpendicular to the fiber axis but does not permit the kind of continuous sequence of

```
- N - H---O = C -
              |
              N - H---O = C -
```

hydrogen bonds found in chitin (9) and N-acetyl-α-D-glucosamine (10,11).

A second view of the two conformational variants (Figure 4) reveals several interesting differences. First, the chains of the compact 4-fold form are highly interdigitated and there are opportunities for direct intermolecular hydrogen bonding involving each chain with its four nearest neighbors and even with its four next nearest neighbors at the adjacent unit cell corners. In the 3-fold forms the individual chains are tightly wound about their respective helix axes and each successive disaccharide acts as a donor in only one direct intermolecular hydrogen bond. All of the hydrophilic groups are distributed near the molecular periphery and under high humidity conditions (ambient relative humidity greater than 70%) hydrogen bonding between chains proceeds largely through water bridges.

The second major difference between these two structures involves the position and orientation of the carboxylate groups together with their associated counter-cation. In the 4-fold form (6) the charged groups are all located less than 0.15nm from the helix axis. This geometry, in addition to providing maximum separation of similarly charged groups, also results in the formation of charge stacks along the helix axis. With the 3-fold forms the radial coordinates of

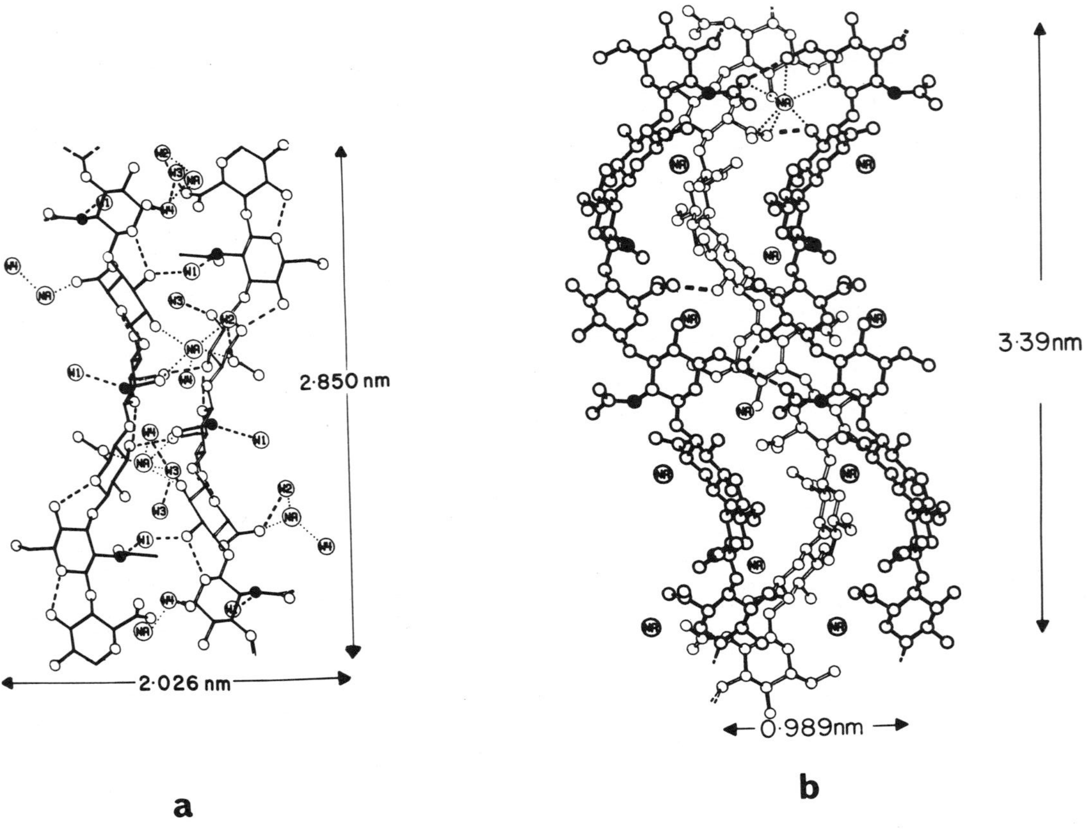

Figure 3. The (a) regular three-fold and (b) compressed regular four-fold hyaluronates viewed in a direction normal to the helix axis

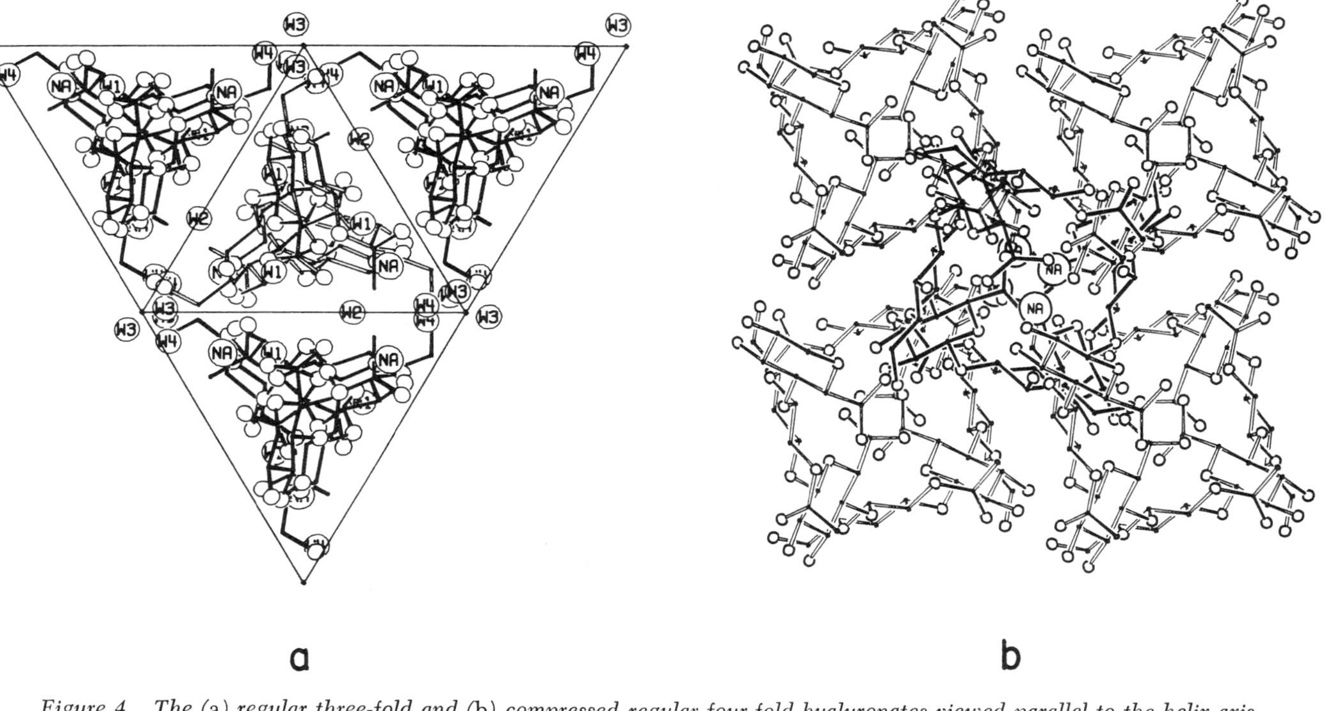

Figure 4. The (a) regular three-fold and (b) compressed regular four-fold hyaluronates viewed parallel to the helix axis

carboxylate oxygens increased by more than 0.06nm and the C-O vectors are directed outward from the helix axis. Nearest neighbor carboxylate oxygens are now only 0.5nm apart. This is exactly the geometry one would expect if the two neighboring carboxylates were sharing a divalent cation such as calcium. Detailed refinement of the calcium hyaluronate structure (Figure 5) indicates only small changes in the conformation and cooperative binding of one calcium by two disaccharides from different chains in exactly the expected manner (8).

Cation and Water Binding by Hyaluronates

In condensed systems such as oriented films and fibers, hyaluronate can bind monovalent or divalent ions to specific sites. Experimentally such sites are usually readily apparent in a 3-dimensional Fourier synthesis using phases derived from a refined model of the polyanion alone and amplitudes equal to difference of the observed and calculated structure amplitudes (Figure 6). In every case to which this procedure has been applied, we observed strong positive peaks within 0.3nm of a carboxylate oxygen. Subsequently these structures of the ion binding region and the ion located could be optimized subject to two criteria:

1) an improved agreement between calculated and observed structure amplitudes and
2) a reasonable set of metal-oxygen interactions coupled with the absence of short non-bonded contacts.

Our procedures for satisfying these requirements have been discussed elsewhere (6,12) and we shall not belabor the point here. Similar procedures have been employed in locating ordered water molecules.

Our conclusions that the cations and at least some water molecules are specifically site bound are reinforced by two important experimental observations. First and most significant is the appearance of additional layer lines in diffraction patterns obtained from hemi-protonated materials. This coupled with the continued presence of 0,0,3n meridional reflections clearly demonstrates that the repeating unit is no longer a disaccharide but a tetrasaccharide of sequence -(AH-B-A^-M^+-B)-. If the ions were randomly distributed through the unit cell or even over specific sites, the repeating unit seen by X-ray diffraction would have remained a disaccharide plus a (now fractional) metal ion.

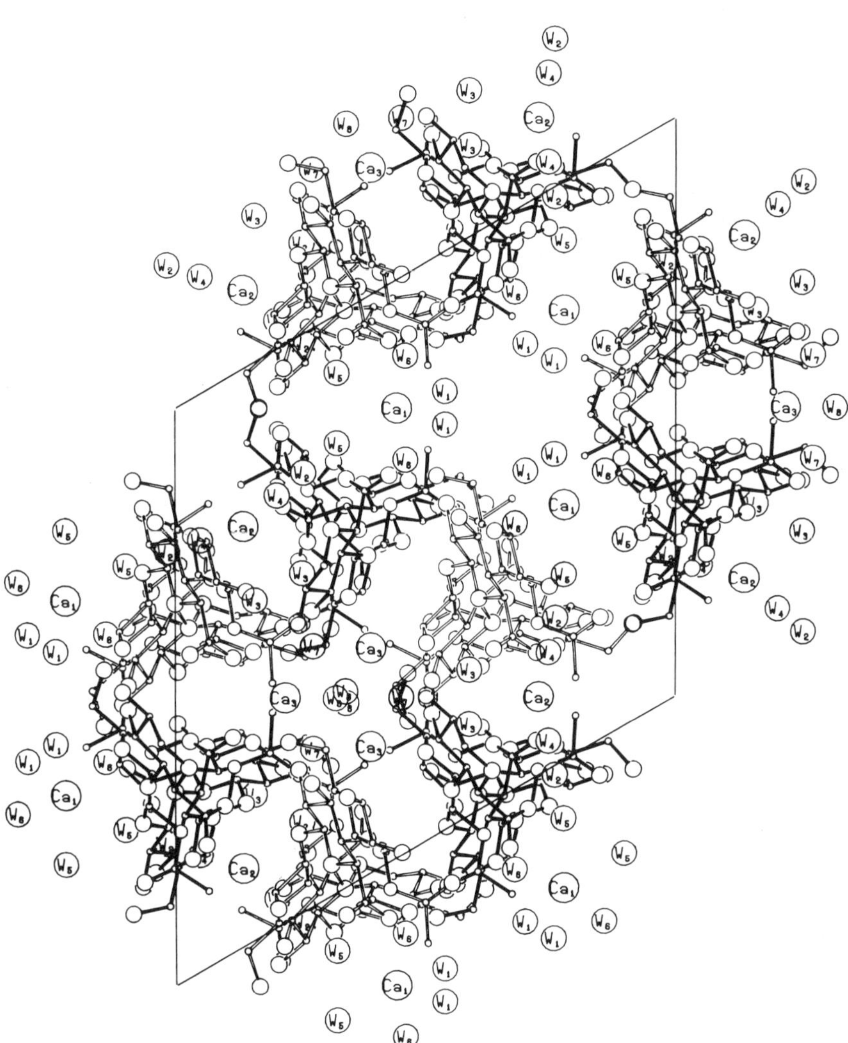

Figure 5. Calcium hyaluronate unit cell viewed parallel to the helix axis. Ca1, Ca2 and Ca3 are independent calcium ions; W1-W7 are water molecules all of which participate in calcium coordination.

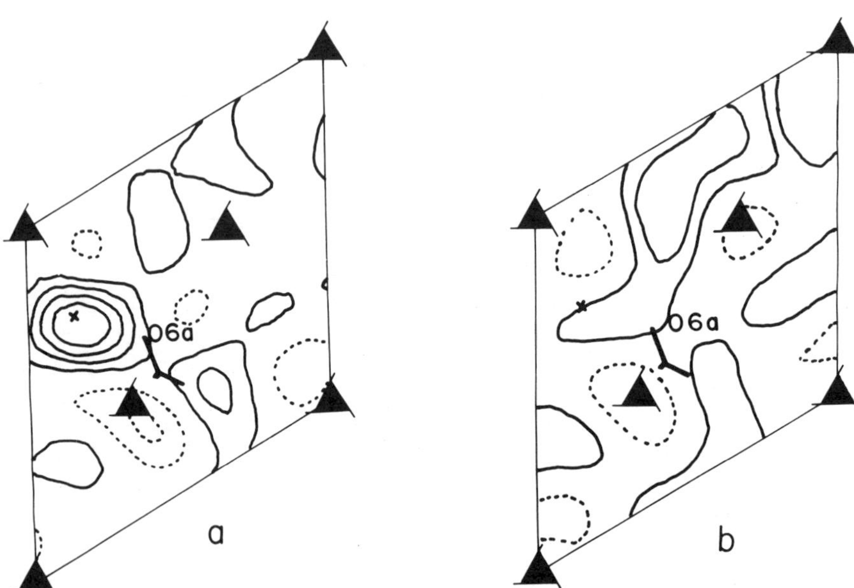

Figure 6. The z = *0.033 section of Fourier difference syntheses for three-fold sodium hyaluronate (a) before, and (b) after location of the sodium ion. The coefficients* $F_o - F_c$ *are the difference between observed and calculated structure amplitudes, and the phases are those of the* F_c *model. Solid contour lines denote regions of positive electron density,* i.e., *regions where the data suggest the presence of additional scattering material not defined in the model. Dotted lines denote regions of negative electron density. The final position of the sodium (x) is at* z = *0.056, and the carboxylate group to which it binds is shown in heavier black line.*

The regular removal of every second cation coupled with some shift in the carboxyl orientation at those sites does account for the apparent doubling of the helix pitch.

Secondly, oriented hyaluronates exhibit, upon drying, dramatic changes in the observed intensities far in excess of the changes predicted by recalculation of structure amplitudes for the shrunken cell. These changes together with the constancy of the helix parameters *n* and *h* are exactly what one would expect from a system where the solvent filled interchain gaps rather than participate in the molecular structure.

The geometry of cation binding in hyaluronates is largely predictable from crystal structures of small molecules. Sodium ions occupy central locations in distorted octahedra where the Na^{+}...O distances are all 0.25 ± .02nm. In all cases at least one carboxylate oxygen is involved in cation binding. Additionally O(2) of the uronic acid and the acetamido carbonyl oxygen are involved either directly or through water bridges in binding univalent metal ions.

In calcium hyaluronate the calcium ions occupy special positions on diad axes where each ion interacts in an identical manner with two polyanion chains. In the highly hydrated form we have studied the two carboxylate groups contribute one oxygen each to the calcium binding shell. Six water molecules act as bridges between each calcium ion and the polyanions and complete a distorted square antiprism with calcium at the center.

Induced Conformational Changes

The appearance of particular hyaluronate conformations appears to be regulated by the presence of certain conformation directing factors in the local milieu. Protonation of the uronic acid, for example, (unlikely *in vivo*) induces the formation of extended 2-fold conformations in hyaluronic acid (13), chondroitin-4-sulfate (14), and dermatan sulfate (15, 16).

Highly purified hyaluronate which is free of divalent cation typically exists as a 4-fold helix with *h* = 0.850nm (6). It is possible to induce a second 4-fold conformation with *h* = 0.93nm by addition of 10% chondroitin-6-sulfate to the hyaluronate solution followed by film casting and orientation in the usual manner (22). In this instance the presence of a minor but conformationally restricted component induces a conformational change in the major component.

For some time now we have been disturbed by our inability to induce transitions between the 3-fold and the compact 4-fold (h = 0.85nm) forms. We are now of the opinion that, in the condensed state, it is not possible to induce such changes in the direction from 3-fold to 4-fold forms and that the reverse can be done only by addition of calcium salts. Similarly, the condensed state conformation for any given sample depends largely on the specific techniques employed during its isolation and purification. Our first clue to the importance of removing all calcium in producing the more compressed forms was that solutions of hyaluronate dialyzed against calcium salts invariably resulted in 3-fold helical structures. Secondly, when we obtained what was at first considered to be a 3-fold sodium hyaluronate, elemental analysis indicated the continued presence of a small quantity of calcium in the sample (7).

The significance of calcium in inducing the 3-fold conformation is apparent from our knowledge of the cationic binding sites. Only the 3-fold form permits two carboxylates to participate cooperatively in the binding of a single ion.

The question then arose as to why calcium persisted in some samples of hyaluronate but not others. Careful examination of the chemical pedigree in every case where complete preparative details were available pointed to a single factor, the use of cetylpyridinium chloride (CPC) precipitation. Without exception every sample where this technique was employed resulted in the 4-fold conformer and those where it was not employed yielded 3-fold conformers.

Aldrich (17) noted that calcium persisted in hyaluronate preparations even after 12 successive dialyses *vs.* 0.2 *M* NaCl and that the last vestiges of calcium were removable only by CPC precipitation. Swann (19) using gentle mechanical and extractive procedures isolated a highly viscous hyaluronate from rooster comb. Again CPC precipitation resulted in a 4-fold decrease in limiting viscosity of this material without decreasing the molecular weight determined by sedimentation equilibrium. In our own experience (7) the presence of less than 3% of the available cation as divalent ion is sufficient to induce the formation of the 3-fold conformation.

From these experiments it seems quite clear that the nature and distribution of other ionic and polyionic species exerts considerable conformational control on the polyanion.

Whether or not any of these conformations or the factors inducing transitions between them are biologically significant remains speculative. It is clear, however, that in those tissues where calcium is associated with hyaluronate, the 3-fold conformations offer the best models of any *in vivo* conformational ordering.

Acknowledgements

We wish to thank Dr. M.B. Mathews and colleagues of the University of Chicago and Dr. E.A. Balazs, Columbia University, New York, for their generous supplies of material. The work at Purdue was supported by grants to one of us (SA) from the National Science Foundation (7420505) and National Institutes of Health (GM 20612). One of us (WTW) was a fellow of the Arthritis Foundation.

Abstract

Hyaluronic acid is a regularly alternating copolymer of N-acetyl-β-D-glucosamine (A) and β-D-glucuronic acid (B) where the linkages A-B and B-A are (1→4) and (1→3) respectively. Examination of oriented films of a variety of hyaluronate salts by X-ray fiber diffraction reveals a broad spectrum of conformation and packing motifs. Using the linked-atom least-squares procedure in which molecular and crystal models are refined simultaneously against X-ray intensity and stereochemical data, we have obtained detailed models for regular 3- and 4-fold sodium hyaluronate and calcium hyaluronate. In each case it has been possible to define the helical conformation, side chain orientations, location of counter cations and the nature of their interaction with the polyanion chains and to propose reasonable intra- and intermolecular hydrogen-bonding. In some cases we have also located small quantities of ordered solvent. The models are described in some detail and with considerable emphasis on the effects of ionic environment and hydration on conformation and association. It is concluded that divalent ions provide the driving force for inducing 3-fold conformations.

Literature Cited

1. Toole, B.P., Jackson, G. and Gross, J. Proc. Natl. Acad. Sci. USA (1972) 69,1384-1386.

2. Balazs, E.A. and Gibbs, D.A. in Chemistry and Molecular Biology of the Intercellular Matrix, Balazs, E.A., Ed., New York, NY (1970).
3. Swann, D.A. in Chemistry and Molecular Biology of the Intercellular Matrix, Balazs, E.A., Ed., New York, NY (1970).
4. Darke, A., Finer, E.G., Moorhouse, R. and Rees, D.A. J. Mol. Biol. (1975) 99,477-486.
5. Ogston, A.G. in Chemistry and Molecular Biology of the Intercellular Matrix, Balazs, E.A., Ed., New York, NY (1970).
6. Guss, J.M., Hukins, D.W.L., Smith, P.J.C., Winter, W.T., Arnott, S., Moorhouse, R. and Rees, D.A. J. Mol. Biol. (1975) 95,359-384.
7. Winter, W.T., Smith, P.J.C. and Arnott, S. J. Mol. Biol. (1975) 99,219-235.
8. Winter, W.T. and Arnott, S. J. Mol. Biol. (1977) (in press).
9. Gardner, K.H. and Blackwell, J. Biopolymers (1975) 14,1581-1595.
10. Johnson, L.N. Acta Crystallogr. (1966) 21,885-891.
11. Mo, F. and Jensen, L.H. Acta Crystallogr. sect. B (1976) 31,2867-2873.
12. Smith, P.J.C. and Arnott, S. Acta Crystallogr. (1977) (submitted).
13. Atkins, E.D.T., Phelps, C.F. and Sheehan, J.K. Biochem. J. (1972) 128 ,1255-1263.
14. Isaac, D.H. and Atkins, E.D.T. Nature New Biol. (1973) 244,252-253.
15. Arnott, S., Guss, J.M., Hukins, D.W.L. and Mathews, M.B. Biochem. Biophys. Res. Commun. (1973) 54,1377-1383.
16. Atkins, E.D.T. and Isaac, D.H. J. Mol. Biol. (1973) 80,773-779.
17. Aldrich, B.L. Biochem. J. (1958) 70,236-244.
18. Sheehan, J.K., Atkins, E.D.T. and Nieduszynski, I.A. J. Mol. Biol. (1975) 91,153-163.
19. Swann, D.A. Biochem. Biophys. Res. Commun. (1969) 35,571-576.
20. Dea, I.C.M., Moorhouse, R., Rees, D.A., Arnott, S., Guss, J.M. and Balazs, E.A. Science (1973) 179,560-562.
21. Atkins, E.D.T. and Sheehan, J.K. Science (1973) 179,526-564.
22. Arnott, S. and Winter, W.T. Fed. Proceed. (1977) (in press).

8

Studies on the Crystalline Structure of β-D-(1→3)-Glucan and Its Triacetate

TERRY L. BLUHM and ANATOLE SARKO

Department of Chemistry, State University of New York, College of Environmental Science and Forestry, Syracuse, NY 13210

Polysaccharides containing the β-(1→3) linkage are found widely distributed in nature. Many of these molecules contain glucose as the principal sugar, and many are nearly completely linear homopolymers of (1→3)-linked β-D-glucose. Typical examples are laminarin of brown seaweeds (1), paramylon of green algae (2), curdlan produced by bacteria (3), along with pachyman (4), and lentinan (5), found in fungi. The function of these glucans appears to be one of reserve polysaccharides. However, a number of β-(1→3)-glucans have been associated with structural functions, for example, a cell wall component isolated from yeasts (6); pachyman; and a fibrillar polysaccharide found in the pollen tube wall of the lily, *Lilium longiflorum* (7). Other examples of this polysaccharide, with as yet undetermined functions, are callose (8) of plants and laricinan (9), found in compression wood.

In view of this diversity of functions, it became of interest to us to determine the crystalline structure of β-D-(1→3)-glucans from different sources. It had previously been found that this polysaccharide does exist natively in crystalline form, as for example, in the lily pollen tube wall (7) and some fungal cell wall tissue (10). Because of the similarity of x-ray diffraction patterns of the glucan with those of β-D-(1→3)-xylan (11), whose structure has been determined to be a triple-stranded helix, it has been suggested that the crystalline structure of the glucan may also be triple helical (10). The same conclusion was suggested by earlier computer model building (12).

Because our attempts to obtain good quality x-ray fiber patterns from both pachyman and lentinan were not successful (probably because of the branching present in these molecules), our initial studies were directed to the determination of the crystal structure of pachyman triacetate which gave acceptable x-ray diffraction diagrams. From this structure we hoped to obtain sufficient molecular information that would help in the characterization of the parent polysaccharide. Information of this type has previously been found valuable, as for example, in the study of V-amylose which followed that of amylose triacetate.

Experimental

Preparation of pachyman triacetate and suitable film samples for x-ray diffraction have been previously described (13). Two crystalline polymorphs were obtained, depending both on the degree of stretching of the film samples and the temperature of stretching. The low stretch polymorph (PTA I) was obtained at 125°C and with degree of stretching not exceeding 50%, while the high stretch polymorph (PTA II) was obtained after stretching PTA I to 300% extension at 215°C. Their x-ray diffractograms will be shown elsewhere (14).

Attempts to convert fibrous specimens of PTA to the corresponding crystalline parent polysaccharide by solid state deacetylation, failed. Similarly, attempts to orient film samples of pachyman did not result in acceptable x-ray diffraction diagrams. The experiments performed on lentinan, a more linear β-D-(1→3) glucan of the fungus *Lentinus elodes*, were more succesful, particularly so in preparing fibrous specimens from the gel state. The latter was obtained by first dissolving 2% (w/w) of lentinan in 1 N NaOH overnight at room temperature. The alkaline solution was dialyzed two days against distilled water. A small amount of the resulting gel was then placed between two glass beads, which were immediately separated a short distance in a fiber stretching clamp, and left to dry. The resulting fiber diagram (shown elsewhere (15)) was consistent with the hexagonal unit cell $a=b=15.8$Å, c (fiber repeat) = 5.95Å, previously proposed for the β-D-(1→3)-glucan of *A. mellea* (10).

Methods of Conformational Analysis

In the case of pachyman triacetate for which good quality diffractograms were available, the structure analysis took the form of a stereochemical conformation and chain packing refinement, combined with the analysis of diffraction intensities. The method of stereochemical refinement has been previously described in detail (16) and its application to the structure solution of pachyman triacetate has likewise been described in a preliminary publication (13). Details of the crystal structure of both PTA I and PTA II and the stereochemical refinement method will be described in a forthcoming publication (14). Briefly, both the conformation and the unit cell packing of pachyman acetate chains were determined by refining a variable chain model of the molecule against acceptable stereochemical criteria, using as variables all bond lengths, bond angles and conformational angles of the monomer residue and both intra- and intermolecular nonbonded contacts. The criteria were average values for the above variables as determined from known carbohydrate crystal structures

(16, 17). The best models thus determined were further refined against x-ray diffraction intensities.

Because of relatively poor quality of the lentinan diffractograms, similar analysis of its structure could not be undertaken. However, it was attempted, by single chain conformational analysis, to predict likely chain models of lentinan whose conformation would be in agreement with the available x-ray data.

The conformational analysis was essentially of the ϕ-ψ type, in which all possible chain conformations were created by rotations about the two bonds to the glycosidic oxygen. The torsional angles ϕ and ψ refer to rotations about the bonds C(1)-O and C(3')-O, respectively.* The conformation of the glucose residues was assumed to be in the C1 chair, and was left invariant in the process. The O(6) hydroxymethyl group was placed in the gg rotational position** and was also left invariant. The coordinates used were those of the average β-D-glucose residue of Arnott and Scott (17) and the bond angle at the glycosidic oxygen was fixed at 118°. Each conformation was evaluated by calculating its nonbonded conformational potential energy (18). The details of the method will be described elsewhere (15).

Results and Discussion

The crystal structures of both PTA polymorphs revealed a common chain conformation of the molecule in that both structures were based on a right-handed, sixfold helix. The differences between them in terms of chain conformation were slight, residing mainly in the axial repeat per residue, h, with 3.73 Å for PTA I and 3.10 Å for PTA II. This difference was sufficient to change the packing and the dimensions of the unit cell. Although in both polymorphs the chains were nearly hexagonally close-packed, as expected, the polarity of chains was antiparallel in PTA I (again, as expected from a solution crystallized sample), whereas in PTA II it was an equal mixture of parallel and antiparallel packing. Because the latter polymorph was a high extension sample, it was thought probable that the reason for this change in polarity was the severe extension of an already crystalline structure. This, in turn, would lead to molecular extension and consequent parallel packing.

In other respects, the chain conformation of PTA was analogous to other known polysaccharide acetates, such as amylose triacetate (19), xylan diacetate (20) and mannan triacetate (21). In all of these structures, the rotational positions of the acetate substituents were such as to nearly eclipse the carbonyl double bond with the C-H bond of the substituted ring atom. The O(6) acetates showed more rotational freedom, as in PTA I the position

*For definition of $(\phi,\psi) = (0,0)$ see caption of Fig. 3.

**This position results when the C(6)-O(6) bond is *gauche* to both the C(5)-O(5) and C(5)-C(4) bonds.

was _tg_ and in PTA II it was either _tg_ or _gg_ depending on packing polarity. From the point of view of information most useful for the analysis of the structure of the parent glucan, the right-handedness of the PTA helix may be the most important feature.

The details of the structures of PTA I and PTA II will be described in a separate publication (14). The chain conformation and nearest neighbor chain packing for PTA I are illustrated in a stereo view in Fig. 1.

As indicated in the Experimental Section, the analysis of the structure of lentinan was approached in a different manner. This was necessitated by the lack of resolution in the x-ray diffraction diagram. The x-ray diagrams were similar under all conditions of relative humidity that were tested, i.e. under vacuum and at all relative humidities between 50% and 100%. The diagrams were not as sharp as, but were very similar to, the diffractograms obtained by Jelsma and Kreger from oriented fungal tissue of _A. mellea_ and were in agreement with the hexagonal unit cell proposed for it (10). The diffractograms contained two well oriented reflections on the equator at _d_-spacings of ~ 14.5 Å and ~ 6.7 Å, and a spread of intensity on the first layer line at approximately _d_=4.5 Å. The _A. mellea_ diagram showed a resolution of the 6.7 Å equatorial reflection into a doublet and a resolution of the first layer intensity into four distinct reflections (10). From lentinan diffractograms a fiber repeat of approximately 6 Å was measured. The lentinan diffractograms were also similar to those obtained from curdlan by Scott and Rees (12) and from _Lilium longiflorum_ by Herth _et al_ (7). It thus appears that all of the β-(1→3) glucans that have been studied to date crystallize in the same hexagonal unit cell.

The volume of this unit cell and the observed density of 1.54 g/cc are in agreement with six glucose residues per cell which results in a calculated density of 1.23 g/cc. The addition of 12 water molecules would bring the calculated and observed densities into good agreement.

The detailed results of the conformational calculations for single and multiple helices of the β-(1→3)-glucan molecule, in terms of potential energy maps, will be shown elsewhere (15). In these calculations, it was found that acceptable conformations for lentinan existed as single helices, parallel and antiparallel double helices as well as triple helices. Whether any of the allowed helices correspond to the observed fiber repeat is determined by the helix parameters _n_ and _h_. Further, any of the conformations satisfying the fiber repeat must also satisfy the observed density. For a single helix the fiber repeat calculated from _n_ and _h_ should agree with the 6 Å repeat measured from the x-ray diffractograms. If the structure were to be composed of double helices with parallel strands, rules of symmetry specify that the odd order layer lines from the single helices would be absent. Therefore, the molecular fiber repeat would be

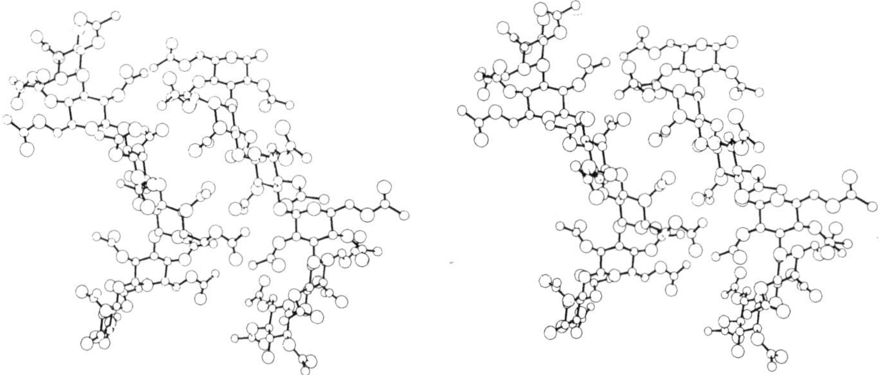

Figure 1. Stereo view of the two chains of the unit cell of PTA I

twice the apparent crystallographic repeat, i.e. ~ 12 Å. In double helices with antiparallel strands, the odd order layer lines would be weakened or unobservable, since the electron density in parallel and antiparallel polysaccharide chains is similar. If the odd order layer lines were indeed unobserved, the actual fiber repeat would be twice that measured from the diffractogram. For this reason, conformations resulting in a fiber repeat of ~ 6 Å as well as ~ 12 Å must be examined. For triple helices with all parallel strands only every third layer would be present and the molecular repeat would be three times the observed crystallographic repeat, i.e. ~ 18 Å. The application of this criterion of fiber repeat coupled with the density calculation eliminated most of the conformations in the allowed regions for all types of helices. Five conformations survived, of which one was a single helix, one a double parallel stranded helix, one double antiparallel stranded helix, and two were triple helices.

The allowed single helix is right-handed and is described by helical parameters _n_=6 and _h_=0.98 Å, therefore, its repeat is 6 x 0.98 = 5.88 Å. Its conformation is stabilized by inter-residue O(2)---O(2') hyrdrogen bonds and there are no short non-bonded contacts present within the helix. It packs reasonably well into a hexagonal unit cell with side dimensions of 15.8 Å.

The allowed double parallel stranded helix is left-handed with helical parameters _n_=6 and _h_=2.0 Å for each strand thus giving a molecular repeat of 12 Å. It contains a rather long inter-residue O(2)---O(2') hydrogen bond of 2.98 Å but no inter-strand hydrogen bonds are present. It does not pack well into the unit cell as the single helix.

The case of the double antiparallel stranded helix is similar to that of the parallel double helix. Only one conformation in the energetically allowed region was found to fit the fiber repeat and density data. In that conformation each strand is right-handed and has helical parameters _n_=6 and _h_ = 1.83 Å, i.e. molecular repeat ~ 11 Å. No interstrand hydrogen bonds exist but, again, an O(2)---O(2') intrachain hydrogen bond 3.06 Å in length can be formed between contiguous residues. However, this double helix could not be packed into the unit cell without difficulty.

Of the two allowed triple helices, one is right-handed and has helical parameters _n_=6 and _h_=2.9 Å, i.e. molecular repeat of 17.4 Å. The strands are held together by a network of O(2)---O(2) interchain hydrogen bonds of length 2.85 Å between equivalent residues. The other allowed helix is left-handed and is characterized by parameters _n_=6 and _h_=3.3 Å (repeat = 19.8 Å). The strands are held together by the same interchain hydrogen bond network as in the right-handed helix. Both of these models pack extremely well into the unit cell.

Of the above five allowed conformations only the double helices could be eliminated by packing restrictions, leaving one single and two triple helices as probable models of the structure.

Further characterization of these models was undertaken by the calculation of the Fourier transform of the helix, using the equation:

$$F(R,\psi,Z) = \sum_{n}\sum_{j} f_j J_n(2\pi R r_j)\exp\{i\,[n(\psi-\phi_j+\pi/2) - 2\pi Z z_j]\} \quad (1)$$

where $F(R,\psi,Z)$ is the Fourier transform at the point in reciprocal space with cylindrical coordinates R,ψ,Z, f_j is the atomic scattering factor and r_j, ϕ_j, z_j are the cylindrical polar coordinates of the jth atom. $J_n(X)$ is the Bessel function of order n and argument X. The transform was calculated for the rotational coordinate from 0° to 60° at 10° increments and for the Z coordinate of all required layer lines. Even though the double helices were previously eliminated from consideration on the basis of packing difficulties, their transforms were also calculated and are shown along with the transforms for the other helices in Fig. 2. (Only the transforms with the ψ coordinate in best agreement with experimental intensities are shown).

For the single helix model, no agreement could be found between the location of possible scattering intensity on the first layer line and the intensity observed on the x-ray diffractogram. In the latter, intense scattering occurs on the first layer line at a position of R=0.22 $Å^{-1}$. The helix transform showed only the possibility of very weak intensity at that location for all rotational positions of the single helix. It is also clear from Fig. 2 that both double helices show serious disagreements between the calculated transforms and the experimentally observed intensity distribution. On the other hand, both left- and right-handed triple helices gave rise to Fourier transforms that were in excellent agreement with the observed intensity distribution on both the equator and the first layer line.

It is thus apparent that the only probable conformation for the crystalline β-(1→3)-glucan of lentinan, curdlan, *A. mellea*, *Lilium longiflorum*, and perhaps others, is a triple helix, much like the corresponding structure of β-(1→3)-xylan. Because of the limited x-ray data available, the chirality of the helix cannot be determined, but both left- and right-handed triple helices are surprisingly alike. Both are stabilized by an interstrand network of O(2)---O(2) hydrogen bonds, almost identical with those found in the structure of β-(1→3)-xylan. The additional hydroxymethyl group of the glucan appears to have no influence on the chain conformation as it lies entirely on the outside of the helix. Because the β-(1→3)-xylan helix is right-handed, as is pachyman triacetate, it is probable that the triple helix of the glucan is also right-handed. A projection of the right-handed helix is shown in Fig. 3.

SINGLE 6_1 HELIX

L=1

L=0

0.45

R(Å^{-1})

DOUBLE -6_1 2 HELIX (PARALLEL)

L=1

L=0

0.45

R(Å^{-1})

DOUBLE 6_1 2 HELIX (ANTIPARALLEL)

L=1

L=0

0.45

R(Å^{-1})

TRIPLE -6_1 3 HELIX

L=1

L=0

0.45

R(Å^{-1})

TRIPLE 6_1 3 HELIX

L=1

L=0

0.45

R(A^{-1})

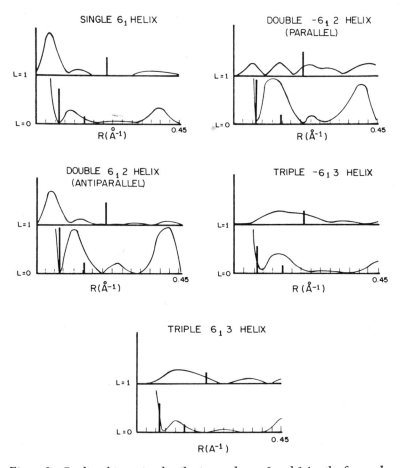

Figure 2. Predicted intensity distribution on layers 0 and 1 for the five probable conformations of β-D-(1 → 3)-glucan, as calculated from Equation 1. The ψ coordinate for all cases is 0°. The vertical lines indicate the position and relative magnitude of observed intensities.

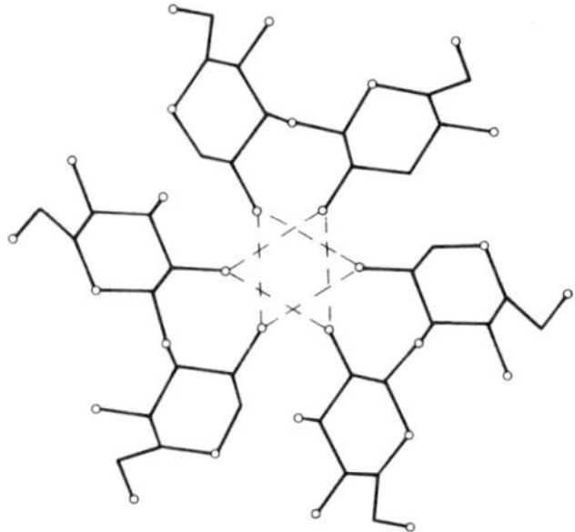

Figure 3. Projection in the x–y plane of the first two residues of each strand of the right-hand triple helical glucan

Abstract

The crystal structure of lentinan, a fungal β-D-(1→3)-glucan, was studied by conformational analysis coupled with limited x-ray diffraction data, and with the aid of x-ray diffraction analysis of pachyman triacetate. The latter is a fully acetylated derivative of another fungal β-D-(1→3)-glucan. Two crystalline polymorphs of pachyman triacetate were obtained, both based on right-handed sixfold helical conformations. The polymorphs differed in fiber repeats but exhibited typical features of polysaccharide acetates. Based on the analysis of ϕ,ψ maps of single, double-and triple-stranded helices and the calculation of Fourier transforms, it was concluded that the only probable structure of crystalline lentinan would be a triple-stranded helix. In that respect the structure would be similar to the corresponding xylan which has previously been determined to be a triple helix. It is likely that all β-D-(1→3)-glucans studied to date possess the same crystal structure.

Acknowledgment.

This work has been supported by the National Science Foundation grant No. MPS7501560. We thank J. Hamuro of Ajinomoto Co., Inc., Kawasaki, Japan, for the sample of lentinan and Dr. T. E. Timell of this College for the sample of pachyman.

Literature Cited

(1) Aspinall, G. O. "Polysaccharides", pp. 74-80, Pergamon Press, Oxford (1970).
(2) Percival, E. and McDowell, R. H. "Chemistry and Enzymology of Marine Algal Polysaccharides," pp. 196-197, Academic Press, London (1967).
(3) Harada, T. Misaki, A. and Saito, H. Arch. Biochem. Biophys. (1968) 124, 292-298.
(4) Hoffman, G. C., Simson, B. W. and Timell, T. E. Carbohyd. Res. (1971) 20, 185-188.
(5) Chihara, G., Maeda, Y., Sasaki, T., Fukuoka, F. and Hamuro, J. Nature (1969) 222, 687-688.
(6) Manners, D. J. and Masson, A. J. Fed. Bur. Biochem Soc. Lett. (1969) 4(2), 122-124.
(7) Herth, W., Franke, W. W., Bittiger, H., Kuppel, A. and Keilich G. Cytobiologie (1974) 9, 344-367.
(8) Aspinall, G. O. "Polysaccharides," p. 74, Pergamon Press, Oxford (1970).
(9) Hoffmann, G. C. and Timell, T. E. Svensk Papperstidn. (1972) 75, 135-145.
(10) Jelsma, J. and Kreger, D. R. Carbohyd. Res. (1975) 43, 200-203.

(11) Atkins, E. D. T. and Parker, K. D. J. Polym. Sci., Part C (1969) 28, 69-81.

(12) Rees, D. A. in "M. T. P. International Review of Science: Organic Chemistry, Series 1, Vol. 7, Carbohydrates," Aspinall, G. O. (ed.), p. 251, Butterworths, London (1973). Scott, W. E. and Rees, D. A., private communication.

(13) Bluhm, T. L. and Sarko, A. Biopolymers (1975) 14, 2639-2643.

(14) Bluhm, T. L. and Sarko, A. Biopolymers (to be published).

(15) Bluhm, T. L. and Sarko, A. Can. J. Chem. (to be published).

(16) Zugenmaier, P. and Sarko, A. Biopolymers (1976) 15, 2121-2136.

(17) Arnott, S. and Scott, W. E. J. Chem. Soc., Perkin Trans. II (1972) 2, 324-335.

(18) Blackwell, J., Sarko, A. and Marchessault, R. H. J. Mol. Biol. (1969) 42, 379-383.

(19) Sarko, A. and Marchessault, R. H. J. Amer. Chem. Soc. (1967) 89, 6454-6462.

(20) Gabbay, S. M., Sundararajan, P. R. and Marchessault, R. H. Biopolymers (1972) 11, 79-94.

(21) Bittiger, H. and Marchessault, R. H. Carbohyd. Res. (1971) 18, 469-470.

9

Conformation and Packing Analysis of Polysaccharides and Derivatives

Detailed Refinement of Trimethylamylose

P. ZUGENMAIER, A. KUPPEL, and E. HUSEMANN[1]

Institut fur Makromolekulare Chemie der Universität Freiburg, D7800 Freiburg i.Br., Germany (BRD)

Trimethylamylose (TMA) is an α-(1→4) linked methylated D-glucan that can be prepared by homogeneous methylation of commercially available amylose (1). It is known that the α-(1→4) linkage in the polysaccharide backbone results in a flexible chain conformation that gives rise to different crystalline polymorphs. For example,amylose, the unsubstituted polysaccharide, crystallizes in at least three polymorphs, known as A-, B-, V-amyloses. Of these, only the structure of V-amylose (2)is characterized in detail at the present time. It was originally thought that V-amylose crystallizes as a sixfold helix with the same symmetry. Not until recently, however, was a twofold screw axis along its chain established, thus indicating that three chemically identical residues comprised one crystallographic asymmetric unit, but with nonidentical conformations. Nonetheless, the conformation of the V-amylose backbone can be described as an approximate sixfold helix. Another polymer of the α-(1→4) linkage type which has been characterized in moderate detail, is amylose triacetate (ATA)(3). Its chain conformation can best be described by a 14_{11} helix, that is, a left handed helix with 14 residues in 3 turns. The rise per residue for V-amylose and for ATA are quite different, as are the conformation angles responsible for their helical structures. Therefore, it was of interest to us to determine the chain conformation of TMA, and especially the influence of the methyl substituents on the conformation of its backbone.

Although excellent fiber X-ray diagrams of TMA have been obtained in this laboratory (4), it was not until recently that an attempt could be made to solve the crystal structure. First, difficulties were encountered in establishing a unique unit cell and its symmetry. These difficulties were overcome only when electron diffraction patterns became available from TMA single crystals. Secondly, it is only recently that a sophisticated modeling method became available to deal simultaneously with conformation and packing of polysaccharide derivatives in which

[1] Deceased November 9, 1975.

simultaneous rotation of all substitutent groups could be handled. This was necessary in order to obtain a phasing model for refinement against X-ray data.

Experimental

Trimethylamylose was prepared by homogeneous methylation of commercial AVEBE amylose. Suitable fibers for X-ray measurements were obtained by stretching TMA films, cast from methylene chloride, about 300 % in a stream of hot air at 60°C. Such fibers showed perfect orientation but low crystallinity. Annealing at 220°C for about 10 minutes improved crystallinity dramatically but did not affect the orientation of the crystallites. A typical X-ray fiber diagram of an annealed sample taken with a cylindrical camera of 5.73 cm radius is shown in Fig. 1. In this diffraction photograph the meridional reflections are scarcely detectable due to the nearly perfect orientation of the crystallites. When the fiber was tilted into the correct positions, all even numbered meridional reflections up to $\underline{l}$ = 10 were observed while all odd numbered were absent. The fiber repeat distance $\underline{c}$ was determined from this pattern but no unique solution for the other unit cell parameters was found. Therefore, attempts were made to solve this problem by electron diffraction of polymer single crystals. The latter were obtained by two methods. First, TMA was dissolved in diethyleneglycol (5g/l), heated to 240°C and then slowly crystallized at 180° - 150°C. In the second method, TMA was dissolved in a solvent-nonsolvent mixture of dioxane (2g/l) and 60 % octane and then crystallized at room temperature. The single crystals obtained by these two methods showed different morphologies (cf. Fig. 2a, b). Crystallization at room temperature resulted in spherulitic crystals. The outer parts of the spherulites showed single crystals represented in Fig. 2a. Crystallization at elevated temperatures resulted in lamellar type crystals which are shown in Fig. 2b. The corresponding X-ray powder diagrams are shown in Fig. 2c and d, respectively. Only the single crystals grown at elevated temperatures showed a diffraction intensity distribution which corresponded to the fiber X-ray pattern of Fig. 1. TMA single crystals grown at room temperature gave a quite different pattern and it has been shown (4) that this results from dioxane having been incorporated in the crystal lattice.

Electron diffraction patterns obtained from TMA single crystals grown at elevated temperatures are schematically represented in Fig. 2e.

The unit cell parameters $\underline{a}$, $\underline{b}$ and $\gamma = 90^{o}$ could now be absolutely determined with the help of an internal standard. The limiting planes of the lamellae of Fig. 2b could now also be determined and indexed with (100) representing the width of the crystals while the two planes representing the front edges in some of the lammelae (e.g. lower left corner) are ($\bar{1}$10) and (110). With $\underline{a}$ and $\underline{b}$ lying in the depicted plane it became clear that the chains of

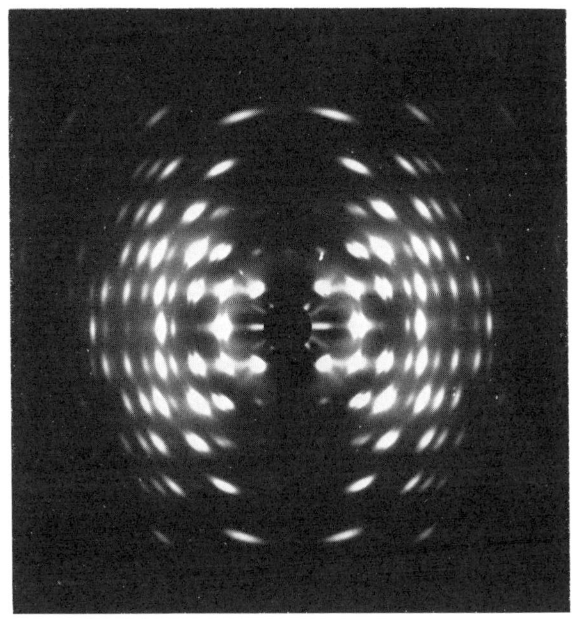

Figure 1. Fiber diffraction pattern of TMA taken with a cylindrical camera of 5.73-cm radius

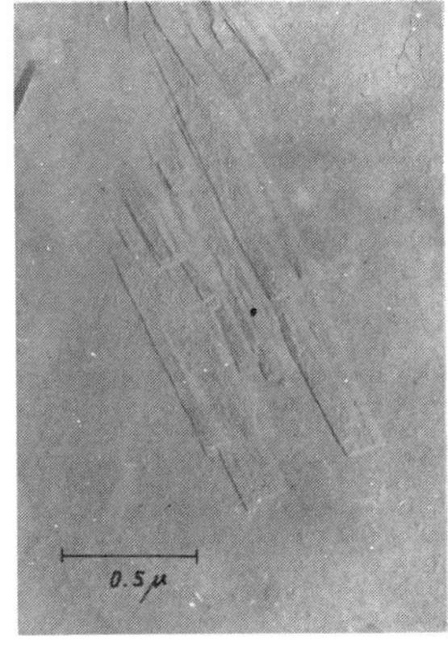

Figure 2a. Electron micrograph of single crystals of TMA grown at room temperature in a solvent–nonsolvent mixture of dioxane and 60% octane

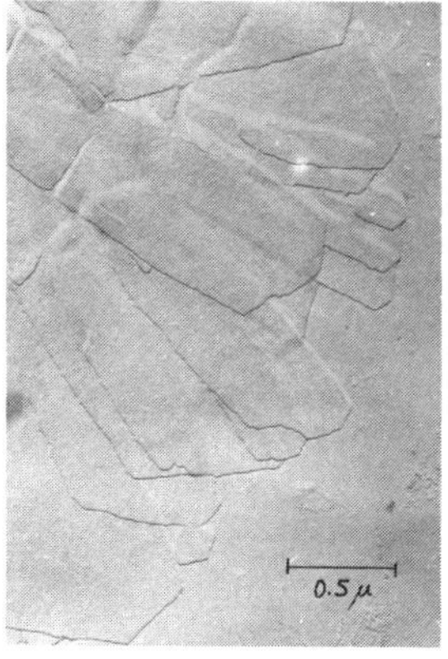

Figure 2b. Electron micrograph of single crystals of TMA grown at 180°–150°C in diethyleneglycol

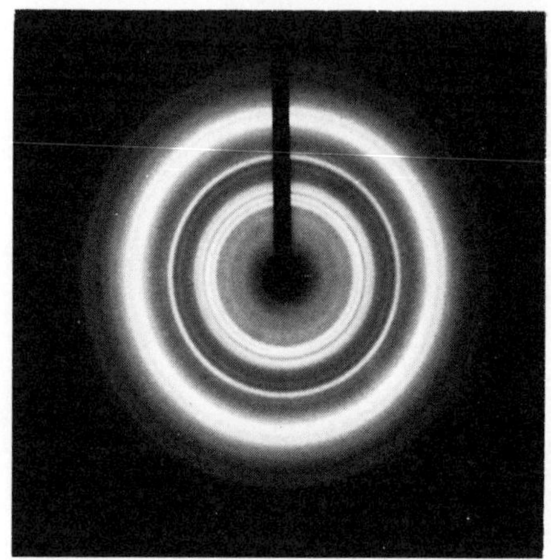

Figure 2c. Powder x-ray pattern of single crystals of Figure 2a on a flat film

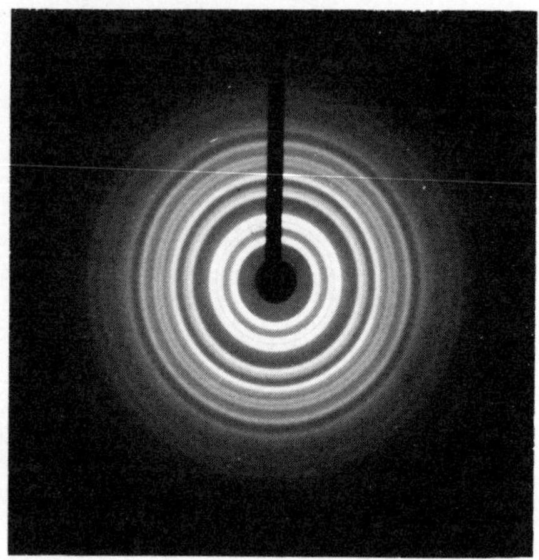

Figure 2d. Powder x-ray pattern of single crystals of Figure 2b on a flat film

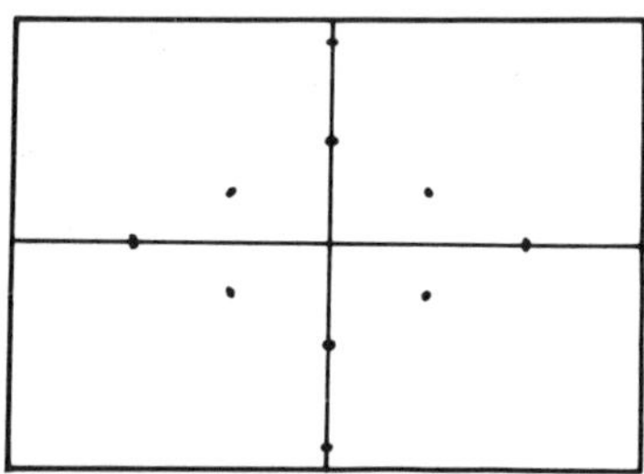

Figure 2e. Schematic representation of the electron diffraction pattern obtained from a single crystal shown in Figure 2b

TMA are perpendicular to that plane. The electron diffraction patterns also suggested that twofold screw axes were present along both $\underline{a}$ and $\underline{b}$. With values of $\underline{a}$ and $\underline{b}$ from electron diffraction and $\underline{c}$ from X-ray fiber diagrams, the reflections of the fiber pattern of Fig. 1 could now be indexed with an orthorhombic unit cell. Its parameters were refined by a least squares procedure and are $\underline{a}$ = 17.24 ± 0.01 Å, $\underline{b}$ = 8.704 ± 0.009 Å, $\underline{c}$ (fiber repeat) = 15.637 ± 0.008 Å at room temperature. The measured density is ρ = 1.20 g/cm^3. With twofold screw axes in all three directions $\underline{a}$, $\underline{b}$ and $\underline{c}$ the most probable space group is $\underline{P2_12_12_1}$. Therefore, the unit cell contains segments of two chains running through it in an antiparallel fashion, as demanded by space group symmetry, with four trimethylglucose monomer units per fiber repeat.

The relative X-ray reflection intensities were obtained from cylindrical film diagrams taken in an evacuated camera. The intensity was recorded along each layer line with a Joyce-Loebl recording densitometer. Areas of individual peaks were measured by planimetry and used as a measure of uncorrected relative intensity. Intensities thus obtained were corrected for Lorentz (5) and polarisation factors, arcing of reflections, unequal film-to-sample distances of diffracted rays and were then converted to relative structure amplitudes. In those instances where observed intensities were actually a composite of contributions from two or more unique diffraction planes, the value of the calculated structure amplitude was taken as

$$|F_c| = (\sum_i m_i F_{ci}^2)^{1/2} \quad (1)$$

with $\underline{m_i}$ the multiplicity and the summation being over all planes contributing to the composite. The structure amplitudes of unobserved reflections were assigned one half of the minimum observable intensity in the corresponding region of diffraction angle.

Results and Discussion

Stereochemical Model Analysis. The method of generating models of helical structures has been described in previous papers (5,6). A flexible ring conformation was introduced which allowed variation, when necessary, in bond lengths , bond and conformation angles, using as the refinement criterion the optimization of the function (2):

$$Y = \sum_{i=1}^{N} STD_{oi}^{-2} (A_i - A_{oi})^2 + W^{-2} \sum_{\substack{i=1 \\ j=1}}^{n} w_{ij} (d_{ij} - d_{oij})^2 \quad (2)$$

The first term in this function represents the bonded and the second term the nonbonded interactions, with:

A_i any calculated bond length, bond or conformation angle of the molecule;

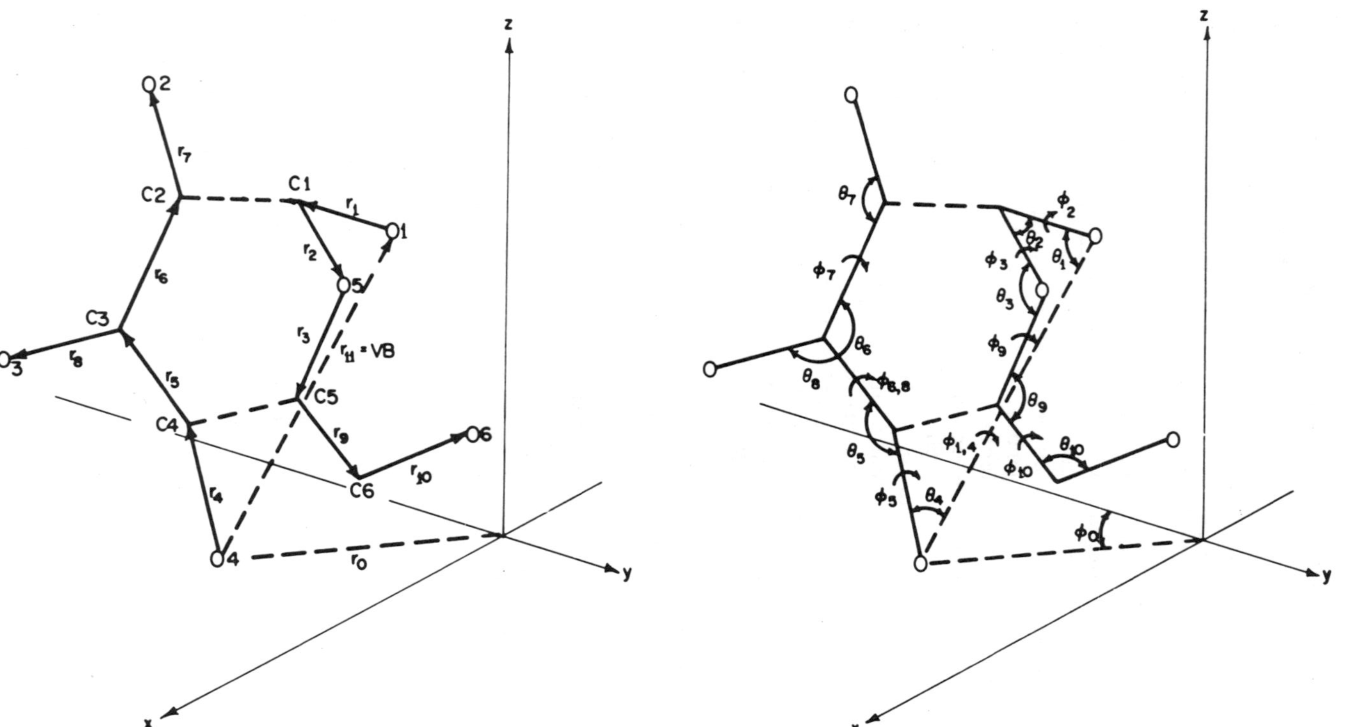

Figure 3. Bond lengths (r_i), bond angles (θ_i), and conformation angles (ϕ_i) required for the description of the α-D-glucose residue. Hydrogen atoms are not shown. The angle ϕ_o measures rotation of the entire helix about its axis, and the distance r_o is constant for a given helix and virtual bond length.

A_{oi} average or standard value of A_i;
STD_{oi} weight or standard deviation for the average value A_{oi};
N number of optimization parameters selected;
d_{oij} nonbonded equilibrium distance between atoms *i* and *j*;
d_{ij} nonbonded distance between atoms *i* and *j* (only repulsive interactions are used, that is, the second term is set to zero if $d_{ij} > d_{oij}$);
w_{ij} the weight factor for the atom pair *i*, *j*;
W overall weight factor of nonbonded interactions;
n number of nonbonded contacts considered.

The actual constants used for this calculation are summarized in reference 2.

The strategy used was to establish first a suitable chain conformation to be refined later for optimal packing. For the description of the chain two sequences of atoms, as shown schematically in Fig. 3, are needed. The open bonds and angles, i. e. the bond C(1)-C(2) and the angles C(3)-C(2)-C(1) and C(2)-C(1)-O(5) are constrained to stay within given limits. As a trial model for the α-(1→4) linked glucan any α-D-glucose residue in the Cl chair conformation may serve, as the latter is thus far the only conformation found in amylose and its derivatives. Pendant atoms and branches are added to the monomer residue to complete the chain description. The model is then refined by minimizing the function *Y*. Models with high values of *Y* and with many short contacts below the established limits are disregarded.

For TMA with four residues per fiber repeat different fourfold helices were considered; however, all but a 4_3 helix had to be disregarded. The O(6) substituent could be placed either in the vicinity of *gt* or *tg* rotational position but not near *gg*. (For O(6) rotation nomenclature see Ref. 8). As the glycosidic angle between adjacent residues is varied by turning the complete residue around the virtual bond (cf. Fig. 3), more than one solution for a certain range of the glycosidic angle can exist for every type of the helix, as shown in Fig. 4.

Two rotational positions ϕ_4 for the vector r_4 (see Fig. 3), can be found which result in a glycosidic angle θ of about 120° for 4_3 helix of TMA. One solution of the two can be ruled out because of short nonbonded contacts between adjacent residues. Of all the fourfold helices which were tested in this manner only one solution was found for the backbone conformation. Even when the fourfold symmetry was removed leaving only a two fold screw axis as indicated by the X-ray data, the minimum energy conformation stayed very close to a 4_3 helix (9).

In the second refinement step, the allowed chain conformation with the substituents in all possible rotational positions were packed within the unit cell, in agreement with space group $P2_12_12_1$. This symmetry limited the possible chain arrangements considerably, by imposing antiparallel packing of chains and twofold screw axes in all three directions of the unit cell. The best packing of the chains was sought by first rotating and translating a fixed

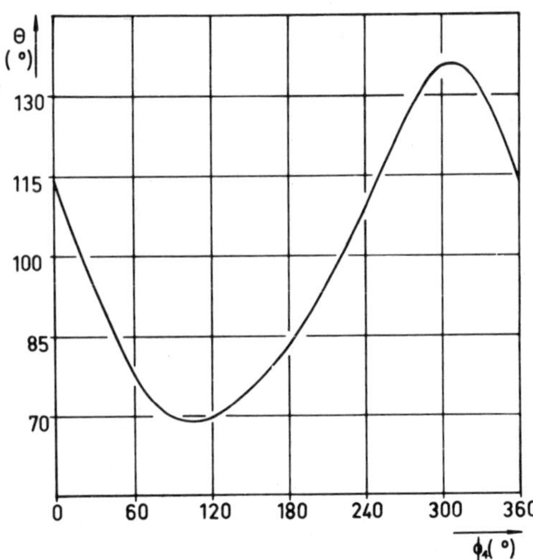

Figure 4. Glycosidic angle θ (C(1) — O(4)2 — C(4)2) as a function of rotation of the complete residue (ϕ_4) around the virtual bond vector of a 4_3 helix of TMA

chain backbone with rotational refinement of the substituents, but omitting methyl hydrogens at this stage. Very little further change was observed in the chain backbone when the conformation was subsequently simultaneously refined with packing.

In the final stage of packing analysis, the TMA structure was refined by optimizing the virtual bond length (i.e., the distance between two adjacent glycosidic oxygens O(4)...O(4)2) and by introducing the methyl hydrogens. The ether bridge angle and oxygen-carbon distances of the substituents were kept at constant values. The optimal virtual bond length was found to be 4.45 Å.

The completed analysis of TMA revealed almost identical rotational positions for the C(2M) and C(3M) methyls of two adjacent residues, but differences appeared in the rotational positions of the O(6) groups. The O(6) of one residue was near the tg position whereas the O(6) of the second residue appeared near gt. The packing analysis favored only this one unique solution.

X-ray Intensity Analysis. The starting model for the structure refinement against X-ray data was the model obtained from the above conformation and packing analysis. The same variable parameters were used as in stereochemical refinement, but the function to be minimized was now the crystallographic disagreement index

$$R = \frac{\sum \left| |F_o| - |F_c| \right|}{\sum |F_o|} \qquad (3)$$

where $|F_o|$ and $|F_c|$ are the observed and calculated structure amplitudes, respectively. In addition, an anisotropic temperature factor was introduced in this stage of refinement. The strategy of refinement was the same as previosly described (2). Only minor differences between the starting model and the refined model resulted, but a stereochemically unacceptable rotational position of the methyl carbons, which were rotated within a 10° range away from the best packing model, was one of the differences.

Therefore, in order to maintain the stereochemical criteria for all rotatable substituents, the final packing model was kept fixed and only the helix was allowed to shift along the c axis and to rotate around it during the final R-index refinement. The coordinates of one asymmetric unit (two residues) of the resulting best model are listed in Table I and the model is shown in Fig. 5. Because a 4_3 helix but with different positions of substituent atoms had been assumed as a model in order that it could be handled in the present stage of calculation, the coordinates of the backbone have to be considered as averaged values with respect to two consecutive residues. The polar coordinates are also listed in Table I, according to the residue definition of Fig. 3. Significant differences in the rotational positions of the two residues are only observed for the O(6) substituents and for the rotation of the methyl hydrogens. The corresponding bond lengths, bond angles and conformation angles are shown in Table II. The deviation from the average values as a result of the refinement are shown in parenthesis. None of the bond lengths exceeded one standard deviation and only three bond angles exceeded the standard deviation of 1.5^{o}, which included the glycosidic bond angle of 121.5^{o}. Major changes for conformation angles occurred with respect to the position of two adjacent residues as compared with V-amylose (2). The shortest inter- and intramolecular contacts are listed in Table III and all are larger than the established limits (11).

The observed and calculated structure amplitudes for TMA are listed in Table IV. The R-index obtained with the coordinates of Table I was 0.32 with all reflections included and 0.31 with only the observed reflections. The estimated intensities of the observed even numbered meridional reflections agreed with the calculated ones.

Conclusion

It became clear during the refinement procedure that even though the TMA backbone structure can be approximated by a 4_3 helix, packing of the chain superimposes a twofold axis only. The effect of this is a rotation of consecutive O(6) substituents into two different positions. The same was found to be the case in the structure of V-amylose (2), in which the backbone could be approximated by a 6_5 helix but the three O(6) groups of consecutive residues were all found to be in different rotational positions. Here again a twofold axis is superimposed but three residues form the asymmetric unit. This resulted in both the better

Table I:

Coordinates of the asymmetric unit for trimethylamylose (VB=4.45 Å). The virtual bond vectors are approximated by a 4_3 helix. The polar coordinates of the pendant atoms are different for the two residues, Best <u>R</u> value is obtained for a rotation by 2° and a shift of 1.73 Å along the <u>c</u>-axis. The asymmetric unit has to be shifted by 1/4 <u>a</u> for space group $\underline{P2_12_12_1}$.

	X	Y	Z	r_i	ϕ_i	θ_i	i
		(Å)		(Å)		(deg.)	
O(4)	0.000	-1.503	0.000				
C(1)	-2.540	-0.138	2.949	1.420	-1.7	67.7	1
C(2)	-2.672	-1.612	2.585	1.524	169.1	111.4	6
C(3)	-1.476	-2.055	1.752	1.523	-58.9	105.8	5
C(4)	-1.277	-1.151	0.542	1.431	0	44.2	4
C(5)	-1.190	0.289	1.030	1.430	63.7	113.9	3
C(6)	-1.075	1.267	-0.125	1.518	-176.7	107.6	9
O(2)	-2.717	-2.359	3.796	1.423	-172.1	109.7	7
O(3)	-1.683	-3.395	1.300	1.429	-70.2	108.9	8
O(5)	-2.348	0.649	1.788	1.416	-59.8	113.7	2
O(6)tg	0.278	1.474	-0.552	1.434	151.0	113.0	10
C(2M)	-4.000	-2.311	4.437	1.435	78	113	
C(3M)	-1.093	-4.376	2.165	1.435	144	113	
C(6M)	0.372	2.212	-1.779	1.435	170	113	
H(1)	-3.408	0.179	3.448				
H(2)	-3.555	-1.760	2.037				
H(3)	-0.610	-2.041	2.346				
H(4)	-2.095	-1.252	-0.109				
H(5)	-0.332	0.406	1.623				
H(6A)	-1.500	2.186	0.151				
H(6B)	-1.649	0.922	-0.934				
H(2M1)	-4.716	-1.948	3.761	1.05	15	109.5	
H(2M2)	-3.950	-1.675	5.271				
H(2M3)	-4.271	-3.275	4.751				
H(3M1)	-0.803	-3.920	3.066	1.05	12	109.5	
H(3M2)	-0.252	-4.796	1.698				
H(3M3)	-1.793	-5.133	2.365				
H(6M1)	-0.024	3.174	-1.640	1.05	63	109.5	
H(6M2)	-0.168	1.715	-2.530				
H(6M3)	1.379	2.286	-2.067				
O(4)2	-1.503	0.000	3.909				
C(1)2	-0.138	2.540	6.858				
C(2)2	-1.612	2.672	6.495				
C(3)2	-2.055	1.476	5.661				
C(4)2	-1.151	1.277	4.451				
C(5)2	0.289	1.190	4.939				
C(6)2	1.309	1.052	3.823	1.519	-174.3	107.1	9

(Table I continued)

O(2)2	-2.356	2.747	7.710	1.427	-172.6	107.8	7
O(3)2	-3.394	1.700	5.224	1.426	-70.9	109.5	8
O(5)2	0.649	2.348	5.697				
O(6)2gt	2.611	0.737	4.313	1.426	67.0	112.4	10
C(2M)2	-2.299	4.043	8.324	1.435	78	113	
C(3M)2	-4.372	1.075	6.067	1.435	142	113	
C(6M)2	3.048	-0.578	3.940	1.435	112.8	113	
H(1)2	0.179	3.408	7.357				
H(2)2	-1.760	3.555	5.946				
H(3)2	-2.041	0.610	6.255				
H(4)2	-1.252	2.095	3.800				
H(5)2	0.406	0.332	5.533				
H(6A)2	1.349	1.945	3.271				
H(6B)2	0.997	0.306	3.153				
H(2M1)2	-2.897	4.714	7.782	1.05	74	109.5	
H(2M2)2	-1.304	4.384	8.329				
H(2M3)2	-2.650	3.979	9.311				
H(3M1)2	-4.442	1.601	6.973	1.05	73	109.5	
H(3M2)2	-4.086	0.083	6.259				
H(3M3)2	-5.305	1.080	5.586				
H(6M1)2	3.463	-0.549	2.976	1.05	84	109.5	
H(6M2)2	2.229	-1.235	3.949				
H(6M3)2	3.771	-0.914	4.623				

Polar coordinates missing in the second residue are the same as in the first one.

Table II

Bond lengths, bond angles and conformation angles for the asymmetric unit of Table I. Differences from the average residue values (10) are shown in brackets.

Parameter	First residue		Second residue	
Bond Lengths (Å)				
O(4)-C(4)	1.431	(0.005)		
C(4)-C(3)	1.523	(0.000)		
C(4)-C(5)	1.523	(-0.002)		
C(1)-C(2)	1.524	(0.001)		
C(1)-O(5)	1.416	(0.002)		
C(1)-O(4)2	1.420	(0.005)		
C(3)-C(2)	1.524	(0.003)		
C(5)-O(5)	1.430	(-0.006)		
C(2)-O(2)	1.423	(0.000)	1.427	(0.004)
C(3)-O(3)	1.429	(0.000)	1.426	(-0.003)
C(5)-C(6)	1.518	(0.004)	1.519	(0.005)
C(6)-O(6)	1.434	(0.007)	1.426	(-0.001)
Bond Angles (deg.)				
O(4)-C(4)-C(3)	105.8	(0.3)		
O(4)-C(4)-C(5)	107.6	(-1.0)		
C(3)-C(4)-C(5)	108.3	(-2.0)		
C(4)-C(3)-C(2)	111.4	(0.9)		
C(4)-C(5)-O(5)	111.2	(1.2)		
C(3)-C(2)-C(1)	110.1	(-0.4)		
C(5)-O(5)-C(1)	113.9	(-0.1)		
C(2)-C(1)-O(5)	110.7	(1.5)		
C(2)-C(1)-O(4)2	108.6	(0.2)		
O(5)-C(1)-O(4)2	113.7	(2.1)		
C(3)-C(2)-O(2)	109.7	(-1.1)	110.8	(0.0)
C(1)-C(2)-O(2)	107.9	(-1.4)	107.8	(-1.5)
C(4)-C(3)-O(3)	108.9	(-0.9)	108.5	(-1.2)
C(2)-C(3)-O(3)	109.4	(-0.2)	109.5	(-0.1)
C(4)-C(5)-C(6)	111.7	(-1.0)	113.9	(1.2)
O(5)-C(5)-C(6)	107.6	(0.7)	107.1	(0.2)
C(5)-C(6)-O(6)	113.0	(1.2)	112.4	(0.6)
C(1)-O(4)2 -C(4)2	121.5			

(Table II continued)

Conformation Angles (deg.)

O(5)-C(1)-C(2)-C(3)	54.1	(-1.9)	
C(1)-C(2)-C(3)-C(4)	-53.4	(-0.2)	
C(2)-C(3)-C(4)-C(5)	54.0	(1.0)	
C(3)-C(4)-C(5)-O(5)	-56.3	(-0.9)	
C(4)-C(5)-O(5)-C(1)	60.7	(-0.4)	
C(5)-O(5)-C(1)-C(2)	-58.8	(3.4)	
O(4)-C(4)-C(5)-O(5)	-170.2		
O(4)-C(4)-C(3)-C(2)	169.1		
O(4)2-C(1)-C(2)-C(3)	-71.3		
O(4)2-C(1)-O(5)-C(5)	63.7		
O(5)-C(5)-C(6)-O(6)	151.0		67.0
O(5)-C(1)-O(4)2-C(4)2	56.7		
C(2)-C(1)-O(4)2-C(4)2	-179.6		
C(1)-O(4)2 -C(4)2-C(3)2	88.2		
C(1)-O(4)2-C(4)2-C(5)2	-156.2		
C(4)2-O(4)2-C(1)-H(1)	-62.8		
C(1)-O(4)2-C(4)2-H(4)2	-33.3		

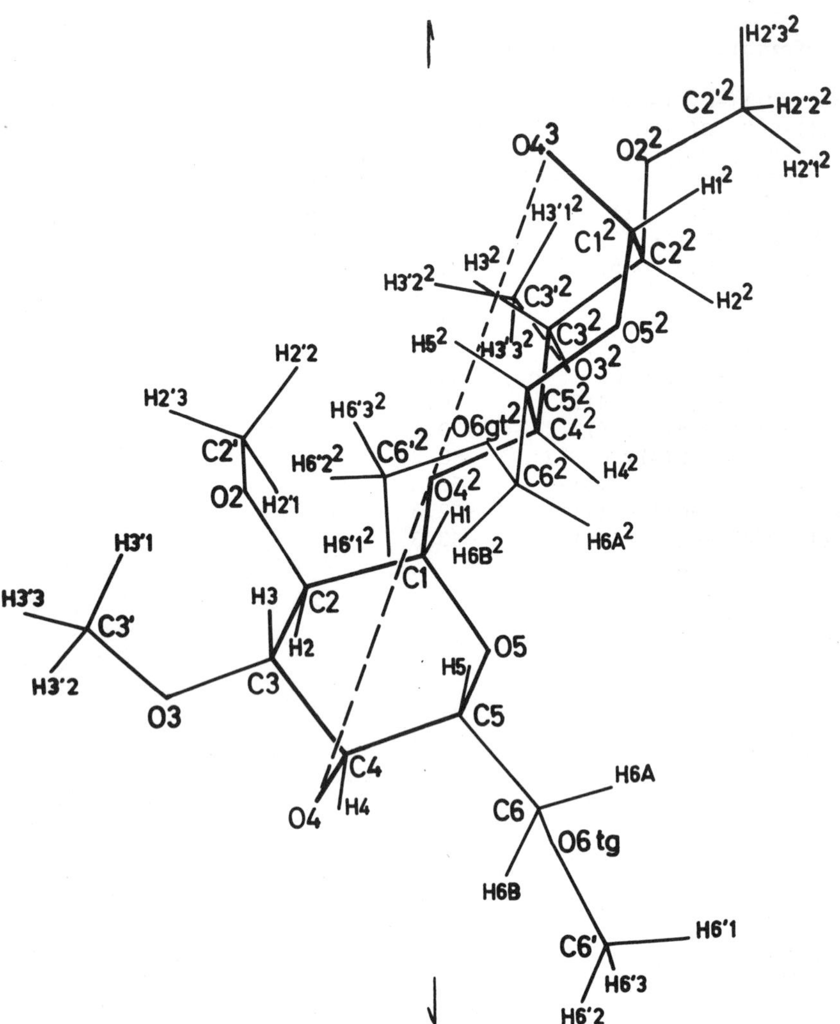

Figure 5. Representation of the asymmetric unit of TMA with all hydrogen atoms included, projected onto y–z plane. Note the different positions for O(6) in the two residues. (The sign ′ is used as abbreviation for M.)

Table III

Short contact table. a) shortest intermolecular contacts between atoms of the asymmetric unit of trimethylamylose. b) shortest contacts between atoms of consecutive residues of the trimethylamylose chain. All distances in Å.

a)	C(2M)	-- O(5)	c	3.24
	H(2M1)	-- O(5)	c	2.66
	H(3M1)2	-- O(6)2	b	2.68
	C(2M)2	-- O(6M)2	b	3.56
	C(3M)	-- H(4)2	a	2.77
	H(3M3)	-- H(4)	a	2.22
	H(2M1)	-- H(6M3)	b	2.24

---

a	x, y-1, z
b	x-1/2, -y+1/2, z+1
c	-x, y-1/2, -z+1/2

b)	O(5)	--	C(4)2	2.99
	C(1)	--	O(3)2	3.05
	H(1)	--	O(3)2	2.34
	C(1)	--	C(3)2	3.19
	H(1)	--	C(4)2	2.70
	C(1)	--	H(4)2	2.71
	H(5)	--	H(6B)2	2.03
	H(2M2)	--	H(3M2)2	2.02

H(5)2	--	H(6B)3	2.22
H(2M2)2	--	H(3M2)3	2.00

Table IV

Observed and calculated structure amplitudes for trimethylamylose with coordinates of Table I.
(B_x=5.5, B_y=4.0, B_z=5.8; temperature factor = $\exp.(-\underline{B}\cdot\sin^2\Theta)$)

hkl	$\lvert F_{obs}\rvert$	$\lvert F_{calc}\rvert$	hkl	$\lvert F_{obs}\rvert$	$\lvert F_{calc}\rvert$
200	11.4	8.9	711	6.6	4.1
110	11.3	10.2	531	+	0.9
210	+	1.2	041,141		
310	4.4	2.1	801	4.4	4.2
400,020	15.7	6.8	721,241		
120	+	2.7	811	4.6	5.2
410,220	+	3.7			
320	3.1	2.2	102	1.3	2.4
510	+	1.9	012,202	4.9	10.0
420,600			112	6.2	5.4
130	8.9	4.3	212	14.2	8.4
230,610			302	6.3	8.1
520	3.3	4.6	312	3.1	3.9
330	3.5	3.3	022,402	13.0	16.7
430,620			122	+	3.6
710	4.5	2.7	222,412	8.4	11.4
530,040			502,322	7.5	6.0
140,800	+	3.5	512,422	+	3.1
720,240			032,602		
810	5.9	3.8	132	+	3.1
			232,612		
101	5.8	9.3	522	7.1	7.0
011,201	11.8	10.9	332	5.0	2.0
111	2.5	4.1	702,432		
211	10.9	7.9	622,712		
301	6.5	8.0	532	+	5.1
311	5.1	7.4	042,142		
021,121			802,722	8.5	6.1
401	12.0	8.1	242,812		
221,411	5.2	7.1	632,342	4.2	8.0
501,321	8.1	5.7			
421	8.5	6.8	103	0.9	4.5
511	2.8	2.9	013,203	2.4	6.0
031,601			113	3.4	7.8
131	4.3	5.7	213	8.1	7.1
231,611			303	7.1	7.3
521	3.3	5.1	313	1.9	1.3
331	4.7	4.1	023,403		
701	+	2.0	123	9.6	9.7
431,621			223,413	6.8	7.7

(Table IV continued)

323,503	6.2	7.4
423,513	3.6	2.7
033,113		
603	5.4	8.5
233,613		
523	5.6	4.1
333	3.8	1.4
703,433		
623,713	5.7	5.7
104	1.0	0.6
014,204	+	0.8
114	6.6	8.3
214	2.4	0.8
304	+	1.4
314	7.9	5.1
024,404		
124	5.8	5.5
224,414	5.8	7.7
324,504	4.7	3.7
514	5.0	3.7
424	2.4	2.2
034,604		
134	5.6	4.0
234,614		
524	3.5	4.9
334	4.4	6.2
704	+	3.1
434,624		
714	8.4	8.1
534	3.7	3.7
044,804		
724,144	5.2	5.1
015,205		
115	7.1	6.4
305,215	1.3	2.7
315	3.2	2.8
025,125		
405	9.2	6.0
225,415	5.2	6.7
325,505	2.6	5.3
425,515	4.7	6.0
035,605		
135	3.2	3.8
235,615		
525	9.6	9.5
335	3.0	2.1
016,206	+	2.1
116	2.2	4.0
216,306	3.8	3.6
116	3.6	1.2
026,406		
126	2.9	4.6
226,416	5.1	4.4
326,506	4.0	2.9
516	2.2	3.7
426	2.3	0.6
036,606		
136	8.0	8.3
236,616		
526	3.7	2.3

+) not observed

hydrogen bonding and better packing.

Comparing the TMA structure with other known α-(1→4) linked D-glucan structures, it is found that the 4_3 helix is both in conformation and in rise per residue closest to the 14_{11} helix of amylose triacetate (3), while the conformation of V-amylose is quite different. These differences can be attributed to the O(2)...O(3)2 distances in the two polymers. In V-amylose, this contact is hydrogen bonded while in TMA and ATA it has to be lengthened to a normal van der Waals contact.

Acknowledgements

This work was supported by a grant from Deutsche Forschungsgemeinschaft. The computations were carried out at the Computing Center of the University of Freiburg.

Abstract

The crystal structure of trimethylamylose was solved by conformation and packing analysis and by refinement against X-ray fiber data. The unit cell is orthorhombic, space group $P2_12_12_1$, with a = 17.24 ± 0.01 Å, b = 8.704 ± 0.009 Å, and c (fiber repeat) = 15.637 ± 0.008 Å, as determined by electron diffraction and fiber X-ray patterns.Two chains with four residues each per fiber repeat are packed in an antiparallel fashion in the unit cell. The conformation of the chain can best be described as an approximate 4_3 helix with a superimposed twofold screw axis. Two non-identical residues form the asymmetric unit of the space group.

Literature Cited

1. Keilich, G., Salminen, P. and Husemann, E. (1971) Makromol. Chem. 141, 117-125.
2. Zugenmaier, P. and Sarko, A. (1976) Biopolymers 15, 2121-2136.
3. Sarko, A. and Marchessault, R. H. (1967) J. Amer. Chem. Soc. 89, 6454-6462.
4. Kuppel, A. and Husemann, E. (1974)Colloid & Polymer Sci. 262, 1005-1007.
5. Cella, R. J., Lee, B. and Hughes, R. E. (1970) Acta Cryst. A 26, 118-124.
6. Zugenmaier, P. and Sarko, A. (1973) Biopolymers 12, 435-444.
7. Zugenmaier, P. (1974) Biopolymers 13, 1127-1139.
8. Sarko, A. and Marchessault, R. H. (1969) J. Polymer Sci., Part C 28, 317-331.
9. Zugenmaier, P. unpublished results.
10. Arnott, S. and Scott, W. E. (1972) J. Chem. Soc. Perkin II, 324-335.
11. Ramachandran, G. N., Ramakrishnan, C. and Sassisekharan, V. (1963) "In Aspects of Protein Structure". Edited by G. N. Ramachandran p. 121. Academic Press, London and New York.

10

Solid State Conformations and Interactions of Some Branched Microbial Polysaccharides

RALPH MOORHOUSE, MALCOLM D. WALKINSHAW, WILLIAM T. WINTER, and STRUTHER ARNOTT

Department of Biological Sciences, Purdue University, West Lafayette, IN 47907

In recent years the primary structures of a number of bacterial polysaccharides have been determined (1,2,3,4). These molecules are, in general, more complex than polysaccharides from algal or mammalian sources. Repeating units of between 2 and 6 sugar residues have been observed, often having a branch of one or more sugars and having acetal-linked pyruvic acid on terminal sugar residues (2,5).

We shall consider two branched extracellular polysaccharides, one from a mutant M41 of *Escherichia coli* serotype 29 and the other from *Xanthomonas* species.

The primary structure of the wild-type *E. coli* serotype 29 capsular polysaccharide (the receptor of *E. coli* K phage 29) has recently been reinvestigated by Choy *et al.* (4) and found to consist of hexasaccharide repeating units (I)

→2)-α-D-Man*p*-(1→3)-β-D-Glc*p*-(1→3)-β-D-GlcA*p*-(1→3)-α-D-Gal*p*-(1→

| (on GlcA*p*) (4 ← 1)

β-D-Glc*p*-(1→2)-α-D-Man*p*

(Glc*p* 4,6-linked to C(H_3C)(CO_2H))

(I)

Pyruvate is found (on average) attached to every terminal D-glucose residue, no O-acetyl residues have been detected. It is believed that the polysaccharide from mutant 41 of *E. coli* serotype 29 (hereafter referred to as *E. coli* M41) has essentially the same primary structure as that from the parent.

The primary structure of the extracellular polysaccharide from *Xanthomonas campestris* has also been recently reinvestigated by Jansson *et al.*, (3) and by Melton *et al.*, (6), who found it to consist of pentasaccharide repeating units (II)

```
        →4)-β-D-Glcp-(1→4)-β-D-Glcp-(1→
                                |
                                3
                                ↑
                                1                    (II)
                                |
β-D-Manp-(1→4)-β-D-GlcAp-(1→2)-α-D-Manp-6-OAc
   4  6
    \/
    C
   / \
H3C   COOH
```

Only about one-half of the terminal D-mannose residues carry acetal-linked pyruvic acid. When previously detected in bacterial polysaccharides, pyruvic acid has usually been observed in every repeating unit, as for example in *Klebsiella* type 21 capsular polysaccharide (7) or the *E. coli* K29 polysaccharide already mentioned. However, the closely related polysaccharides from various *Xanthomonas* species show differing pyruvic acid contents (8).

Using the combined techniques of X-ray diffraction and computer aided modeling (9,10), we have investigated the conformation and packing of these microbial polysaccharides.

Structure of the *E. coli* M41 Capsular Polysaccharide

Diffraction Pattern and Unit Cell Properties. Variation of the relative humidity over a large range during crystallisation, orientation and X-ray examination, resulted in fiber diffraction patterns showing essentially the same distribution of intensities and differing slightly in that an overall sharpness of the diffraction pattern correlated with increasing relative humidity (Figure 1). All showed Bragg reflections with layer line spacings of 3.044nm and meridional (*00ℓ*) reflections for ℓ = 2n. The reflections were indexed on the basis of an orthorhombic lattice with *a* = 2.030 nm, *b* = 1.178nm, *c* (fiber axis) = 3.044nm. The systematic absences on the equator, where only even order (*00ℓ*) and (*0k0*) reflections are observed, together with the even order meridional (*00ℓ*) reflections, suggest 3 sets of mutually perpendicular 2-fold screw axes consistent with the orthorhombic space-group $P2_12_12_1$ and a 2-fold helical conformation. There was no evidence of even very weak additional reflections on the equator that would suggest a lower symmetry space-group. This space group and the measured relative density of 1.48 is consistent with a unit cell containing 2 antiparallel, 2-fold helical, polyhexasaccharide chains with about 30 water molecules associated with each hexasaccharide unit.

When samples are examined in a dry helium atmosphere after prolonged (100h) drying over silica gel, the *a* and *b* dimensions are found to have shrunk about 14% to 1.73nm and 1.02nm respectively but no change occurs in *c* or in the systematic absences. The pattern is however, more diffuse, suggesting that while the overall molecular conformation and symmetry are not being appreciably altered upon drying, some disorder is becoming apparent. This also suggests that much of the solvent is located between the polyanion chains.

Rehumidification to 92% relative humidity gives the original pattern (Figure 1).

The diffraction pattern from the wild-type *E. coli* K29 polysaccharide is the same as that from the mutant. The distribution of intensities is the same although the pattern does not show the same degree of orientation exhibited by the mutant M41 polysaccharide. It therefore seems likely that the two strains produce chemically identical (on average) polysaccharides.

Structure Determination. A survey of the literature did not produce a crystal structure of a 4,6-*O*-(1-carboxethylidene)-β-D-glucose or similar molecule. A model residue (for the fused pyruvate part) was constructed using a linked-atom description with an average set of bond lengths, and angles derived from a survey of crystal structures (e.g. 11). The pyruvate carboxyl was placed in the axial configuration at the quaternary carbon atom as proposed for the pyruvic acid ketal that occurs in the polysaccharide from *Xanthomonas campestris* (12).

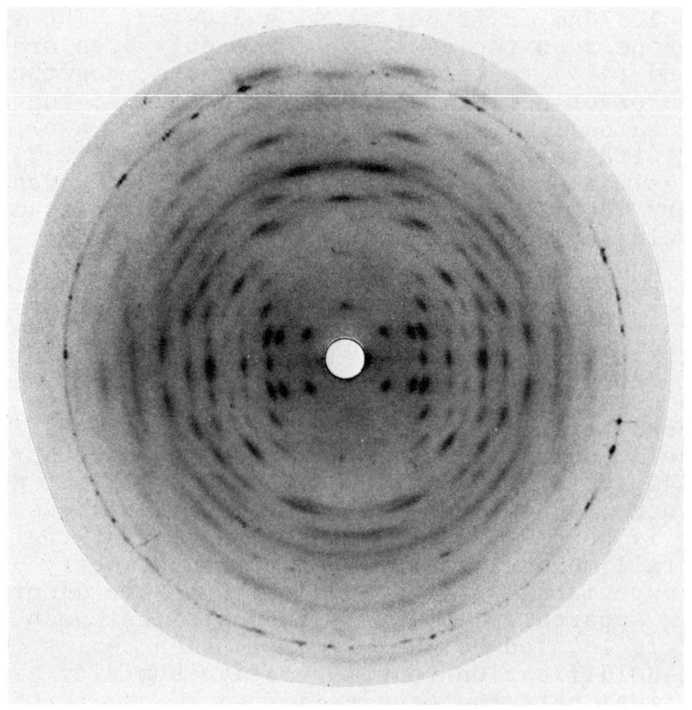

Figure 1. Diffraction pattern from two-fold E. coli *M41 polysaccharide. Two chains pass through the orthorhombic unit cell with dimensions* a = *2.03,* b = *1.178,* c *(fiber axis) = 3.044 nm. The meridional direction is vertical, and the sample was tilted from a direction normal to the beam by 9°.*

The 1,3 dioxane 'pyruvate' ring was refined using the linked-atom least-squares program to minimise steric strain with the bond angles and conformation angles as explicit variables. The bond angles were tied to their initial values to preserve values typical of related small molecules; no explicit restrictions were placed on the values of the conformation angles.

Single-stranded molecular models were generated having the hexasaccharide repeat (Figure 2) and refined using the linked-atom least-squares method (10). The variable parameters in the refinement were, the 12 conformation angles about the glycosidic linkages, 7 side group conformation angles (hydroxymethyl, carboxyl, methyl), the X-ray scale factor and the orientation of the molecules in the unit cell. Water-weighted scattering factors were used (13). The models were constrained to maintain regular 2-fold symmetry and an axial pitch of 3.044nm (Figure 3).

The model refined to give a conventional crystallographic residual of R = 0.25. Particularly good agreement between the observed and calculated structure factors is shown on the equator (R_{hk0} = 0.09). It was of interest to note during the final stages of the refinement, that resetting the hydroxymethyl groups to the *gauche-trans* position and the carboxyl O(6a) on the glucuronic acid to the single crystal setting in which it eclipses O(5), resulted in a further 5% R-factor reduction. Further refinement resulted in the carboxylate orientation returning to a position similar to that observed in various forms of hyaluronate with O(6a) almost eclipsing H(5) (10,14). It will be apparent from the view of the cell down the molecular axis (Figure 4) that very little interdigitation of the helical chain occurs. This is emphasised in Figure 5. Contrary to what might be expected, for such an apparently loose structure, the sharpness of the diffraction pattern, absence of layer line streaks and small spot breadth all indicate that there is very little conformational or packing variability for this structure.

Hydrogen Bonding and Water Location. One would expect a molecule of this complexity, which nevertheless maintains its general molecular conformation over a wide range of humidities, to be stabilised by intra- and inter-molecular forces such as hydrogen bonds. Only 3 hydrogen bonds appear to occur within each hexasaccharide unit, with 2 additional between chains. Thus, only 5 of the available 15 hydroxyl groups are acting as hydrogen bond donors contrary to the

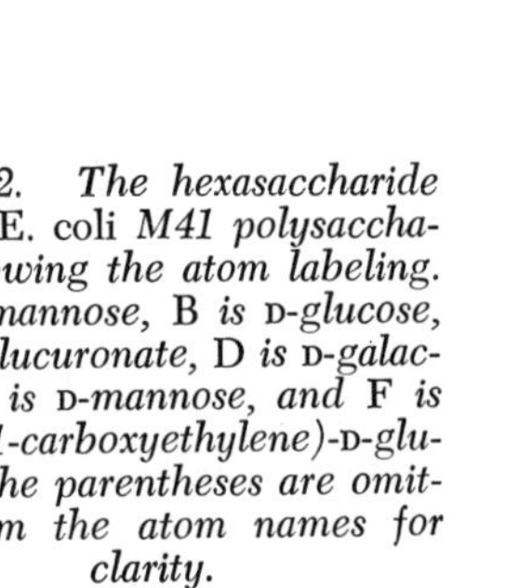

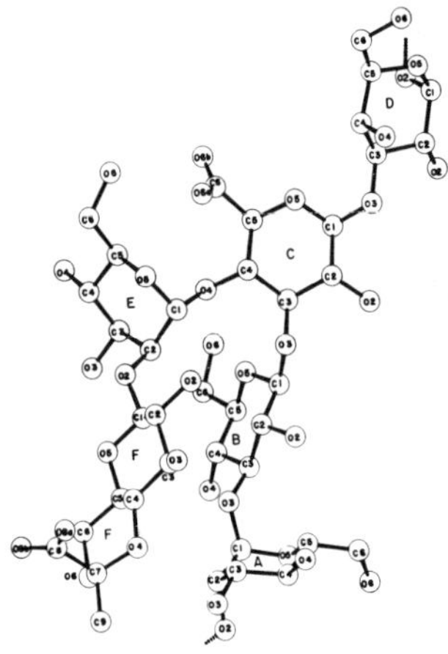

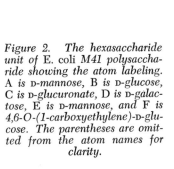

Figure 2. The hexasaccharide unit of E. coli *M41 polysaccharide showing the atom labeling.* A *is* D-*mannose,* B *is* D-*glucose,* C *is* D-*glucuronate,* D *is* D-*galactose,* E *is* D-*mannose, and* F *is 4,6-O-(1-carboxyethylene)-*D-*glucose. The parentheses are omitted from the atom names for clarity.*

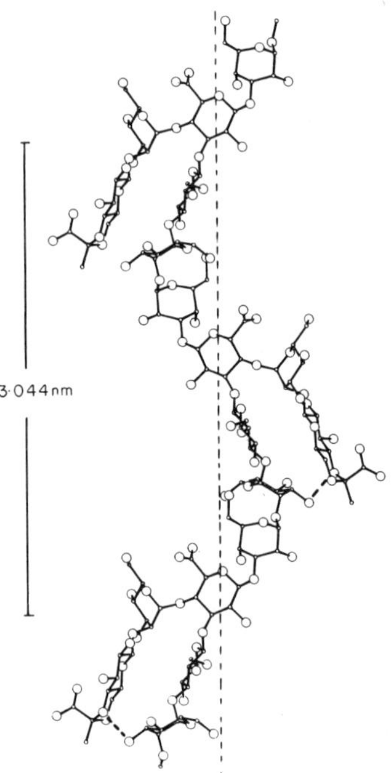

Figure 3. Molecular conformation of the E. coli *M41 polyanion*

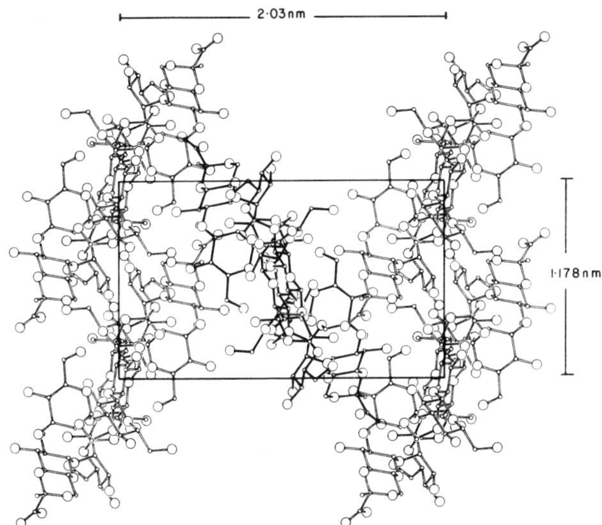

Figure 4. View along the helix axis of the orthorhombic cell

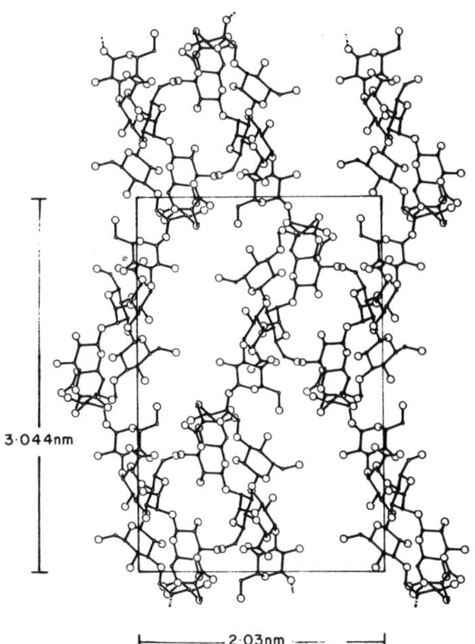

Figure 5. View normal to the helix axis of the orthorhombic cell

situation in other polysaccharides, where most if not all are involved in ion binding hydrogen bonding either directly to sugar acceptor atoms or via water. Of course we already know that 2 counterions and about 30 water molecules are associated with each asymmetric unit and some or all of the molecules/atoms clearly must be involved maintaining the molecular conformation of the molecule. However, the location of all of these presents a considerable problem since the X-ray data to variable ratio would soon approach or exceed 1:1. We have 68 X-ray data (48 excluding those spots systematically absent or unobservable) and 25 variables to define the existing model, each additional individual molecule requires 3 variables to define it. This was the reason we chose initially to use water weighted scattering factors (13) in the refinement. However, using phases derived from this and non-water weighted scattering factors in the model, we have now succeeded in locating some of the more obvious 'water' molecules from Fourier difference maps.

We have therefore a simple hydrogen bonding scheme that is, in part, responsible for stabilising the molecular conformation of the model together with overall packing arrangement (Table 1).

TABLE 1. Attractive interactions in the *E. coli* M41 polysaccharide.

Hydrogen bonds	O---O Distance (nm)	Type
O[3]A --→ O[4]F	0.30	intra
O[4]B --→ O[6]D	0.27	intra
O[3]E --→ O[5]F	0.31	intra
O[4]D --→ O[3]F	0.28	inter
O[4]E --→ O[4]D	0.26	inter

Additional stabilisation probably arises from the ordered water/ion bridges within or between polyanion chains. It is interesting to note that the most intense Fourier difference peak occurs in the region where the backbone glucuronic acid carboxyl on say a corner molecule lies adjacent to a pyruvate carboxyl on say a center antiparallel chain separated by 0.4nm.

Inclusion of four water molecules in this region results in a slight decrease in the residual, R, to 0.24 after a few cycles of refinement. There is a suggestion from the number of hydrogen bonds involved that at least one of these waters could be a cation (15).

Xanthomonas Polysaccharides: Preliminary Studies

Diffraction Patterns. A large number of samples of *X. campestris* and *X. phaseoli* have been surveyed by our X-ray diffraction techniques. All produce diffraction patterns having essentially the same distribution of intensity, exhibiting orientation but not crystallinity, those from *X. phaseoli* being the most oriented (Figure 6).

Although the pattern is confined to layer lines and, therefore, indicates that the molecules are oriented with their screw axes approximately parallel, the intensity distribution along each layer line is, for the most part, continuous with local maxima. This shows that the parallel molecules have random translations along and rotations about their axes and are not packed together so that their repeat units fit into a well-developed crystal lattice. However, destructive interference has occurred near the center of the equator, leaving one broad Bragg reflection of spacing 1.85nm, the array of molecules therefore has some order when viewed down a molecular screw axis at sufficiently low resolution. This effect is to be expected in an array having equally spaced helical molecules with their helix axes parallel but with different translations along and rotations about these axes (16). Apart from this small region of the equator of the diffraction pattern where destructive interference occurs, the calculated intensity distribution is proportional to the square of a cylindrically averaged Fourier transform (17,18).

The layer line spacing is consistent with a helix of pitch 4.70nm, the meridionals on every 5th layer line (00ℓ = 5n) suggest a five-fold helix. This gives a rise-per-*backbone* disaccharide of 0.94nm (Figure 7), which results in an extended single-chain conformation; multi-chain helices being stereochemically unreasonable.

Molecular Models of Xanthan. *A priori* we could have no preference for the four 5-fold helical symmetry possibilities, the 5/1 and 5/4, right and left-handed respectively, single turn-per-helix pitch or the 5/2 or 5/3 right and left-handed two turns-per-helix pitch.

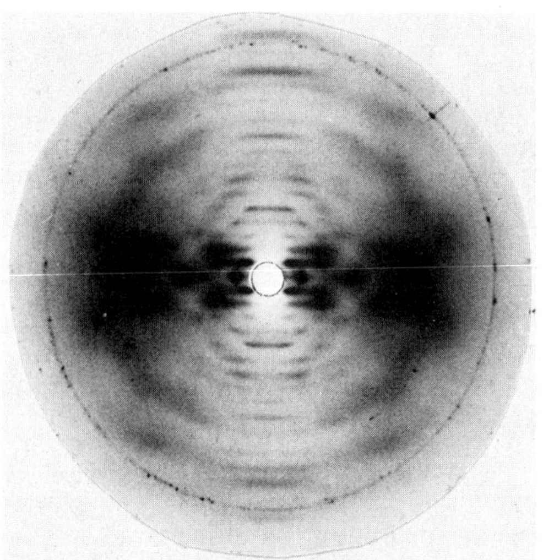

Figure 6. Diffraction pattern typical for both Xanthomonas campestris *and* Xanthomonas phaseoli *showing five-fold helical symmetry. The "sharp" Bragg reflection on the equator has a spacing of 1.85 nm.*

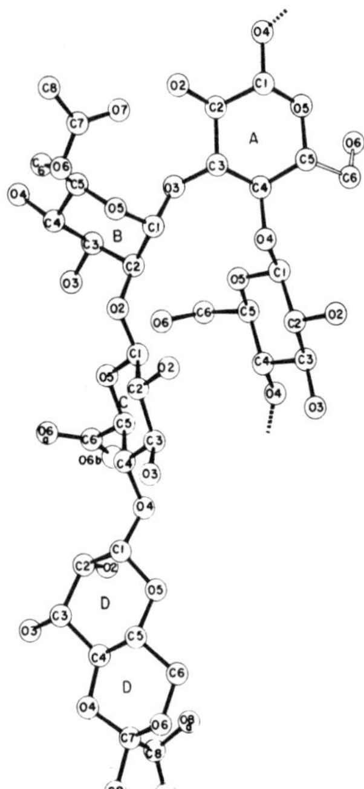

Figure 7. The pentasaccharide repeating unit of X. campestris *showing the atom labelling. The unlettered residue and residue* A *are* D*-glucose,* B *is* D*-mannose,* C *is* D*-glucuronate, and* D *is 4,6-*O*-(1-carboxyethylene)-*D*-mannose.*

Molecular models of the four possibilities in isolation were constructed, maintaining a helix pitch of 4.70nm. Purely on a 'hard-sphere' interatomic contact basis the right-handed 5/1 and 5/2 helices were most favoured having glycosidic conformational angles within the 'normal' range (Table 2), the 5/4 helix being close behind while the 5/3 helix had some unacceptably short interatomic contacts. Of course in isolation there is no driving force to hold the side chains close to the backbone as in the case of the *E. coli* M41 polysaccharide (Figure 8,9). However, we already know from the diffraction pattern that the helical molecules are spaced approximately 1.85nm apart, while the isolated molecular models have a diameter of about 3.8nm. We have therefore undertaken a preliminary packing study of the various models.

TABLE 2. Comparison of backbone conformation angles in the isolated and 'packed' 5/1 and 5/2 helical models.

Parameter	Range	Isolated helices 5/1	Isolated helices 5/2	'Packed' helices 5/1	'Packed' helices 5/2
θ_1	-100 → -161	-136	-121	-148	-119
θ_2	-78 → -98	-63	-64	-76	-30
θ_3	-100 → -161	-111	-99	-98	-97
θ_4	-78 → -98	-92	-22	-81	-61

where, using the atom notation in Figure 7:

$\theta_1 = \theta[C_{(1)A}, O_{(4)}, C_{(4)}, C_{(5)}]$; $\theta_2 = \theta[O_{(5)A}, C_{(1)A}, O_{(4)}, C_{(4)}]$;

$\theta_3 = \theta[C_{(1)}, O_{(4)A}, C_{(4)A}, C_{(5)A}]$; $\theta_4 = \theta[O_{(5)}, C_{(1)}, O_{(4)A}, C_{(4)A}]$.

Two packing modes are consistent with the single Bragg reflection of 1.85nm; a tetragonal cell of side $a = b = 1.85$nm, c (fiber axis) = 4.70nm, and a hexagonal cell having $a = b = 2.13$nm, $c = 4.70$nm, each containing one polysaccharide chain.

Stereochemically both the 5/4 and 5/3 helices are unlikely, having an unacceptable number of intramolecular overshort contacts (19) thus reinforcing the previous results. Both the 5/1 and 5/2 helices are

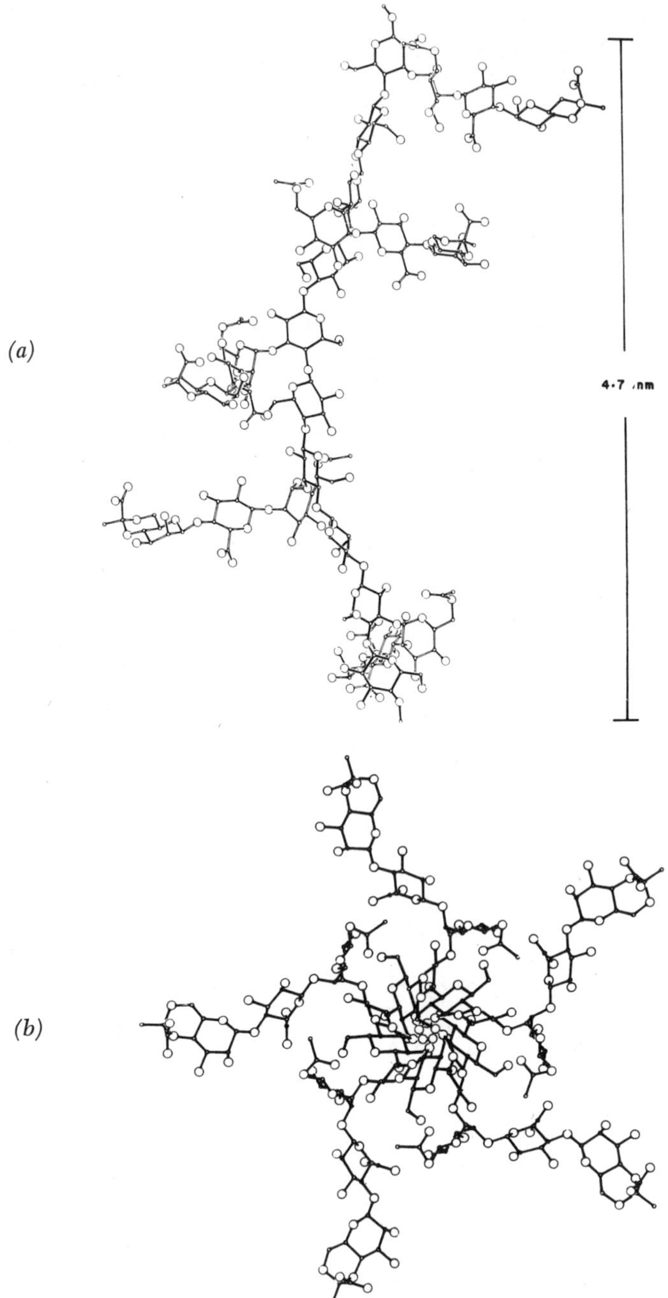

Figure 8. The isolated 5/1 helix viewed (a) *perpendicular and* (b) *down the helix axis*

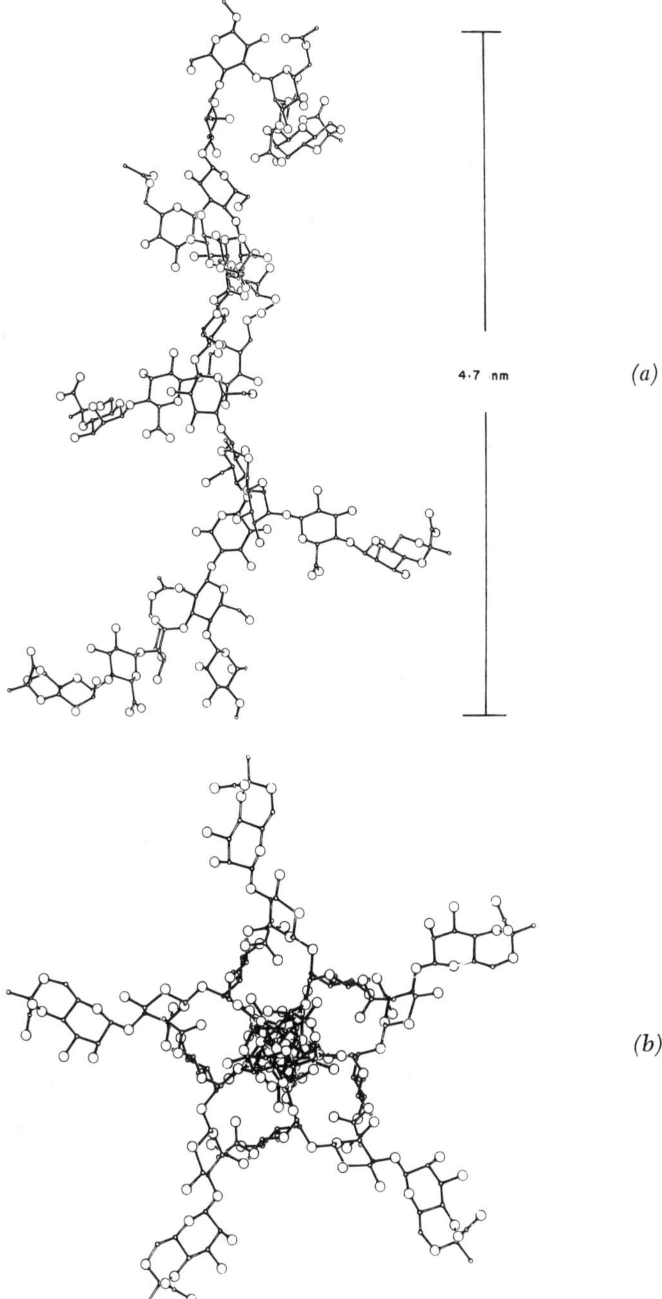

Figure 9. The isolated 5/2 helix viewed (a) *perpendicular and* (b) *down the helix axis*

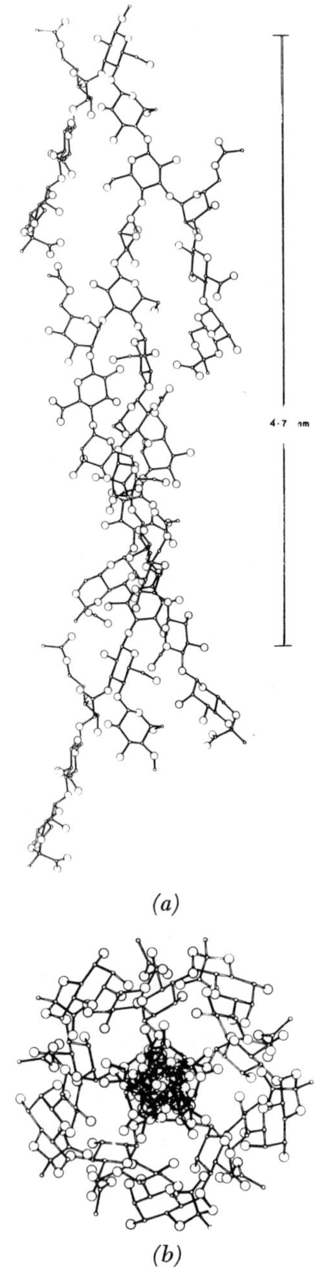

(a)

(b)

Figure 10. The contracted "tetragonal" 5/2 helix viewed (a) *perpendicular and* (b) *down the helix axis*

stereochemically good (Figures 10,11) in the tetragonal cell. No appreciable change in the conformation angle of both models were apparent when the tetragonal models (maintaining any hydrogen bonds) were placed in the larger hexagonal cell. Relaxing the hydrogen bond constraints similarly maintained the molecular conformations. It was only when considerable perturbations were made to the side-chain conformation angles that any overall change occurred in the backbone causing a poorer agreement for the 5/2 helix with the model angles shown in Table 2.

Overall the hydrogen-bonded 5/1 helix in either the tetragonal or hexagonal packing model is, we believe, the most likely at this stage since all the backbone conformation angles are within model single crystal ranges (Table 2). Additionally the 5/1 helix has no overshort contacts and several hydrogen bonds (Table 3) that could stabilise the backbone and side chain (Figure 12).

TABLE 3. Attractive interactions in the *X. campestris* 5/1 helix.

Model	Overshort contacts (nm)	Hydrogen Bonds
5/1	none	$O_{(3)}$---→$O_{(5)A}$
		$O_{(2)}$---→$O_{(8a)D}$
		$O_{(6)}$---→$O_{(5)C}$
		$O_{(2)A}$---→$O_{(7)B}$
		$O_{(3)B}$---→$O_{(6)}$
		[or $O_{(3)B}$---→$O_{(5)C}$]
		$O_{(2)D}$---→$O_{(6b)C}$
5/2	$O_{(5)A}$•••$H_{(4)A}$	$O_{(3)}$---→$O_{(5)A}$
	(0.195)	$O_{(6)A}$---→$O_{(5)}$
		$O_{(3)B}$---→$O_{(5)C}$
		$O_{(3)C}$---→$O_{(5)D}$

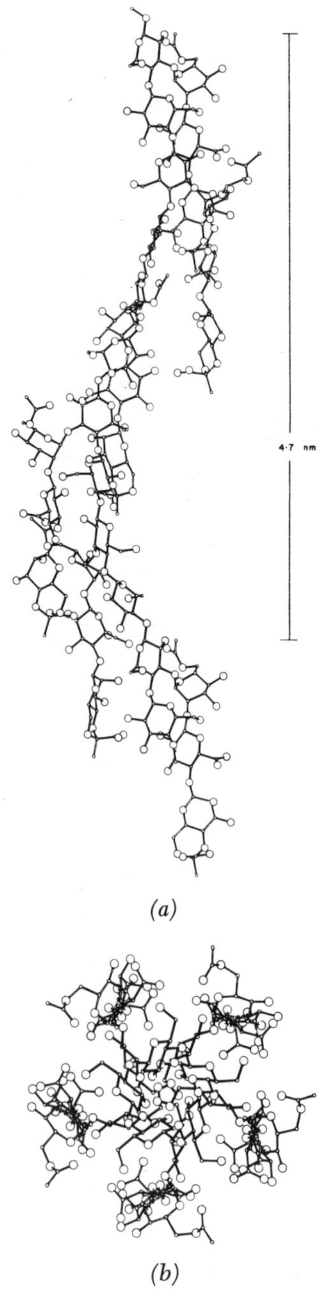

Figure 11. The contracted "tetragonal" 5/1 helix viewed (a) *perpendicular and* (b) *down the helix axis*

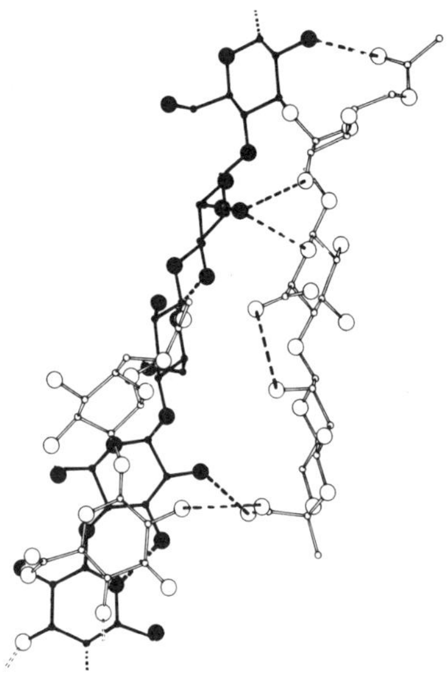

Figure 12. Possible hydrogen bonds holding the side chain close to the backbone in the 5/1 helix. Some adjoining residues are omitted for clarity. The backbone has solid bonds and atoms.

Discussion

E. coli M41 polysaccharide. It is perhaps surprising that a heteropolysaccharide of this complexity can form such crystalline and oriented fibers. However, we have been able to report here the first fiber diffraction and packing study on a branched heteropolysaccharide. The good agreement between observed and calculated structure factors reflected in the relatively low R factor (R = 0.24) suggests that while we have included a hypothetical acetal-linked pyruvic acid residue, the overall molecular conformation is correct. That the molecular conformation and packing is relatively insensitive to hydration implies that the molecule is stabilised intramolecularly by a combination of hydrogen bonds that must also involve water although we have not been able to specify many of these. We do not intend to speculate on the biological significance of this structure since there is no complementary solution conformational data although we would expect that it would persist *in vivo* due to the high water content of the fiber samples studied.

Xanthomonas polysaccharide. This polysaccharide has been shown to have an ordered aggregated conformation in solution which reversibly 'melts' out on heating (20). This 'helix coil' transition is not concentration dependent and therefore is a unimolecular process. It is not known if the 5/1 helical conformation presented here for the solid state persists in solution although past experience suggests that it does (9,21).

The side chains which drape around the backbone like a discontinuous second strand are stabilised and reinforce the stability of the backbone by a series of hydrogen bonds (Table 2). However, in the simple packing study we have undertaken, no direct interchain hydrogen bonding is apparent although there are several hydroxyl groups on the outside of the helix that could participate in a hydrogen bonding scheme. Since 2/3 of the charge density is on the outside of the helix, ionic forces probably play a major role in stabilising the packing of this molecule. Indeed it has been shown that the addition of excess inorganic salts enhances the viscosity of this polysaccharide (22) and causes a shift in the melting temperature of the ordered conformation to higher values (20). Water presumably also has a role in linking the helices via water bridges. Such a network may be complex because of the apparent random orientation and translation of the molecules.

It is interesting to note that the 5/1 helix presents 2 distinct faces -- one containing the side chains and charged groups, the other essentially the cellobiose backbone. Since Xanthan gum can bind cooperatively to polysaccharides having unbranched sequences of β-1, 4 linked D-mannose, D-glucose or D-xylose, possibly mimicing the cell-cell recognition between the bacteria and plant wall, it may be that this cellobiose face is the recognition site for the polymannan chain to slot into.

Acknowledgements

We wish to thank Dr. M.E. Bayer, Institute for Cancer Research, Philadelphia (*E. coli* polysaccharides) and Drs. A. Jeanes and P.A. Sandford, U.S.D.A., Peoria (Xanthomonas polysaccharides), for their generous supplies of material.

Abstract

The extracellular microbial polysaccharides from the *E. coli* K29 mutant M41, *Xanthomonas campestris* (Xanthan gums) and *Xanthomonas phaseoli* have been examined by X-ray diffraction and computer-aided molecular modeling.

A detailed model has been obtained for the *E. coli* polysaccharide in which it has been possible to define both the molecular conformation and the crystal packing. The structure suggests possible intra- and intermolecular attractive interactions possibly involving countercations and hydrogen bonding through water molecules.

Preliminary molecular conformations for the *Xanthomonas* polysaccharides are discussed.

Literature Cited

1. Björadal, H., Hellergvist, C.G., Lindberg, B. and Svensson, S. Angew. Chem. Int. Ed. Engl. (1970) 9,610-619.
2. Lindberg, B., Lönngren, J. and Thompson, J.L. Carbohyd. Res. (1973) 28,351-357.
3. Jansson, P.E., Kenne, L. and Lindberg, L. Carbohyd. Res. (1976) 45,275-282.
4. Choy, Y.M., Fehmel, F., Frank, N. and Stirm, S. J. Virology. (1975) 16,581-590.
5. Thurow, H., Choy, Y.M., Frank, N., Niemann, H. and Stirm, S. Carbohyd. Res. (1970) 41,241-255.

6. Melton, L.D., Mindt, L., Rees, D.A. and Sanderson, G.R. Carbohyd. Res. (1976) 46, 245-257.
7. Choy, Y.M. and Dutton, G.G.A. Can. J. Chem. (1973) 51,198-207.
8. Orentas, D.A., Sloneker, J.H. and Jeanes, A. Can. J. Microbiol. (1963) 9,427-430.
9. Arnott, S., Guss, J.M., Hukins, D.W.L., Dea, I.C.M. and Rees, D.A. J. Mol. Biol. (1974) 88,175-185.
10. Guss, J.M., Hukins, D.W.L., Smith, P.J.C., Winter, W.T., Arnott, S., Moorhouse, R. and Rees, D.A. J. Mol. Biol. (1975) 95,359-384.
11. Arnott, S. and Scott, W.E. J. Chem. Soc. Perkin Trans. II (1972) 324-335.
12. Gorin, P.A.J., Ishikawa, T., Spencer, J.F.T. and Sloneker, J.H. Can. J. Chem. (1967) 45, 2005-2008.
13. Arnott, S. and Hukins, D.W.L. J. Mol. Biol. (1973) 81,93-105.
14. Winter, W.T., Smith, P.J.C. and Arnott, S. J. Mol. Biol. (1975) 99,219-235.
15. Moorhouse, R., Winter, W.T., Arnott, S. and Bayer, M.E. J. Mol. Biol. (1976) (in press).
16. Arnott, S. Trans. Amer. Crystallogr. Asso. (1973) 9,31-56.
17. Cochran, W., Crick, F.H.C. and Vand, V. Acta Crystallogr. (1952) 5,581-586.
18. Klug, A., Crick, F.H.C. and Wyckoff, H.W. Acta Crystallogr. (1958) 11,199-213.
19. Rees, D.A. J. Chem. Soc. (1969) B,217-226.
20. Morris, E.R., Rees, D.A. and Walkinshaw, M.D. J. Mol. Biol. (1977) (submitted).
21. Arnott, S., Fulmer, A., Scott, W.E., Dea, I.C.M., Moorhouse, R. and Rees, D.A. J. Mol. Biol. (1974) 90,269-284.
22. Jeanes, A., Pittsley, J.E. and Senti, F.R. J. App. Polymer Sci. (1961) 5,519-526.

11

Changes in Cellulose Structure during Manufacture and Converting of Paper

E. L. AKIM

Leningrad Technological Institute of Pulp and Paper Industry, Leningrad, U.S.S.R.

A characteristic feature of development of the pulp and paper industry during the last decades was a wide use of achievements of chemistry, physico-chemistry and technology of polymers and close relationship between the technology of paper manufacture and converting and the technology of polymer materials in general and of chemical fibers and films in particular. The principal trends in this field were: first, the preparation of composite materials "paper-synthetic polymer", such as "paper-film" complexes and laminated plastics, secondly, the manufacture of synthetic papers which include papers from synthetic fibers, synthetic paper films and papers from the so-called "synthetic pulp" (i.e., a synthetic material imitating fibrous structure of cellulose) and, finally, chemical modification of paper. As a result, the pulp and paper industry of today, apart from traditional processes and equipment, is characterized by processes and equipment similar to those in chemical industry, such as manufacture of photographic and motion picture film and magnetic tape, manufacture and processing of chemical fibers and plastics.

Fibrous and non-fibrous synthetic and natural half-finished products used for paper manufacture as well as methods of treatment and converting of paper are very varied. Nevertheless, they are all based on chemistry and physicochemistry of polymers. The progress in this field is based on the investigation and solution of relatively few fundamental problems. One of them is the strength development in a paper sheet.

Irrespective of the type of the half-finished

products used (natural or synthetic), paper (with the exception of synthetic paper film) is a fibrous porous material in which not only fibers themselves, but also fiber-to-fiber bonds ensure physicomechanical properties of the material. Paper strength is mainly determined by the strength of fiber-to-fiber bonds and therefore their nature should be considered in detail. For papers from natural cellulose fibers, taking into account their long successful manufacture, this problem is mainly theoretical. However, for papers from synthetic fibers or "synthetic cellulose" this problem is of primary practical importance, since its solution determines successful solution of purely technological problems. It is advisable to begin the consideration of this problem with ordinary paper.

For several decades the development of paper strength was of great interest for scientists of many countries. Various aspects of this problem have been investigated more or less thoroughly and the existing theories have been considered in detail in many monographs. Nevertheless, one aspect of this problem has hardly been dealt with for a long time. This is the problem of changes in the physical (relaxational) state of cellulose in the course of paper manufacture.

For historic reasons some problems of physical chemistry of polymers virtually have not been considered for cellulose. This was caused by an artificial division of physico-chemistry and polymers in general and that of cellulose in particular. Many papers and books openly or in a veiled form adhered to the idea that cellulose is an exceptional polymer synthesized by nature itself and that theoretical concepts developed in detail for synthetic polymers are inapplicable to it. All this certainly concerns the problems of changes in the physical (relaxational) state of cellulose, its glass transition temperatures and possibilities of its transition into a highly elastic state. Nevertheless, before considering problems related to the physical state of cellulose, we should briefly deal with its phase state. At present there are several viewpoints on this problem. According to one of them, cellulose is regarded as a faulty crystal (in this case it is impossible to speak about any physical transitions in cellulose). According to the most widely held and experimentally confirmed viewpoint cellulose is a crystallizing polymer with comparatively well defined crystalline and amorphous regions. The presence of amorphous regions permits us to speak about the glass transition temperature of cellulose and about a possibility of its transition into a highly elastic state.

There are very few works dealing with the glass transition temperature of cellulose. Nevertheless, all service properties of polymer materials and their behavior in the process of mechanical, physico-chemical and chemical treatment are closely related to temperature ranges of physical states. Consequently, the determination of these ranges, the investigation of possibilities of transition of cellulose from one physical state into another is of decisive importance for cellulose materials also.

In 1949 in the work of Kargin and co-workers (1) it was shown that the glass transition temperature of cellulose is 220° C. This figure was obtained in an indirect way by extrapolating the dependences of glass transition temperatures of cellulose samples on amounts of plasticizer to its zero amount. Later this value has been confirmed by a direct method (2). The results obtained by Kargin suggested that under normal conditions cellulose is in a glassy state and since the glass transition temperature is higher than the temperature of degradation of cellulose materials in air, it follows that it is impossible to transform it into a highly elastic state without degradation. However, if it is impossible to achieve this by heating, in principle it is possible to do this by plastification. One of the methods of plastification is to bring the polymer sample into contact and into a sorption equilibrium with the plasticizing liquid medium. Thus, Thus, Bryant and Walter have shown that for cellulose materials in water the values of elastic modulus correspond to values characteristic of polymers in a highly elastic state (3).

At the end of the sixties we have investigated the effect of liquid media on the temperature of transition from a glassy to a highly elastic state for a number of polymers (4). For cellulose these investigations have been carried out in collaboration with

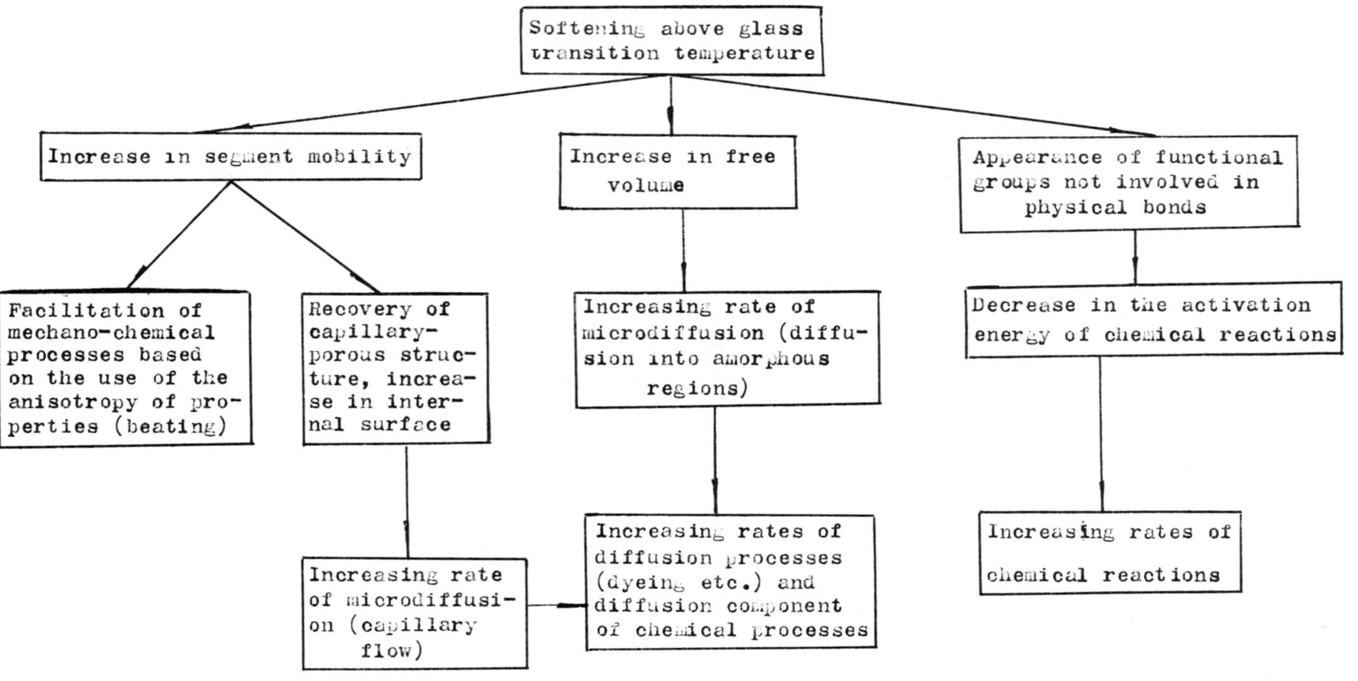

Figure 1. Schematic of effect of transition of the polymer from a glassy to a highly elastic state on its behavior in chemical and other processes

N. J. Naimark, B. A. Fomenko and others (5, 6). These investigations made it possible to analyze the role of the highly elastic state of polymers in mechano-chemical, physico-chemical, and chemical processes (4) and to formulate the peculiarities of these processes for a highly elastic state of polymers (Fig. 1 and 2). In particular, it has been shown that one of the indispensable conditions of dyeing of textile materials is a transition of a fiber forming polymer from a glassy state into a highly elastic state (7). An indispensable condition of chemical reactions with a polymer in a condensed state is also its softening above the glass transition temperature. For example, it has been proved experimentally that the mechanism of such a technological operation as activation by treating cellulose with acetic acid before its acetylation is the transition of cellulose from a glassy to a highly elastic state owing to the plasticizing effect of acetic acid (8).

This paper is concerned only with those aspects of the effect of a transition of polymers from a glassy to a highly elastic state which dominate in the processes of paper manufacture and converting; it summarizes the results of our recent investigations.

Natural fibrous half-made products used for the paper and board manufacture contain three main components: cellulose, hemicelluloses, and lignin. Each of them plays its part in the development of paper strength. Fibrous structure of cellulose is the main framework forming the fibrous porous structure of paper and board whereas lignin, hemicellulose, and low molecular weight fractions of cellulose ensure the formation of fiber-to-fiber bonds in this structure. This paper is mainly concerned with the role of cellulose, since changes in lignin are considered in some other papers, in particular in Goring's works (9).

Thus, for dry cellulose in air the glass transition temperature is about 220° C. In liquid media (Table I) this temperature decreases. In particular, in water it is below room temperature (in later works (10) it has been shown that water decreases the glass transition temperature of native cellulose to -45° C. and that of regenerated cellulose to -25° C.).

For cellulose and materials based on it the

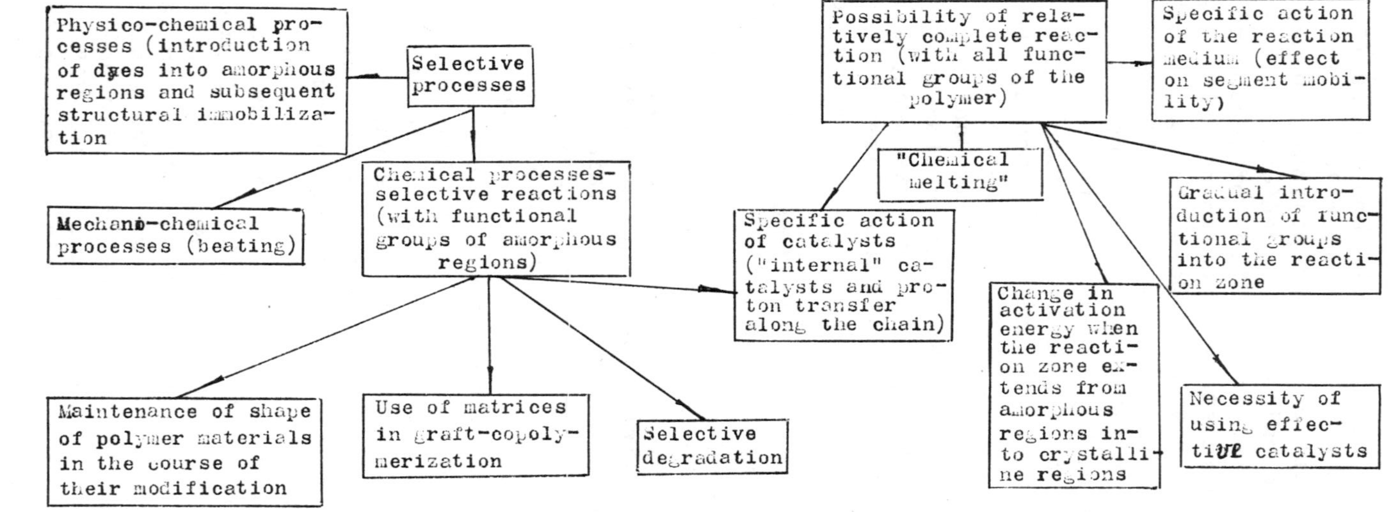

Figure 2. Schematic of peculiarities of processes for polymers in a highly elastic state

transition: wet cellulose - dry cellulose is of particular importance. Wet cellulose is in a highly elastic state whereas dry cellulose is in a glassy state. Hence, a change in the physical state of cellulose occurs in the course of its drying. Conventional methods of cellulose drying combine processes of drainage (water removal) and drying proper (removal of the liquid wetting cellulose). The glass transition of cellulose caused by the removal of plasticizer and water occurs under conditions of a considerable shrinkage stress. In the manufacture of dissolving pulp shrinkage stresses play a negative part leading to the formation of a dense ("hornified") layer on the fiber surface which makes further processing difficult.

Table I

Glass transition temperatures of cellulose preparations in different media

No.	Plasticizing medium	T_g of viscose fiber, °C.	T_c of cotton, °C.
1	air	220-230	220-230
2	water	<20	<20
3	ethylene glycol	60	<20
4	glycerol	90	40
5	methanol	20	<20
6	ethanol	>80	>80
7	butanol	>80	>80
8	formic acid	<20	<20
9	acetic acid	90	<20
10	propionic acid	>140	40-50
11	butyric acid	>140	40-50
12	acetic anhydride	>140	>140
13	dimethyl sulfoxide	60	<20
14	dimethyl formamide	100	60
15	dimethyl acetamide	125	90
16	monoethanolamine	40	20
17	diethanolamine	120	100
18	triethanolamine	140	--
19	diethylamine	>55	--

In paper and board manufacture shrinkage stresses play a positive part. They draw together fibrillar elements of paper during drying up to distances at which all fiber-to-fiber bonds appear to the maximum extent owing to the formation of intermolecular bonds between hydroxyl groups of macromolecules located on the surfaces of fibrillar supermolecular structures. Subsequent glass transition immobilizes this "contracted" sheet structure. Consequently, the aim of the main stages of paper manufacture (beating, formation, pressing, drying) is provide fiber-to-fiber bonds in paper in order to obtain a strong sheet.

Beating of fibrous materials (Fig. 3) is possible only under conditions ensuring softening of the fiber-forming polymer above glass transition temperature. For cellulose materials the medium ensuring this is water. In most non-aqueous media cellulose fibers are dispersed as fragile materials rather than fibrillated. Correspondingly, beating of chemical fibers (in the case of fibrillated fibers) is possible only in media and at temperatures ensuring the transition of the fiber-forming polymer to a highly elastic state (11). Moreover, in beating of cellulose materials favorable conditions exist for extracting hemicelluloses (pentosans and hexosans) from the fiber. In water they partly pass into a viscous state forming a thin film on the surface of fibrils. These fractions play a specific part in paper manufacture ensuring an increase in the surface area of interfibrillar contacts in the drying paper and, hence, an increase in its strength. In paper drying under conditions of high shrinkage stresses cellulose passes from a highly elastic state into a glassy state while its low molecular weight fractions pass from a viscous into a glassy state and the sheet structure is immobilized by intermolecular hydrogen bonds. Thus, it is just hemicelluloses (pentosans and hexosans) - low molecular weight fractions of cellulose, that immobilize the capillary-porous structure of paper and cellulose drawn together during drying. The role of hemicelluloses in the manufacture of a strong paper sheet has been known to papermakers for a long time. Here many well known factors may be mentioned: first, difficulties arising in the manufacture of strong paper from cotton pulp; secondly, decrease in paper quality with increasing alkali concentration in alkaline pulping (12), and, finally,

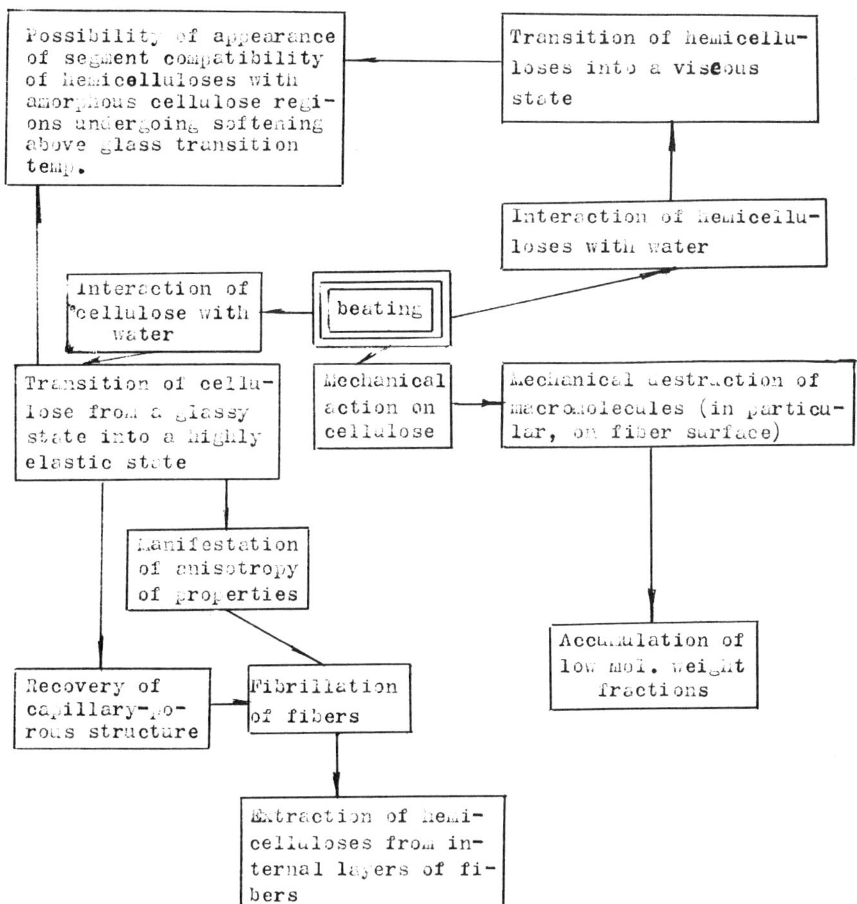

Figure 3. Schematic of main processes taking place in the course of beating

improvement in paper forming properties of pulp when hemicelluloses are reprecipitated in alkaline pulping (13). It is known that in the initial stages of alkaline pulping hemicelluloses pass in solution more or less completely. Their precipitation on hard cellulose material in the final stages of cooking leads not only to an increase in the yield of the pulp but also to an improvement in its paper-forming properties. In contrast, in the preparation of dissolving pulp, reprecipitation of hemicelluloses, i.e., sorption by cellulose of components dissolved in the course of cooking, is very undesirable.

As already stated, beating conditions favor extraction of hemicelluloses from the fiber and their transition into a viscous state. Owing to low mobility of macromolecules, a great part of hemicelluloses passing into a viscous state remains directly on cellulose fibers (moreover, hemicelluloses exhibiting segment compatibility with cellulose, cooperative movement of chain segments, interact with its macromolecules in softened amorphous regions). The part of hemicelluloses which in a dissolved form becomes "abstracted" from the hard cellulose material accumulates in white water and sooner or later also passes into the wet paper. In particular, this accounts for the well known fact that in some cases when fresh water is used instead of white water, paper strength decreases.

In paper drying the role of hemicelluloses in development of paper strength is very great. One of the main components of shrinkage stresses is the process of capillary contraction, i.e., closing ("clapping up") of capillaries when the liquid in them evaporates. Under these conditions precipitation of hemicelluloses leads to increasing forces of capillary contraction and this additionally increases "monolithization" of the material. Hemicellulose macromolecules segmentally compatible with amorphous regions of cellulose in a highly elastic state become partially "jammed" in the course of drying ensuring additional strong contacts between the fiber and interfibrous material.

All these processes are also favored by mechanical destruction of the polymer (cellulose) occurring during beating in the surface layer of the fiber, i.e., in layers directly interacting with knives of beaters or refiners. Low molecular weight fractions formed as a result behave in the course of further processes just as hemicelluloses of the initial cellulose material (it is just they that ensure to a great extent the strength of paper from cotton cellulose).

The transition of low molecular weight fractions into a viscous state in the beating stage is favored by intense mechanical action and in the drying stage by high temperature. Pressing of wet paper web favors denser packing of fibrous material and to a certain extent favors segment compatibility of low molecular weight fractions and fibrous material.

The influence of some factors, in particular, temperature and humidity, on paper properties shows the importance of hemicelluloses in development of paper strength. For hemicelluloses the glass transition temperature is much lower than for cellulose. At the same time small amounts of water are sufficient for the plastification of hemicellulose which leads to plasticization of junctions binding fibrous structure of paper and, hence, to plasticization of the paper itself. Thus, very small amounts of water profoundly affect properties of paper and board.

The picture outlined above is characteristic of papers of cellulose origin which do not contain lignin. When lignin is present in pulps, all the more so when mechanical pulp is used, similar processes seem to occur with lignin which, just as hemicelluloses, undergoes a change in its physical state in the course of paper sheet manufacture. This change has been considered in detail by Goring (9). However, he thinks that these processes dominate in paper strength development. He may be right, and only to a certain extent, in the cases when newsprint is considered, while in papers of higher grades it is just hemicelluloses that play the principal part in the strength development. It is possible to evaluate approximately the contribution of each of these components to paper strength development by using liquid media exhibiting different activity with respect to lignin and to hemicelluloses. On the other hand, by using model systems or investigating paper containing synthetic polymers as adhesives, it is possible to evaluate the contribution of the adhesive to the overall complex of physico-mechanical and elastic-relaxational properties of the material.

In order to evaluate elastic-relaxational properties of paper we have used methods of thermomechanics in liquid media, stress relaxation and some other methods (14). Values of elastic modulus, dynamic shear modulus, deformation rate and logarithms of relaxation time were used as criteria for evaluating the physical state of the polymer.

Data in Table II show that the relaxation time for papers in air is very long; this indicates that the material is in a glassy state. Under the effect of liquid media relaxation time decreases sharply and for relatively active media its values are commensurable with the time of experiment. It is interesting to note that for boards containing fibers of polyvinyl alcohol, the relaxation time logarithm is virtually determined only by the presence of this fiber. This suggests that the nature of the adhexive often determines elastic-relaxational properties of the material.

Table II

Effect of liquid media on relaxation time for different paper grades

Medium	Logarithm of time of complete relaxation (log c.r. in min.) for paper or board			
	Experimental unsized cotton paper	Bag paper from sulfate pulp	Impregnated paper from sulfite pulp	Filter-board with PVC fibers in composition
air	87	240	130	10
xylene	57	11	40	--
heptane	37	40	34	--
benzene	34	--	--	--
methylene chloride	22	163	36	--
isopropanol	18	50	90	--
ethanol	6	--	--	--
acetone	4	9	15	--

In the manufacture of papers from synthetic fibers the physico-chemical nature of paper strength development remains just the same as in the manufacture of paper from cellulose fibers. Nevertheless, taking into account thermoplastic character of many synthetic fibers, in this case changes in the physical state of the paper-forming polymer may be caused by thermal action, such as hot calendering. From the viewpoint of physico-chemical basis of papermaking processes it is interesting to use polyvinyl alcohol fibers which partially dissolve in the stage of paper drying (i.e. pass into a viscous state) and ensure fiber-to-fiber bonds in the sheet. At any rate, irrespective of the use of fibers of natural or synthetic origin for paper manufacture, changes in the structure of the material in various stages of paper manufacture are similar. Fig. 4 shows changes in the physical (relaxational) state of cellulose and hemicellulose in different stages of papermaking. When synthetic fibers are used, the picture is somewhat different.

The physical state of cellulose also plays an important part in the processes of treatment and converting of pulp, paper, and board. In most of them the course of the process itself and its results are determined to a great extent by changes in the physical state during the process. We will consider as an example some trends in the treatment and converting of cellulose material and, just as in the preceding case, we will attempt to show the relationship between the processes of manufacture and converting of ordinary (cellulose) and synthetic papers.

Coating of paper may be regarded as film casting from a polymer solution with a filler or without it on a capillary-porous substrate (Fig. 5). In the course of coating the physical (relaxational) state of all polymer components participating in the process changes. When solvents of relatively high activity with respect to the paper-forming polymer are used (such as water for ordinary cellulose paper), softening of the polymer above the glass transition temperature occurs in the surface paper layers. In the case of cellulose material the capillary-porous structure of cellulose fibers is recovered. This facilitates the flow of the polymer solution into the opening capillaries with subsequent formation of "hook-like" bonds. In the case of using water soluble polymers exhibiting segment compatibility with cellulose the increasing free volume of the polymer creates conditions for the appearance of segment compatibility. In drying of coated paper the polymer

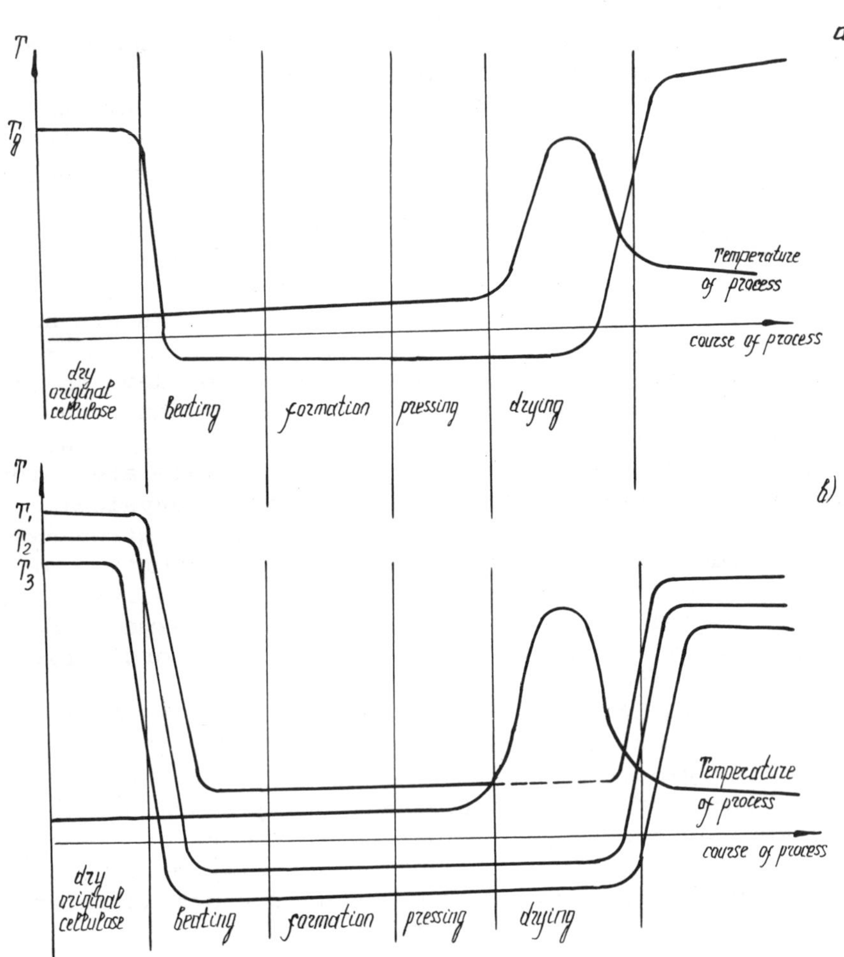

Figure 4. Change in the physical (relaxational) state of (a) *cellulose and* (b) *hemicellulose in the course paper manufacture* (T_{g1} *and* T_{f1}*—glass transition and flow temperatures of high molecular weight hemicellulose fractions, respectively.* T_{f2}*—flow temperature of low molecular weight fractions of hemicellulose).*

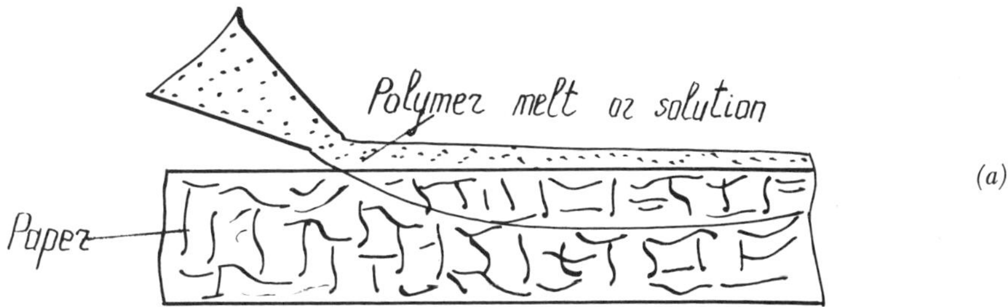

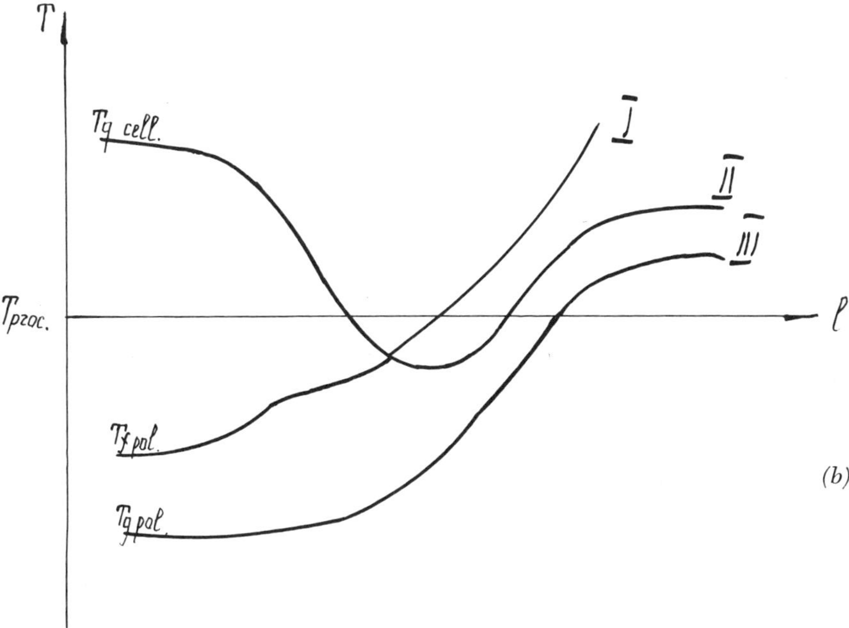

Figure 5. (a) Scheme of paper coating. (b) Change in physical (relaxational) state of polymer components (cellulose and polymer coating) in paper coating. I—change in flow temperature of the polymer. II—change in glass transition temperature of cellulose. III—change in glass transition temperature of polymer (processes are regarded as conditionally isothermal).

contained in the coating passes from a viscous into a highly elastic state and subsequently into a glassy state. Simultaneously the paper-forming polymer undergoes glass transition in the surface layers which earlier have been softened above the glass transition temperature. It is these processes of glass transition that ensure good adhesion of coating to substrate.

Similar phenomena occur in the process of paperization of the surface of synthetic papers of the film type. In this case, however, the formation of "hook-like" bonds is no longer possible and the adhesion of paperized coating to paper may be ensured by the softening of the substrate surface above the glass transition temperature and by using coating components exhibiting segment compatibility with the polymer of the substrate (15).

The physical state of cellulose also plays an essential part in paper (or board) modification. This paper does not deal with problem of chemistry of processes of paper modification. It is sufficient to say that in the first approximation all methods of chemical modification of cellulose materials developed and investigated by Professor Z. A. Rogovin and his pupils (16, 17) can be used for paper modification. As a rule, these processes require preliminary activation of cellulose material but the choice of activation conditions in paper and board modification has some peculiarities which can be considered taking as an example the process of partial acylation, in particular, acetylation of paper (18). As has been shown above, the mechanism of activation of cellulose before acetylation is its transition from a glassy to a highly elastic state. During this transition fiber-to-fiber bonds ensuring paper strength are weakened and after a relatively extended activation they are completely broken. Therefore, it is necessary to choose conditions of paper activation ensuring on the one hand a relatively rapid process of further modification and on the other hand partial retention of fiber-to-fiber bonds. In other words, in contrast to most cellulose reactions, paper activation should ensure incomplete softening of the polymer above glass transition temperature.

Technological conditions of modification should ensure relatively extended modification with a minimum extent of side reactions. One of the principal side reactions is degradation. In the manufacture of some cellulose derivatives degradation is a desirable process ensuring the required viscosity of products, whereas in modification mainly selective degradation in softened amorphous regions of the polymer takes place. Since these regions ensure paper strength, it follows that as a result of degradation paper strength decreases to a much greater extent than it could be expected from changes in degrees of polymerization. Since in paper modification by partial acylation it is necessary to prevent degradation, it is desirable to use basic rather than acid catalysts. However, kinetic factors should also be considered. In the course of acetylation of cellulose materials the reaction rate decreases appreciably (whereas in the acetylation of low molecular weight alcohols the opposite phenomenon is observed - Fig. 6). This decrease is particularly pronounced in the acetylation of preparations of native cellulose (cellulose I). Hence, in the presence of basic catalysts cellulose I is virtually acetylated only to a very small extent. Since in the presence of basic catalysts cellulose II is acetylated much faster than cellulose I (Fig. 7), it is advisable to use cellulose II at least as one of components in the paper composition. The required mercerization of cellulose may be carried out, for example, in the stage of cellulose isolation from plant tissues (19). It is noteworthy that, as has been shown above, in the manufacture of paper pulp the increase in alkali concentration in alkaline pulping is undesirable whereas in this case, i.e., in the manufacture of paper pulp undergoing subsequent modification this increase in alkali concentration is very advantageous.

In modification of paper-forming fibers structural aspects of the problem are also of primary importance (20). In this case it is not advisable to introduce substituents into crystalline regions of cellulose. It is more advantageous to introduce selectively the substituent groups only into amorphous regions of cellulose or into macromolecules located on the surface of supermolecular structures. For instance, in the case of oxyethylation this introduction may be carried out

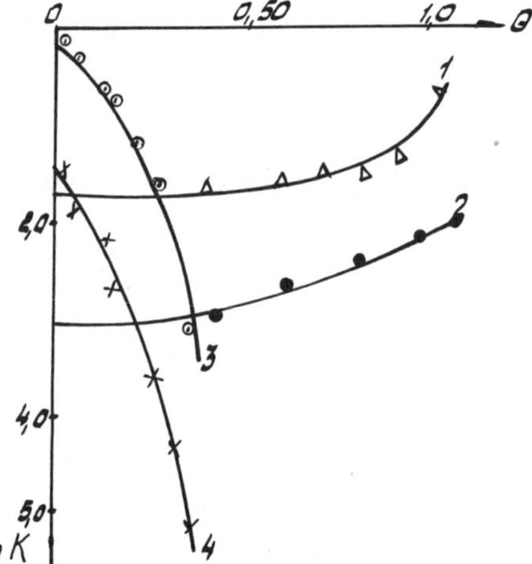

Figure 6. The ln K vs. degree of completion of the process in the case of acetylation of high and low molecular weight alcohols at 70°C (curves 2 and 4) and 90°C (curves 1 and 3). Curves 1,2—1-butanol; curves 3,4—cellulose II (36° SR).

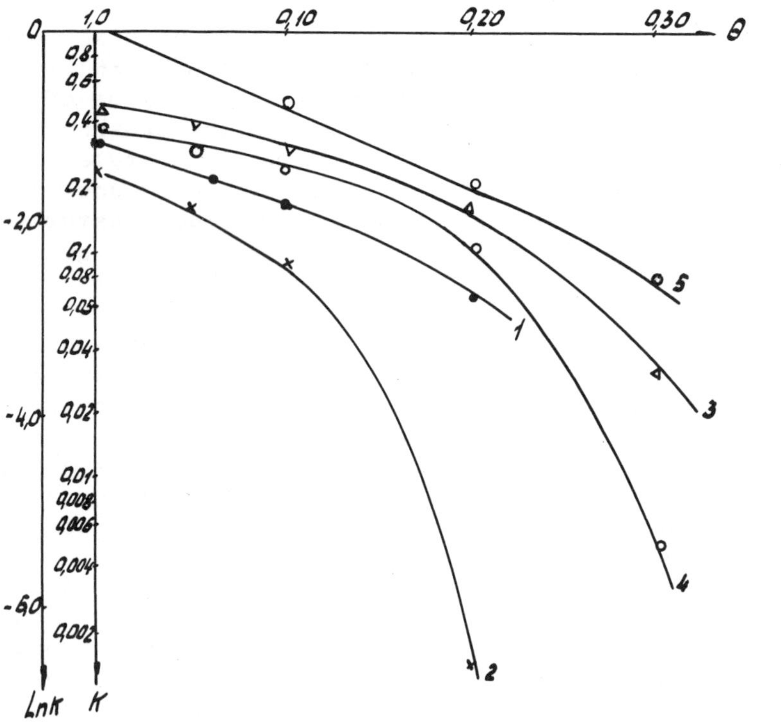

Figure 7. Effect of the type of cellulose material on acetylation kinetics (the initial temperature—85°C); 1—cellulose I; 2—paper from cellulose I; 3—mixture of celluloses I and II (50:50); 4—paper from mixture of celluloses I and II; 5—cellulose II (36° SR).

by conducting the process at low alkali concentration (1-2%). This makes it possible to use for partial oxyethylation alkali treatments which cellulose materials undergo during isolation from plant tissues (21) and to obtain manufacture paper with markedly improved physicochemical properties by using partial modification(20).

Finally, when processes of paper converting are based on the interaction of cellulose with some synthetic polymer, e.g., in obtaining plastic laminates based on cellulose materials, one of the main conditions for the formation of chemical bonds between the cellulose filler and the synthetic polymer (resin) is the coincidence of temperature-time ranges of high elasticity of both components in this composite material.

Abstract. The problems considered above concerning changes in cellulose structure in the course of paper manufacture and converting make it possible to draw an analogy between the technology of the manufacture and converting ordinary (cellulose) paper grades and that of synthetic papers. This consideration shows the similarity of physicochemical phenomena which form the basis of the technology of manufacture and converting of any paper grades irrespective of the type of paper-forming polymer. In this brief survey, it was not possible to deal with many aspects of the problem, in particular, with the question of optimum molecular-mass distribution. Apart from purely structural questions, these aspects of the problem are very important and evidently deserve to be separated and analyzed. The progress in paper industry will doubtless be accompanied by closer relationship between the use of natural and synthetic polymers.

Literature Cited

1. Kargin, V. A., Kozlov, P. V., Nai-Chan, V., Dokl. Akad. Nauk, SSSR (1960) 130, 356-358.
2. Naimark, N. I., Fomenko, B. A., Vysokomolek. Soedin. (1971) 13, 45.
3. Bryant, G. M., Walter, A. T., Text. Res. J. (1959) 29, 211.
4. Akim, E. L., "Issledovanie protsessa sinteza voloknoobrazuyushchikh atsetatov tsellulozy", Thesis, Leningrad, 1971.
5. Akim, E. L., Naimark, N. I., Perepechkin, L. P.,

Fomenko, B. A., "Vliyanie zhidkikh sred na temperaturu perekhoda tsellulozy is stekloobraznogo v vysokoelasticheskoe sostoyania", Program of scientific session dedicated to the discovery of the periodic law by D. I. Mendeleev, Leningrad, 1969, 162-163.
6. Akim, E. L., Naimark, N. I., Vasiliev, B. V., Fomenko, B. A., Ignatieva, E. V., Zhegalova, N. N., Vysokomolek. Soedin. (1971) A XIII, 1.
7. Akim, E. L., Colourage Annual (India) (1970) 20-22.
8. Akim, E. L., Perepechkin, L. P., "Tselluloza dlya atsetilizovaniya i atsetaty tsellulozy", Lesnaya Promyshlennost, Ed., Moscow, 1971.
9. Goring, D. A. I., "Lignins", K. V. Sarkanen, C. H. Ludwig, Eds., Wiley-Interscience, New York, 1971.
10. Naimark, N. J., Fomenko, B. A., Ignatieva, E. V., Vysokomolek. Soedin. (1975) BXVII, 355.
11. Akim, E. L., Chigov, R. A., USSR Patent (1973) 368,366.
12. Connors, W. J., Sanyer, N., TAPPI (1975) 58(2), 80-82.
13. Klevenskaya, V. A., Katkovich, R. G., Gromov, V.S., Khimiya drevesiny (1974) 1, 18-26.
14. Akim, E. L., Erykhov, B. P., Mirkamilov, Sh. M., Uzbekskii Khim. Zhurnal (1976) 1, 40-44.
15. Akim, E. L., Mikhalevich, D. S., USSR Patent(1975) 475294.
16. Galbraikh, L., Rogovin Z., "Cellulose and Cellulose Derivatives", Parts IV-V, N. Bikales, L. Segal, Eds., Mir, Moscow, 1974.
17. Rogovin, Z. A., "Khimiya tsellulozy", Moscow, 1973.
18. Eremenko, Yu. P., Akim, E. L., Yasnovsky, V. M., Transactions VNIIB (1975) No. 65, 107-119; Lesnaya promyshlennost, Ed., Moscow.
19. Akim, E. L., Eremenko, Yu. P., USSR Patent (1972) 344059.
20. Akim, E. L., USSR Patent (1964) 164261.
21. Akim, E. L., Perepetchkin, L. P., USSR Patent (1964) 160266.

12

Characterization of Cellulose and Synthetic Fibers by Static and Dynamic Thermoacoustical Techniques

PRONOY K. CHATTERJEE

Personal Products Co., Subsidiary of Johnson & Johnson, Milltown, NJ 08850

The characterization of materials by acoustical techniques is an old art. Perhaps it dates back hundreds of centuries. There is no record in history when man first learned to differentiate between a rock and a metal by striking them with another object and listening to the characteristic frequency and amplitude of the resulting sound waves. The technique, however, emerged as a science when the subjective listening was changed into the objective measurements by modern instrumentations.

The subject of discussion of this paper is not the differentiation between two dissimilar macro objects but the determination of the changes at a molecular level which takes place due to the environmental changes of a polymeric material, viz, cellulosic and synthetic fibers. More specifically, three individual topics of the acoustical techniques have been briefly discussed, viz. characterization of interfiber bonding in cellulose sheets, water absorption in a paper like non-woven structure, and dynamic thermal analysis of textile fibers including cotton, rayon, and varieties of synthetic fibers. In describing the principles of the technique, mathematical derivations have been avoided as much as possible. The versatility of this novel acoustical technique has been clearly demonstrated by the following examples.

Sonic Pulse Propagation in a Paper Like Structure and the Characterization of Interfiber Bonding

Technique. A number of handsheets were made according to TAPPI standard T205m-58 using fully bleached commercial kraft pulps as listed in Table I. The sonic velocity in the sheets was determined with a pulse propagation meter, PPM-5R, manufactured by H. M. Morgan and Co. The principle of the procedure is described elsewhere (1). The other physical properties of the sheets were measured according to the standard techniques (2).

Table I

Pulp Samples

(Kraft, fully bleached pulp)

Sample Identification	Description
A	80% Cedar, 20% Hemlock
B	Southern Pine
C	Southern Pine
D	Southern Pine
E	Douglas Fir
F	Douglas Fir
G	50% Douglas Fir, 50% Sawdust
H	Southern Pine, Mercerized
I	Southern Pine, Mercerized

Results and Discussion. The structure of a pulp handsheet (or paper) could be defined as a heterogeneous system consisting in part of cellulose unbonded fibers and in part of bonded regions with pores containing air. When a pulse propagates through the sheet (Figure 1), an adiabatic compression takes place in all the constituents of the sheet. However, according to Taylor and Craver (3), the acoustical mismatch of a fiber/air surface is too great to permit direct transmission of the sound through fiber and void space in series. Let it be assumed that the velocity of sound c, is controlled by only two structural constituents: the bonded and the unbonded fiber regions. It is also considered that (1) handsheets are made of randomly oriented fibers which are interconnected by numerous interfiber bonds, and that (2) the bonded regions, the porous spaces, and the unbonded portions of fibers are uniformly distributed in the sheet structure. The bonded region is defined here as the area of intimate contact between two fibers.

A pulse, as it propagates, travels through a series of two different structural constituents as cited above. Based on the above hypothesis, Chatterjee (2) derived the following equation

$$\beta c^2 = \frac{\beta E_1 E_2}{\rho_1 [\beta E_1 + (E_2 - \beta E_2)]} \qquad (1)$$

where $(1-\beta)$ = proportion (by surface area) of unbonded fiber regions, ρ_1 = the density of fiber, and E_1 and E_2 = inverse of compressibility or modulus of unbonded portion of fibers and interfiber bonded regions, respectively.

In the case of unbeaten pulp handsheets, the sheet has an appreciable proportion of unbonded regions, and the modulus of interfiber bonded regions is very low compared to the modulus of a single fiber. By taking into account the equations for mechanical properties of a sheet, equation (1) can be reduced to the following form:

$$dc^2 = k_6 T_\beta \qquad (2)$$

where, $k_6 = \phi(\lambda, d, k', \nu, \rho)$ and λ = ratio of the tensile modulus to the modulus at break, d = apparent density of the sheet, k' = a constant for the basis weight-density relationship, ν = rate of tensile loading factor, ρ = true density of fiber, and T_β = tensile breaking strength measured by instron tester.

A plot of dc^2 versus T_β is shown in Figure 2 along with the plot of ultimate tensile stress and Mullen burst. As expected from equation (2), dc^2 and T_β have a linear relationship.

The theory and the experimental results indicate that the tensile strength of a pulp sheet which is primarily a function of interfiber bonding can be determined by a nondestructive acoustical technique. A similar relationship has also been derived for the modulus of the sheet and for different kinds of paper structure including resin bonded papers (2).

Among many applications, it is worthy of mentioning that the technique has been uniquely applied in characterizing the embossing on fluff pad by Chatterjee (4), on-line measurement of strength characteristics by Lu (5), and in determining interaction between fibers and phenolic resins by Marton and Crosby (6).

Absorption of Water in Cellulose Sheet by Sonic Velocity Response

Technique. The wicking of water in a paper sheet is conventionally done by a technique which is known as the Klemm test. According to this test, a paper strip is hung vertically above a trough filled with water. The strip is lowered slowly until the lower end of it is touched by the water surface. Simultaneously, a series of stopwatches is activated. The water front rises through the paper and when it reaches specified marks, the times are recorded. A plot of wicking height (h) versus time (t) provides the information concerning the rate of fluid wicking in the paper. The rate constant is calculated by the Lucas-Washburn equation (7), which relates the wicking rate and capillary structure of the paper as follows:

$$h = kt^m \qquad (3)$$

where k = constant = $\phi(r, \gamma, \theta, \eta)$ and r = average radius of interfiber capillary, θ = constant angle, η, γ = viscosity and surface tension of water respectively, m = constant (~0.5) and h = wicking height.

A continuous monitoring device for the wicking measurement would help to develop an automatic instrument for testing the rate

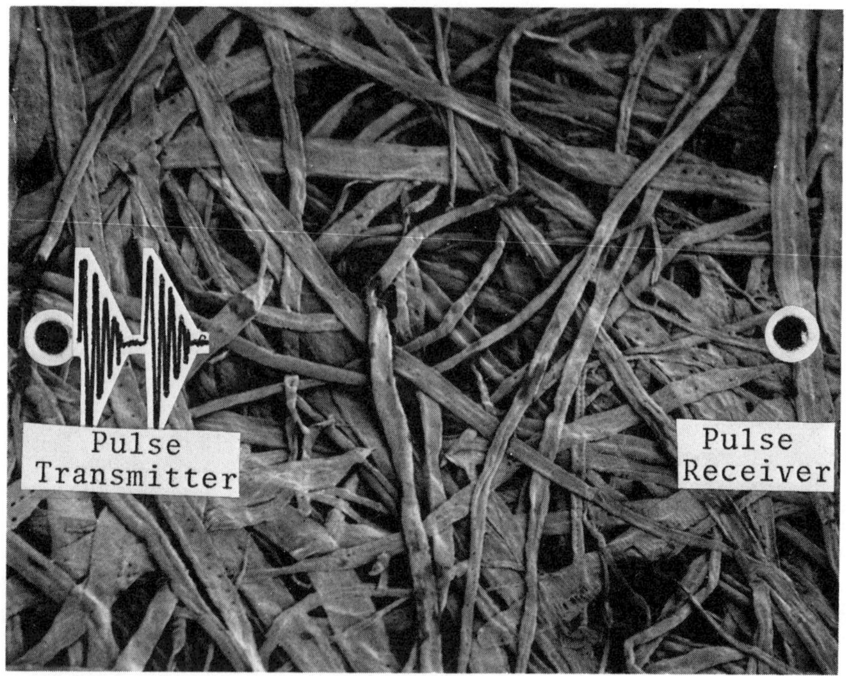

Figure 1. Pulse propagation technique for the characterization of interfiber bonding in pulp sheet (photomicrograph of Southern Pine handsheet, 125×)

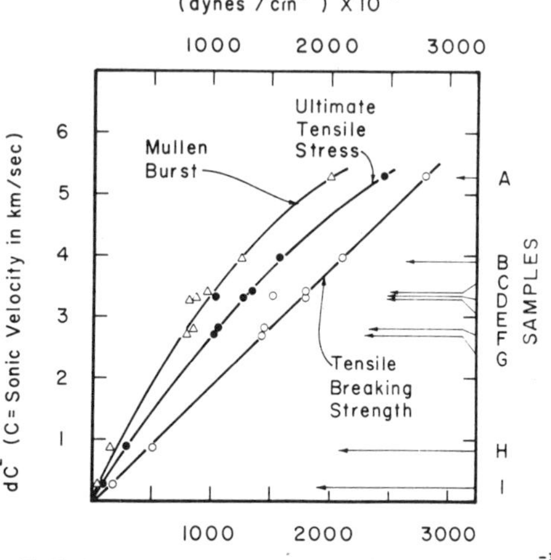

Figure 2. Relationship between the tensile strength properties and dc^2 *(c = sonic velocity in km/sec). (Plots corresponding to individual samples have the same ordinate value and are indicated by an arrow parallel to abscissa.)*

of wicking in paper. In view of this consideration, the acoustical method for the measurement of wicking in a paper like structure has been developed (8).

For the present investigation, standard handsheets consisting of bleached commercial pulps were used. The samples are briefly described in Table II.

A set of piezoelectric transducers of a pulse propagation meter is placed at a fixed distance on the sample sheet as represented in Figure 3. Water is applied by a capillary tubing at the midpoint between the transducers. The sonic pulse propagation time in microseconds is continuously measured during the application of the water. A typical plot of sonic response before and during the wicking process is shown in Figure 4. New coordinates are drawn so as to intersect at the point corresponding to the commencement of the application of water. The converted ordinate would represent the change of the reciprocal velocity of sound due to wicking or $(1/c_w - 1/c_o)$ where c_w is the sound velocity at any time, t, during the wicking and c_o is the sound velocity before the application of water. The abscissa representing the wicking time is calibrated in seconds according to the speed of the chart paper. From the graph, the value of $(1/c_w - 1/c_o)$, is determined where c_w is the sound velocity at any time, t, during wicking and c_o is the sound velocity before the application of water.

According to an earlier publication of Chatterjee (8), the rate equation through acoustical measurement can be expressed as

$$(1/c_w - 1/c_o) = k't^m \qquad (4)$$

where k' is a constant and proportional to k of equation (3).

Results and Discussion. The typical plots of log $(1/c_w - 1/c_o)$ versus log t are shown in Figure 5. The linearity of the plots confirm equation (4). The values of m, k, and k' obtained from Klemm and sonic tests are given in Table III.

The values of m obtained from both techniques are not too far apart from the theoretical value of 0.5. Theoretically, k is defined as the height of the liquid at one second wicking time whereas k' is defined as the reduction of sonic velocity at one second wicking time. Hence, the absolute values of k and k' must not be compared. A fairly good correlation between k and k' is evident from Figure 6. Therefore, it can be inferred that k as obtained from the conventional Klemm test and k' as obtained from the sonic pulse propagation test should represent the rate of wicking in relative terms.

The result indicates that the pulse propagation technique is an excellent automatic method for the study of wicking in a paper like structure. The technique is also very simple and quick.

Table II

PULP SAMPLES

Sample Identification	Apparent Density of handsheet g/cc	Description of Pulp
A	0.35	Ground Wood, Partially Bleached
B	0.65	80% Cedar-20% Hemlock, Kraft
C	0.59	Northern Pine, Sulfite
D	0.59	Hemlock, Sulfite
E	0.60	Southern Pine, Kraft
F	0.56	Redwood, Kraft
G	0.52	50% Douglas Fir-50% Sawdust, Kraft
H	0.54	Douglas Fir, Kraft (sample 1)
I	0.50	Douglas Fir, Kraft (sample 2)
J	0.50	Southern Pine, Kraft (sample 2)
K	0.49	Southern Pine, Kraft Semibleached
L	0.50	Douglas Fir, Kraft (sample 3)

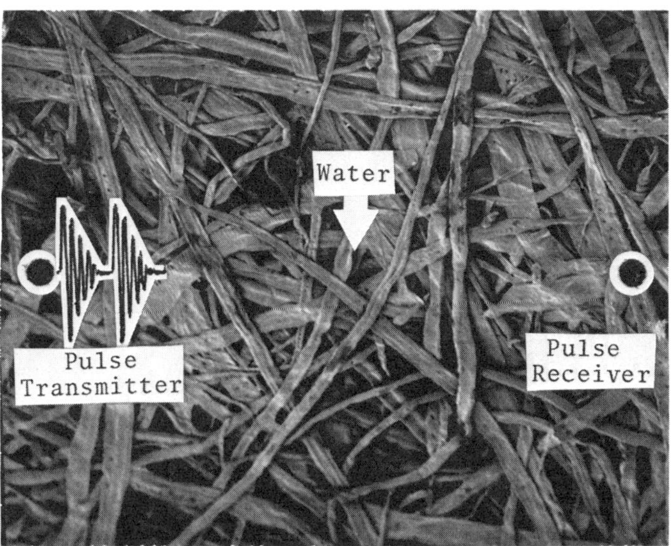

Figure 3. Pulse propagation technique for the measurement of fluid wicking in pulp sheet (photomicrograph of Southern Pine handsheet, 100×).

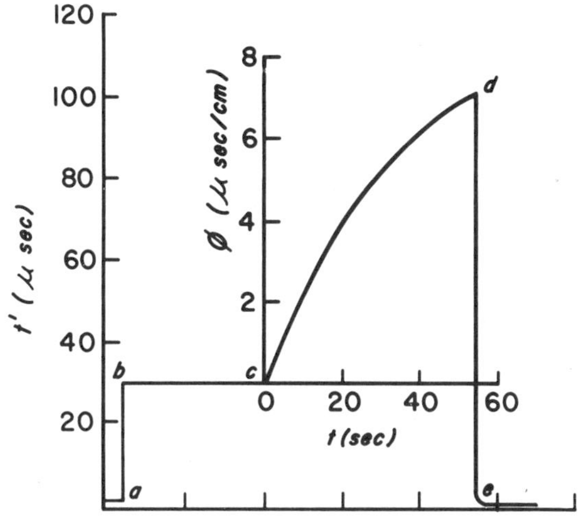

Figure 4. Typical sonic response during wicking. Width of the strip: 15 mm. t′ *= pulse propagation time for 10-cm distance,* $\Phi = [1/c_w - 1/c_o]$, t *= wicking time; sonic response:* bc, *before wicking;* c, *application of water;* cd, *during wicking;* d, *termination of test.*

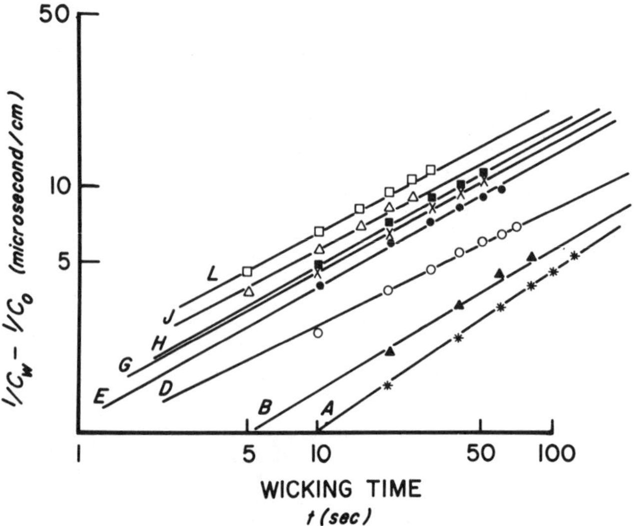

Figure 5. Relationship between sonic velocity and wicking in handsheets. Logarithmic plots of Equation 4.

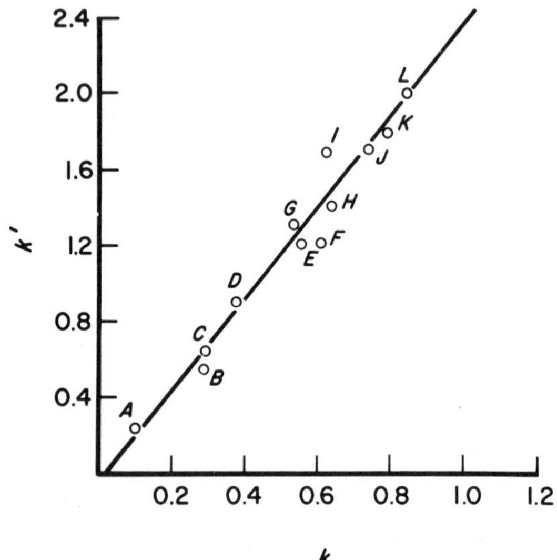

Figure 6. Correlation between the rate constants (k & k′) determined by Klemm and sonic tests respectively. Correlation coefficient is 0.976.

Table III

EQUATION CONSTANT AND RATE PARAMETERS DETERMINED BY KLEMM AND SONIC TESTS

Sample Identification	Equation Constant and Rate Parameters			
	Klemm Test (Equation 3)		Sonic Test (Equation 4)	
	m	k	m	k
A	0.54	0.10	0.55	0.22
B	0.48	0.29	0.52	0.54
C	0.49	0.29	0.54	0.63
D	0.48	0.38	0.49	0.90
E	0.48	0.56	0.54	1.20
F	0.44	0.61	0.50	1.20
G	0.46	0.54	0.54	1.30
H	0.49	0.66	0.55	1.40
I	0.46	0.62	0.56	1.70
J	0.45	0.76	0.56	1.70
K	0.47	0.79	0.51	1.80
L	0.44	0.85	0.52	2.00

Dynamic Thermoacoustical Technique

This section presents an acoustical technique which can be used to determine the changes in relaxation associated with polymers (fibers) under dynamic heating conditions. The principle of the technique with its theoretical derivations was described earlier (9,10).

Technique and Theory. In principle, the method consists of a continuous measurement of the propagation time of sonic pulses of constant frequency (7kc) through the test sample which is being held under light tension and heated at programmed temperature (Figure 7).

The experimental setup is shown in Figure 8. In the figure, A represents the heating block of a DuPont 900 DTA where H represents the heating element and T_2 and T_3 represent the reference and sample wells, respectively. Two additional holes (T_1 and T_4) were drilled all the way from one end to the other through the heating block. The heating block was thoroughly insulated by putting asbestos caps on both ends and covering the rest with asbestos tape. Melting point tubes, open on both ends, were inserted into holes T_1 and T_4. The heating block was mounted horizontally and thermocouples were inserted in T_1, T_2, and T_3. These thermocouples were connected to different terminals of the DuPont DTA cell as shown in Figure 8. The fiber sample was tied with one end to a clamp S, passed under a pulley P_1 and on a notched ceramic piezoelectric crystal transducer Z_1, through hole T_4 and supported by another identical piezoelectric crystal Z_2 and a set of pulleys P_2 and finally terminated at a suspended weight of 5g. The piezoelectric crystals were connected to an electrical pulse generating and recording device, R. The distance between Z_1 and Z_2 was 3.6 cm which was kept constant throughout the experiment.

The sound pulses are transmitted through the fiber sample and the time required for pulses to propagate from Z_1 to Z_2 is recorded. Knowing this propagation time and the distance between Z_1 and Z_2, one can calculate the velocity of sound through the sample. However, in the present technique, it is necessary to record the pulse propagation time only; the velocity conversion is not required.

The sample is heated in air atmosphere at a programmed rate of 20°C per minute. The system temperature is continuously recorded on a DTA chart whereas the sonic response is simultaneously recorded on a time base recorder provided with the pulse propagation device. The abscissa of the original sonic chart is later converted to a temperature scale.

In the case of a simultaneous differential thermal analysis and dynamic thermoacoustical analysis (9), the fibers are cut into small pieces by using a Wiley mill with 60 mesh screen, and 5 mg of this sample was poured into a melting point tube. The tube is

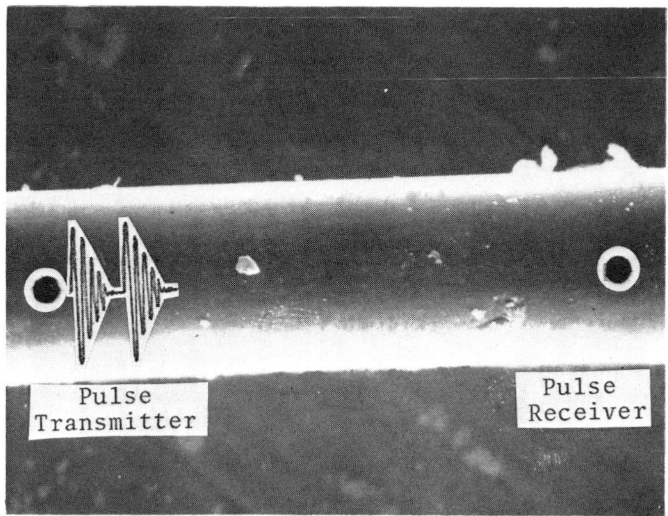

Figure 7. Dynamic thermoacoustical technique for the characterization of textile fibers (photomicrograph of partially drawn polyester fiber, 1000×)

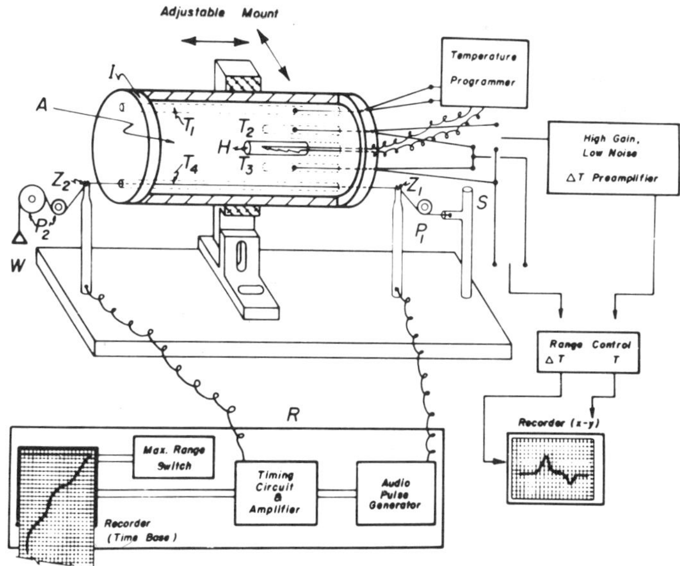

Figure 8. A schematic of dynamic thermoacoustical system

then placed into the sample cavity T_2; and a thermocouple, as shown in Figure 8, is inserted into the sample. Similarly, in reference cavity T_3, a melting point tube containing reference glass beads is inserted. The reference thermocouple is then embedded into the glass beads. For thermoacoustical analysis, the setup is the same as that described in the preceding section. On heating the metal block 'A', the DTA curve of the sample is obtained on the X-Y recorder of the DTA instrument and a thermoacoustical curve of the sample is obtained on the time base recorder.

A sketch of a hypothetical dynamic thermoacoustical curve is shown in Figure 9. The pulse propagation time for a distance of χ cm of the sample at room temperature is represented by the horizontal portion of the curve AB. As long as the distance χ is kept constant, AB remains parallel to the abscissa. The velocity of sound through the material at 25°C is equal to $(\chi/120) \times 10^6$ km/sec. The temperature programming of the DTA apparatus is initiated at B. As long as the sample remains physically and chemically unchanged, the curve continues to indicate the horizontal line. At C the sample begins to transform to a different phase and the curve deviates from the base line. A change towards the upward direction indicates the lowering of the sound velocity. It is known that the velocity of sound is highest in solid, lowest in gas, and intermediate in liquid. Therefore, one may assume that the upward trend of the curve would indicate the change of polymer from compact form to relatively fluid form or, in other words, an increase of molecular motion of polymers. An opposite phenomenon is indicated by the downward trend of the curve FG. The sample at G is certainly in a less fluid state than at F. Again G to H shows no physical or chemical change in the sample. At H the sample reveals the premelting behavior. As the melting starts, there is a sharp upward trend of the curve until the sample breaks at I due to the actual melting. The recorder pen drops immediately to zero, indicating thereby a discontinuity of the pulse propagation path.

It has been shown earlier (9) that the pulse propagation time (μ) in a fiber (or polymer film) can be expressed by the following equation:

$$\mu = 0.82\chi \, k \, \varepsilon_s \qquad (5)$$

where ε_s is defined as the sonic viscoelastic function of the polymer at a constant sonic frequency, k is the molecular orientation factor, and χ is the distance between transducers. Therefore, according to equation (5), the dynamic thermoacoustical curves which will be discussed here represent ε_s as a function of temperature.

Result and Discussion. Thermoacoustical curves of a variety of synthetic fibers are shown in Figure 10. These curves are all

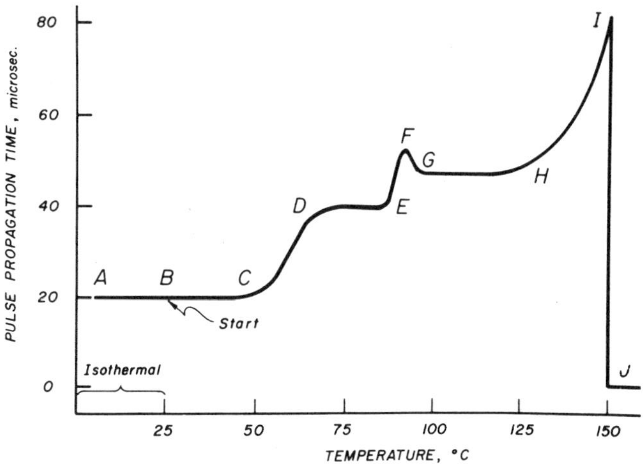

Figure 9. A hypothetical dynamic thermoacoustical curve

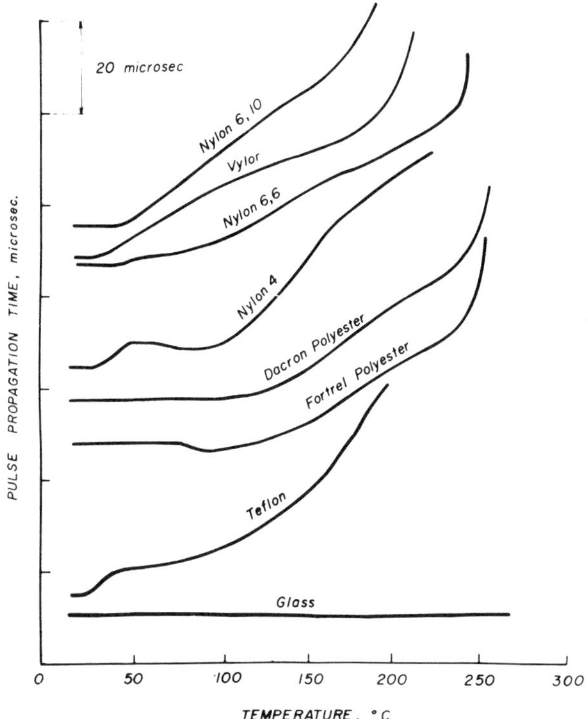

Figure 10. Dynamic thermoacoustical curves of synthetic fibers in air

obtained under the same condition and at the same scale sensitivity. The first curve is for nylon 6,10. The curve shows a distinct upward deviation from the base line at 45°C. This deviation is attributed to the glass transition temperature of nylon 6,10. Premelting behavior of the polymer is revealed by the change of slope of the curve above 150°C. The melting is indicated by the sharp upward trend of the curve and then an instantaneous drop to the zero line (not shown in the chart). It is important to note that above the glass transition temperature, the pulse propagation time increased continuously with the rise of temperature. Prior to melting, however, the propagation time increased at an accelerating rate.

The curves of different fibers all show different characteristic natures. However, as expected, the thermoacoustical behavior of a glass fiber indicates no significant change in the temperature range shown.

The dynamic thermoacoustical curves of cotton and rayon are shown in Figure 11. They are distinctly different, particularly above 250°C. Rayon shows a distinct peak at about 340°C, whereas cotton shows a series of overlapping peaks at higher temperatures. These peaks can be attributed to the decomposition of cellulose fibers. Because of a variety of chemical changes at the decomposition temperature, such as polymer scission, end-group unzipping (11), etc., the velocity of pulses was slowed down resulting in a peak. The right-hand side of the peak indicates the resumption of the original speed as the chemical changes were over and the cellulose molecule was converted to a stable carbonized form. Again, the curve of glass fiber has been included as a reference.

A simultaneous thermoacoustical curve and differential thermal analysis curve were obtained with nylon 6,10 by the technique described earlier. For the thermoacoustical curve, the sample was mounted as shown in Figure 8 and for DTA the sample was cut into small pieces. Both curves were obtained simultaneously as shown in Figure 12.

The dynamic thermoacoustical technique offers an opportunity to develop a new instrumentation in the field of thermal analysis. However, further refinement is required in the experimental technique. More precise measurements can be done by inserting the entire fiber mounting setup and both piezoelectric transducers inside the heating chamber. Atmospheric control and an ability to cool the furnace below room temperature are also essential. In brief, to build a standard instrument, the experimental design shown here would require further modification.

Concluding Remarks

The sonic pulse propagation technique is shown to be an excellent nondestructive method for estimating interfiber bonding in cellulose sheet. The technique also has a unique application for studying the liquid wicking in cellulose sheets. The dynamic

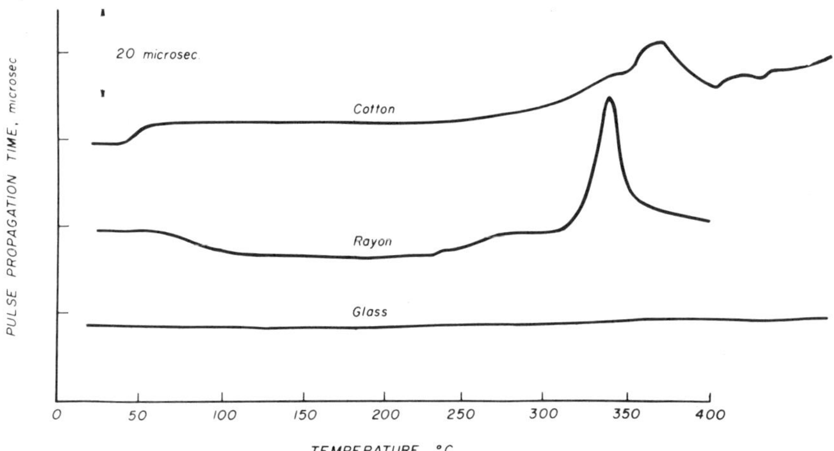

Figure 11. Dynamic thermoacoustical curves of cellulose fibers in air

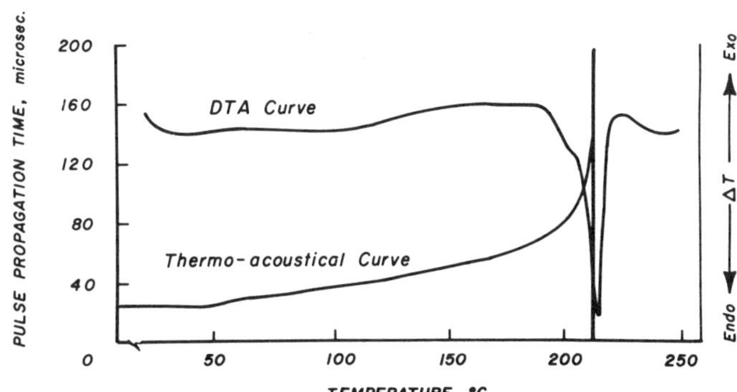

Figure 12. Simultaneous DTA and dynamic thermoacoustical analysis of Nylon 6, 10 in air

thermoacoustical technique described in this paper further expands the thermoanalytical field.

Acknowledgment

The author wishes to acknowledge the management of Personal Products Company, a Johnson & Johnson Company for giving permission to present this paper to the Centennial meeting of the American Chemical Society.

Abstract

The mechanism of sonic pulse propagation in cellulose sheets at isothermal condition and in cotton, rayon and synthetic fibers at dynamic heating conditions have been reviewed.

The velocity of sonic pulse in a dry cellulose sheet is controlled by the proportion, modulus and density of two structural constituents of the sheet: the bonded and unbonded fiber regions. Use of this technique to characterize interfiber bonding and liquid absorption is discussed.

Theory and application of Dynamic Thermoacoustical Analysis are described. The method consists of a continuous measurement of the propagation time of sonic pulses through the sample held under light tension and heated at programmed temperatures. Viscoelastic properties of a variety of synthetic and cellulosic fibers were examined by this technique.

Literature Cited

1. Craver, J. K. and Taylor, D. L., Tappi (1965) 48(3), 142.
2. Chatterjee, P. K., Tappi (1969) 52(4), 699.
3. Taylor, D. L. and Craver, J. K., "Consolidation of the Paper Web", London, British Paper and Board Makers' Association, Trans. Cambridge Symposium, 852 (1965).
4. Chatterjee, P. K., unpublished work.
5. Lu, M. T., Tappi (1975) 58(6), 80.
6. Marton, R. and Crosby, C. M., Tappi (1971) 54(8), 1319.
7. Washburn, E. W., The Phys. Review (1921) 17, 273.
8. Chatterjee, P. K., Sevensk Papperstidning (1971) 74, 503.
9. Chatterjee, P. K., J. Macromol. Sci.-Chem. (1974) A8(1), 191.
10. Chatterjee, P. K., "Thermal Analysis", Vol. 3, Proceedings of the Fourth International Conference on Thermal Analysis, held in Budapest, Hungary, July 8-13, 1974; Akademiai Kiado, Budapest (1975) 835.
11. Chatterjee, P. K. and Conrad, C. M., J. Polymer Sci., Part A-1 (1968) 6, 3217.

13

Heat-Induced Changes in the Properties of Cotton Fibers

S. H. ZERONIAN

Division of Textiles and Clothing, University of California, Davis, CA 95616

Although a considerable number of studies have been made on the effect of heat on the properties of cellulosic materials (1-5), the alterations induced in the cellulose fine structure are incompletely understood. Little is known of the causes of the changes and how the changes affect physical properties of cotton fibers. As noted previously (6), when cellulose is heated in the range 100°C to 250°C some of the observed changes can be explained in terms of alterations in either physical or chemical properties of the material. It is possible also that both can occur simultaneously. In order to obtain clarification, additional studies are needed. Such studies would aid in understanding the mechanism of cellulose pyrolysis.

The work may also have practical implications. It is possible that the physical properties of cotton fibers can be changed advantageously for some end uses by heat treatment. One reason that little consideration appears to have been paid to the utilization of heat for such purposes is the apparent lack of knowledge of the temperature levels before permanent changes occur in the physical properties of the cotton fiber. A number of workers have attempted to determine the glass transition temperature (T_g) of dry cellulose. The values, which have been summarized previously (6), vary from 22 to 230°C. Recently, with the aid of a torsion pendulum, damping peaks were detected on ramie cellulose (6). Two of importance to this paper are at ca 160 and 230°C. Tentative assignments were made for the peaks. One suggestion is that there is a double glass transition in cellulose. Another is that only the damping peak at ca 160°C is associated with T_g for cellulose, and that the damping peak at ca 230°C may be due to release of water during the formation of anhydrocellulose in the amorphous regions of the cellulose. If T_g for cellulose is in the region of 160°C, then it appears that a temperature higher than 160°C is required to bring about, relatively easily, significant changes in the physical properties of dry cotton fibers.

The effect of heat on the stiffness modulus and resiliency (recovery) of dry cotton yarn has been measured by Bryant and Walter (7), and also by Conrad et al. (8). Stiffness modulus was defined by Bryant and Walter as 100 times the stress at 1% elongation. Conrad et al. used essentially the same definition. Recoveries were determined from the relative length of the base lines of stress-strain curves after loading to about 1% extension, and then unloading. Bryant and Walter deduced from their data that T_g for dry cotton is above 240°C. Between 100 and 240°C, the resiliency of their cotton yarn remained roughly constant. However, the stiffness modulus of the yarn remained constant only between 80 and 140°C. Above 180°C it began to decrease relatively rapidly. Bryant and Walter (7) did not comment on this decrease, but it can be postulated that it is an indication of the onset of thermal softening of cotton. In contrast, Conrad et al. (8) found a gradual decrease in stiffness modulus for cotton as the temperature was raised from 100 to 233°C. They stated there was no evidence of a thermal softening temperature. The recovery-temperature curve of Conrad et al. for cotton yarn displayed a continuous increase between 100 and 233°C. Some of the discrepancies between the results of these two groups of workers may have been due to the measured mechanical properties having been affected by yarn structure as well as by the physical properties of the fibers.

Studies have been initiated at Davis to obtain a better understanding of heat-induced changes in the properties of cotton cellulose. In the investigation reported here, the effect of heat on the mechanical properties of cotton yarn in the range of 100 to 240°C was determined to help resolve the discrepancies between the results of Bryant and Walter (7) and those of Conrad et al. (8). Also, the effect of heat in this temperature range on the supramolecular structure of cotton fiber was studied. Depolymerization occurs when cellulose is heated. Thus, the properties of the heat-treated samples were compared with those of acid-hydrolyzed materials to establish if the changes observed in the fine structure of thermally treated fibers could be explained in terms of depolymerization.

Materials

Kier-boiled cotton yarn was used. For mechanical properties, two types were employed, an 80/2's filling twist and a 20's singles mercerizing twist. For the remaining experiments, only the latter yarn was used. Cupriethylenediamine hydroxide (CED) solution was obtained from Ecusta Paper Division, Olin-Matheson Chemical Corp., and SF-96 (50) silicone fluid from Silicone Products Dept., General Electric Co. Other chemicals were reagent grade.

Methods of Treatment

Heat treatments. Cotton yarn was dried for 18 hr at 60°C under vacuum. Phosphorus pentoxide was placed in the vacuum oven to assist the drying. The yarn was then heated in silicone oil in a resin reaction vessel for different lengths of time at various temperatures. Nitrogen was bubbled through the oil continuously throughout the heat treatment. At the end of such treatments, the products were washed with toluene to remove the oil. Then the samples were solvent exchanged with ethanol, followed by ethyl ether. Finally, the samples were exposed to the laboratory atmosphere.

Acid hydrolysis. Cotton yarn (5 g per 250 ml of solution) was hydrolyzed in 2.0 N hydrochloric acid at either 21, 50, or 60°C for different lengths of time, depending on the extent of hydrolysis required. To terminate the reaction, the sample was washed thoroughly with distilled water until it was acid free. To dry the product, the yarn was solvent exchanged with ethanol, followed by ethyl ether. Finally, it was exposed to the laboratory atmosphere.

Level-off degree of polymerization (LODP) samples were prepared in the manner described previously (9).

Characterization of Products

Descriptions have been given previously of the procedures used to determine the following: moisture regains at 59% RH and 21°C (10); infrared spectra (9); and alkali sorption capacity (ASC) (11). Alkali solubility in 0.25 N NaOH was measured by a method essentially similar to that of Davidson (12). Fibers were examined for morphological change with a light microscope at a magnification of 200. The fibers were mounted in 5N NaOH at room temperature. X-ray diffraction measurements were made with a Siemens x-ray diffractometer using a focusing technique in a manner essentially similar to that described by Segal et al. (13). The sample was scanned over the range $2\Theta = 6^{\circ}$ to 30° using K_{α} radiation obtained from a copper target and nickel filter at 35 Kv and 24 ma.

The extent of depolymerization of the degraded samples was determined from viscosity measurements. The samples were sufficiently degraded so that viscosities were determined in CED by ASTM, D1795-62 (14). This method is rapid, and is satisfactory for samples with intrinsic viscosities less than 15. The factor used to convert intrinsic viscosity to DP was 190. The intrinsic viscosity of nondegraded cotton was measured in tris(ethylenediamine)cadmium hydroxide (cadoxen) as described by Henley (15). DP was calculated from the following relation (16).

$$(\eta) = 1.84 \times 10^{-2} \ (DP)^{0.76}$$

Mechanical Properties. Measurements were made with a table model Instron Universal Testing machine equipped with an environmental chamber. Starting with an initial gauge length of 4 inch the yarn, conditioned at 65% RH and 21°C, was extended at a constant rate of 1 inch per min until a load of 40 g had been applied at which time the crosshead was stopped. After 1 min it was returned to the starting position at the same speed used in the elongation phase. After a relaxation time of 1 min, the specimen was extended again at the same rate as before until the specimen became taut. This condition was indicated by the chart pen rising from the zero reading on the chart. The crosshead was again returned to the starting position, and the sample was subjected to the described testing procedure at each temperature as the yarn temperature was raised in five consecutive steps.

Stress decay was determined by measuring the reduction in load as the yarn was held at constant length for 1 min after it had been extended until it bore a load of 40 g, and dividing this value by the initial load (40 g). Recovery was determined by measuring the extension required to increase the load in the yarn from 0 to 40 g (A cm), and also the slack remaining in the yarn after removal of the load followed by 1 min relaxation (B cm). Then,

$$\text{Recovery, \%} = \frac{A - B}{A} \times 100$$

Modulus was measured as the slope of the load-elongation curve at 40 g load. All results are the mean of 6 tests.

The stress decay and modulus data are presented as a fraction of values determined at 100°C.

Results and Discussion

Mechanical properties. The initial portion of the load-extension curve of cotton contains a pronounced toe (Fig. 1), after which the curve appears to straighten. In actual fact, it remains concave. Bryant and Walters (7) did not state whether or not they ignored the toe region in determining the extension applied to their yarn during the measurements relating the effect of temperature on modulus and resiliency of cotton yarn. Conrad et al. (8) included the toe region in extension measurements. With the yarn used in the present work, 1% extension occurred just beyond the toe region. To make measurements unambiguous, it was decided that all testing would be done on yarn extended until it bore a 40-g load. In this manner, measurements were being made in a region well beyond the toe, and thus in an area in which the fibers were in tension and bearing stress. The slope, or modulus, of the load-extension curve at 40-g load was taken as a measure of stiffness, rather than taking the load at 1% extension. As the temperature was raised to 100°C, the yarn shrank in length due to moisture release from the fibers. Shrinkage appeared to be insignificant above 100°C. Thus, only data above 100°C are considered.

To establish if yarn structure was affecting results, two types of yarn were used, namely a low twist singles and an 80/2's yarn. Both yarns have essentially similar results (Fig. 2 and 3 and Table I). All measurements presented were made in air. Pilot measurements were made in a nitrogen atmosphere, but the results were essentially similar to those made in air. The modulus data (Fig. 2) and the resiliency data (Table I) are roughly similar to those of Bryant and Walter (7). In the present case, the modulus of the cotton yarn began to fall at 180°C and had decreased significantly above 200°C with respect to the yarn modulus at 100°C. The stress decay data (Fig. 3) also increased significantly above 200°C. Both the lowering of the modulus and the increase in stress decay are phenomena that could be expected to occur in the region of a glass transition. In our earlier study (6), it was suggested T_g occurs at ca 160°C. It should be noted, however, that the measurements made in the present study would be affected not only by intrafiber properties, but also be interfiber forces. Thus, modulus and stress decay measurements of cellulose in the form of yarn may not be sufficiently sensitive methods for determining T_g. The recovery data (Table I) are scattered and do not indicate a significant trend. Again, these recovery measurements may not be sufficiently sensitive to changes in fiber properties to give a determination of T_g. Back and Didrikson (17) have claimed a secondary transition for cellulose at 175 to 200°C, and a glass transition at 230°C, from measurements made on the modulus of elasticity of paper. Goring (18) has reported that the softening temperature for cellulose ranges between 230 and 250°C depending on the type of cellulose. It should be emphasized, however, that in all the techniques described for

TABLE I

Recovery[a] of two types of cotton yarn progressively heated above 100°C.

Temperature, °C	Yarn Recovery, % 80/2's	20's singles
100	77	72
150	78	79
175	84	79
200	81	78
225	68	75

[a]Recoveries measured after application of 40 g load.

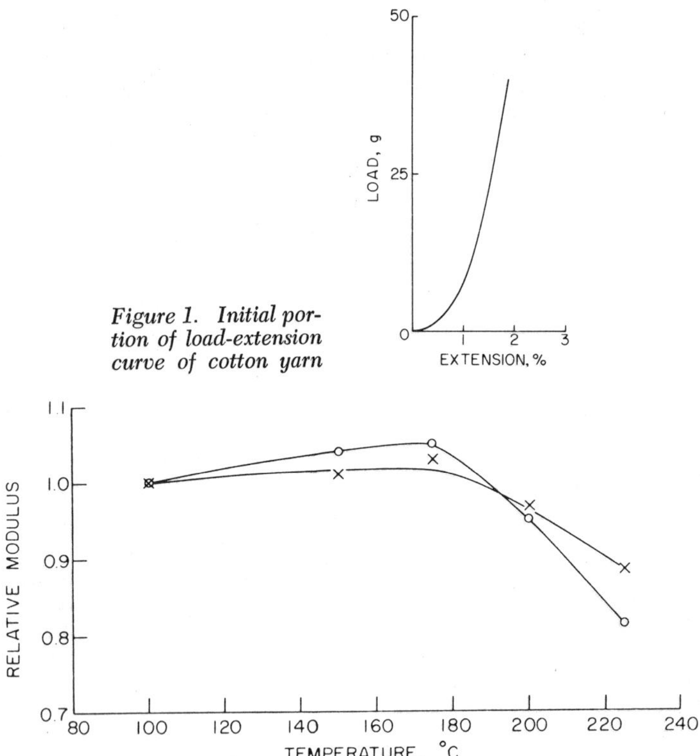

Figure 1. Initial portion of load-extension curve of cotton yarn

Figure 2. Effect of temperature on the relative modulus of cotton yarn. Moduli measured at 40 g load and expressed as a fraction of the modulus at 100°C: (○) 80/2's yarn; (×) 20's singles yarn.

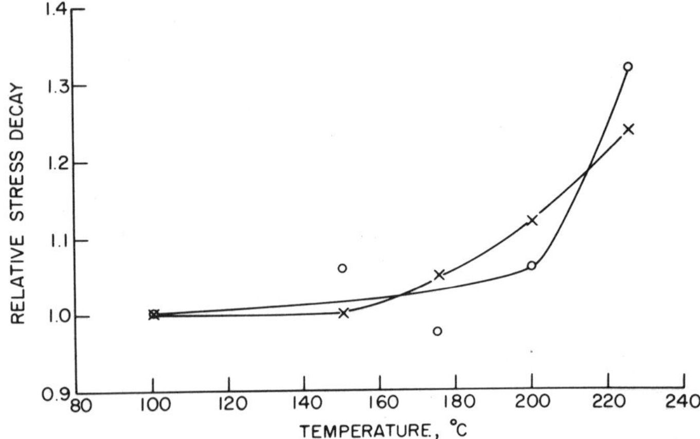

Figure 3. Effect of temperature on the relative stress decay of cotton yarn. Stress decays measured after application of a load of 40 g and expressed as a fraction of the stress decay at 100°C: (○) 80/2's yarn; (×) 20's singles yarn.

measuring transitions namely stiffness modulus, stress decay, elastic modulus, and softening temperature, the data could be affected by the cellulose degradation known to occur during heating. The rate of degradation would accelerate at the higher temperatures. Thus additional work is required before the physical changes described here can be attributed mainly to cellulose being heated above its T_g, and before they can be used to establish accurately T_g for cellulose.

Supramolecular Structure

The results presented in this portion of the paper were obtained with samples heated in silicone oil. The severest thermal treatment used was heating at 240°C for 6 hr. The only obvious change in the infrared spectrum of cellulose caused by this treatment was a shallow absorption band at ca 5.8 μ (Fig. 4). This band has been attributed to C = 0 stretching (acid) (19). Onset of the band occurred at 210°C. In contrast to the white color of the starting cotton, cellulose heated at 190°C for 6 hr had a creamy color. Samples heated for 6 hr at temperatures between 200 and 220°C were beige in color, and those heated at 230 and 240°C were light brown. The crystal lattice of cotton heated at 240°C remained in the cellulose I form (Fig. 5).

Moisture regain can be used as an index of the accessibility of the amorphous regions of cotton. The fraction of amorphous cellulose (F_{am}) can be calculated from Valentine's relation (20), $F_{am} = \frac{SR}{2.60}$. SR is the sorption ratio, which is defined as the ratio of moisture regain of a cellulose to that of nondegraded cotton at the same relative humidity and temperature. Heating for 6 hr reduced the moisture regain of all heat-treated samples (Table II).

It appears that the crystallinity of the fiber was increased by heating. For example, F_{am} decreased from 0.38 to 0.30 by heating at 240°C. The DP of the samples fell during heat treatment indicating that chain scission had occurred (Table II). The ability to crystallize could be increased by chain scission since this would permit the cellulose chains to realign themselves more easily and then crystallize.

In heterogeneous acid hydrolysis, depolymerization occurs within the amorphous regions of cellulose fibers, and it is generally accepted (21, 22) that crystallization can occur due to realignment of the cellulose molecules following chain scission. Thus, to establish the mechanism of crystallization by heat, the moisture regain of cotton cellulose after heterogeneous acid hydrolysis was determined for comparison with heat-treated samples. As the DP decreased, the moisture regain of acid-hydrolyzed samples passed through a maximum, and then fell below the moisture regain of the nonhydrolyzed fiber (Table III). The initial increase in moisture regain may have been due to

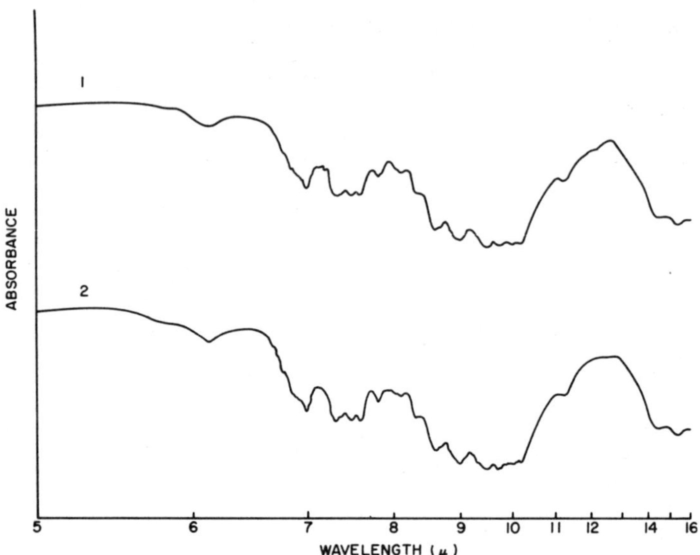

Figure 4. Infrared spectra of (1) cotton cellulose, (2) cotton cellulose heated at 240°C for 6 hr

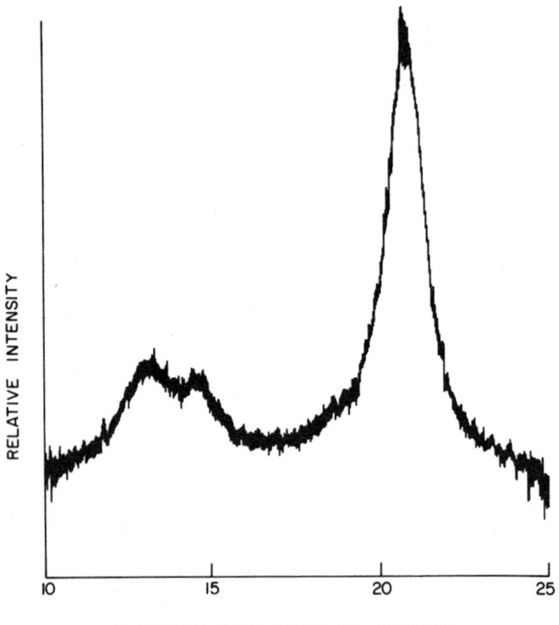

Figure 5. X-Ray diffractogram of cotton cellulose heated at 240°C for 6 hr

TABLE II

Effect of heating for 6 hr in silicone oil at various temperatures above 160°C on degree of polymerization (DP), alkali sorption capacity (ASC), moisture regain (MR) and fraction of amorphous material (F_{am}) of cotton.

Temperature °C	DP[a]	ASC %	MR %	F_{am}[b]
Control	5360	208	6.23	0.38
165	1370	238	5.84	0.36
175	1310	237	5.90	0.36
180	1280	244	5.89	0.36
190	1230	224	5.87	0.36
200	730	302	5.54	0.34
210	910	264	5.64	0.35
220	640	300	5.46	0.34
230	420	334	4.94	0.30
240	320	395	4.85	0.30

[a] DPs measured in CED with the exception of non-heated starting cotton in which case DP was measured in cadoxen.

[b] F_{am} calculated from relation,

$$F_{am} = \frac{\text{sorption ratio}}{2.60}$$

TABLE III

Effect of lowering the degree of polymerization (DP) of cotton by acid hydrolysis[a] on alkali sorption capacity (ASC), moisture regain (MR), and fraction of amorphous material (F_{am}) of cotton.

DP[b]	ASC %	MR %	F_{am}[c]
5360[d]	208	6.23	0.38
1130	226	6.70	0.41
1010	234	6.58	0.41
670	262	6.39	0.39
460	284	6.30	0.39
400	295	5.46	0.34
260	imm[e]	5.25	0.32
210	imm[e]	5.15	0.32

[a] Hydrolysis in 2 N HCl for various lengths of time at temperatures between 20°C and 60°C.

[b] DPs measured in CED with the exception of the nonhydrolyzed starting cotton in which case DP was measured in cadoxen.

[c] F_{am} calculated from relation $F_{am} = \frac{\text{sorption ratio}}{2.60}$

[d] Nonhydrolyzed starting cotton.

[e] Immeasurable.

additional regions within the amorphous areas of the fiber being made accessible to moisture as chain scission occurred. Supporting evidence that such regions exist can be found in acetylation studies (23) where it has been shown that the moisture regain of cotton cellulose is increased by a small amount of acetylation. At DPs lower than 460, the moisture regain fell below that of the starting cotton. In this case, crystallization has been caused by chains within the amorphous regions realigning themselves after chain scission. However, it will be noted that, for a given DP, below a DP of 460, the moisture regain of the hydrolyzed sample remained higher than that of the heat-treated sample (Fig. 6). Thus the heat-treated samples have higher crystallinity, as can be seen by a comparison of F_{am} values, at low DPs, for heat-treated and acid-hydrolyzed samples (Tables II and III). It appears, therefore, that the increased mobility of the cellulose chains at higher temperatures contributes to crystallization. Since the moisture regain of the acid hydrolyzed materials did not fall below that of the non-hydrolyzed material until the DP fell below 460 it is likely that the ability to crystallize by heat treatment is not affected by chain scission until the DP falls below this value. Atalla and Nagel (24), working with cellulose in the form of mercerized cotton and mercerized Avicel, have presented x-ray evidence indicating that heat will induce crystallization, and have suggested that it was due to annealing. However, the present results indicate that, at least for fibrous materials, crystallization is assisted by chain scission as well as by heat when DP is reduced below 460 by heat degradation.

The crystallization due to the heat treatments may result from an increase in the size of preexisting crystallites by, for example, the realignment of cellulose chains on crystallite surfaces, or at their ends. It is possible also that the crystallization occurs by the formation of completely new crystallites within the amorphous areas.

To determine whether existing crystallites had changed in length, LODP of heat-treated samples was measured; and, for comparison, the LODP of acid-hydrolyzed samples was determined also. LODP is a relative measure of crystallite length (9). The LODP of the nondegraded cotton was 198 (Table IV). The severest heat treatment had reduced the LODP by 10%. In contrast, the LODP of hydrocelluloses, with DPs comparable to those of the severely heat-treated samples, was similar to that of the nondegraded sample. It is interesting to note from residue determinations (Table IV) that the severest heat treatments yielded less crystalline product than did either the nondegraded material or samples degraded by acid hydrolysis. It thus appears, from the LODP data that heating at a temperature of 230°C or above damaged the crystallites in the cotton cellulose. This damage could be random chain scission within the crystallites, or it could be confined at the ends of the

TABLE IV

Level-off Degree of Polymerization (LODP) of heat-treated and acid-hydrolyzed cotton cellulose.

Sample	DP[a] of sample	LODP[b]	Residue[b,c] %
Starting cotton	5360	198	96
Heat treated 6 hr at 165°C	1370	194	93
Heat treated 6 hr at 190°C	1230	194	95
Heat treated 6 hr at 230°C	420	178	90
Heat treated 6 hr at 240°C	320	178	90
Acid hydrolyzed 2 hr at 50°C	670	197	96
Acid hydrolyzed 6 hr at 50°C	460	194	95

[a] DPs measured in CED with the exception of starting cotton in which case DP was measured in cadoxen.

[b] LODP and residue determined after 15 min hydrolysis with boiling 2.5 N HCl.

[c] Expressed in terms of dry weights.

crystallites. If random chain scission had occurred, then it is probable that the amount of residue obtained from the preparation of LODP material would have remained similar to that from nondegraded cellulose, while the DP of the product dropped. Thus it is speculated that the thermal damage is mainly confined to the ends of the crystallites, and that it consists of nicks or cracks perpendicular to the b axis of the crystallites. Acid would be able to penetrate these fissures, and thus thermally damaged material could be removed by hydrolysis. However, inner portions of these damaged regions would remain inaccessible to water vapor, and therefore would still be considered crystalline. The fact that heat increased the crystallinity of cellulose, as measured by moisture regain determinations, even though LODP decreases, may be explained by the formation of junction points (25) within the amorphous regions after chain scission occurs. Junction points are defined here as small regions of parallel chains which are inaccessible to water vapor but which would be eliminated in the preparation of LODP material.

It has been stated by earlier workers (5, 26) that when cellulose is heated a breakdown occurs, and the DP approaches the LODP value of the sample. In this respect, it was thought (5), thermal degradation resembles heterogeneous acid hydrolysis, where a LODP value is normally encountered. The present results also indicate heat may reduce the DP of the product to a value close to that of the LODP of the nondegraded starting material. After heating at 240 for 6 hr, the DP of the cellulose was 320, while the LODP of the nondegraded material was 198. However, as the DP of the thermally treated product approaches the LODP of

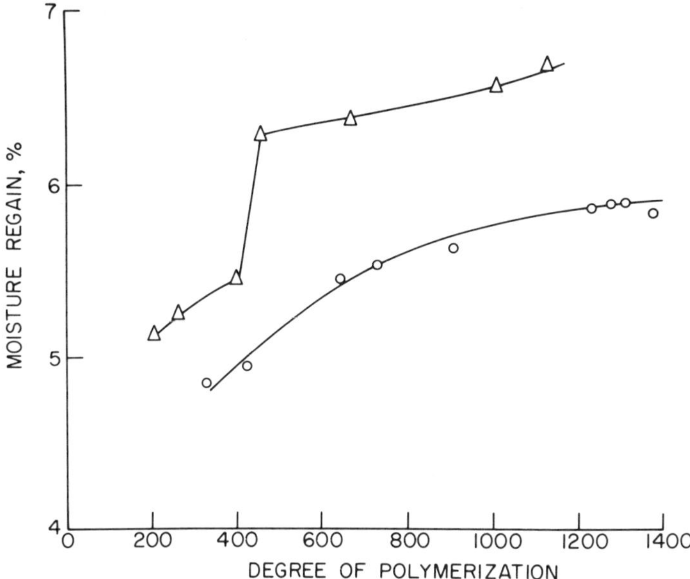

Figure 6. Relation between degree of polymerization (DP) and moisture regain for highly degraded cotton cellulose: (○) heat treated; (△) acid hydrolyzed. Note: DPs measured in CED.

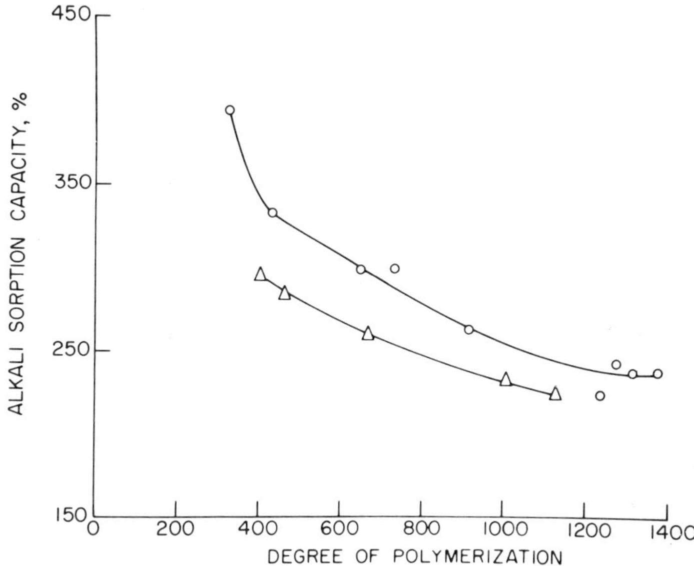

Figure 7. Relation between degree of polymerization (DP) and alkali sorption capacity for highly degraded cotton cellulose: (○) heat treated, (△) acid hydrolyzed. Note: DPs measured in CED.

the nondegraded starting material, the LODP of this degraded product falls, indicating that damage has not been confined to the amorphous regions. In the present example, the LODP of cellulose heated at 240°C for 6 hr was 178. Thus the resemblance between heat degradation and acid hydrolysis may be more apparent than real.

Differences between acid-hydrolyzed and heat-degraded cellulose can also be detected by ASC and alkali solubility determinations. ASC is a measure of the swelling of cotton fiber in 15% NaOH. The concentration of alkali is sufficiently high that intra- as well as inter-crystalline swelling will occur. For a given temperature, the ASC varied as the time of heating increased. At temperatures between 110 and 140°C, the ASC decreased as the time of heating increased. For example, at 130°C the ASC fell from 213 to 199, as the time of heating was raised from 2 hr to 6 hr. In contrast, at 190°C and above, the ASC increased with increasing heating time. For example, at 210°C the ASC increased from 237 to 264 as the heating time was raised from 2 hr to 6 hr.

The ASC of the nonheated cotton was 208. The ASC of samples heated for 6 hr began to increase rapidly above 190°C (Table II). As noted earlier, degradation occurred as the samples were heated, resulting in a decrease of DP (Table II). It appears that the increased swelling ability of the fiber is related to the chain scission which occurs. This is supported by acid hydrolysis data. As the DP of cotton was decreased by acid hydrolysis, ASC again increased (Table III). However, for a given DP, the ASC of the heat-treated samples remained greater than that of the acid-hydrolyzed samples (Fig. 7). The greater swelling of the heat-treated samples at lower DPs may have been due, in part, to the thermal degradation which occurred in the crystalline regions of the fiber, permitting higher overall swelling. The increased swelling could also have resulted from changes that occurred in the morphology of the heat-treated fibers. Using a light microscope, nicks and cracks could be easily detected in fibers heated at 230°C and higher, and then swollen in alkali. Minimal damage to the fiber structure was detected by this method for fibers heated at lower temperatures. Also, no damage could be detected in acid-hydrolyzed fibers of DP of 400 after alkali swelling.

It can be deduced from the alkali solubility determinations that, for materials of low DP, there are chemical differences between acid-hydrolyzed and heat-treated cellulose (Table V). Alkali solubilities were measured after heating the samples in 0.25 N NaOH at 100°C for 6 hrs. Cellulose is attacked at the reducing end by hot alkali (27, 28). Nondegraded cellulose is attacked only very slowly because of the small number of carbonyl groups compared with the number of glycosidic linkages. The alkali solubility of the sample used in the present determinations was 3.1%. This value increased to 9.9% when the DP

TABLE V

Alkali solubility of heat-treated and acid-hydrolyzed cotton cellulose.

Sample	DP[a]	Alkali solubility[b] %
Starting cotton	5360	3.1
Heat treated 6 hr at 230°C	420	9.0
Heat treated 6 hr at 240°C	320	9.7
Acid hydrolyzed 2 hr at 50°C	670	9.9
Acid hydrolyzed 2 hr at 60°C	400	20.0

[a] DPs measured in CED with the exception of nonheated starting cotton in which case DP was measured in cadoxen.

[b] Alkali solubility measured using 0.25 N NaOH at 100°C for 6 hr. Results expressed as percentage loss of weight in terms of dry weights.

of the sample was decreased to 670 by acid hydrolysis. Hydrocelluloses suffer increased degradation owing to the increased number of carbonyl groups. It is interesting, however, that at a given DP for extensively damaged samples, heat degradation did not increase alkali solubility as much as did acid hydrolysis. It appears, therefore, that the heat-degraded materials contain few reducing ends. Other evidence, which supports this, is that cotton cellulose heated at 270°C under nonoxidative conditions contains a minimal number of reducing groups although the DP is very low (29).

In summary, measurements of the modulus and stress decay of cotton yarn at temperatures between 100 and 240°C can be interpreted as showing a glass transition in cellulose at ca 200°C. However, it will be difficult to use such measurements to obtain an accurate determination of T_g since the mechanical properties of cotton yarn are influenced by inter-fiber as well as intra-fiber forces. Also, degradation of the cellulose occurs during heating and this could affect the accuracy of the T_g measurement. Secondly, there are differences in the supramolecular structure of degraded cellulose, depending on whether it is degraded by acid or by heat. However, a comparison of the properties of heat-degraded celluloses and acid hydrolyzed celluloses can be useful in clarifying the changes occurring to cellulose by heat, provided alterations in chemical properties are taken into account.

Acknowledgments

The author is grateful to Mr. K. Alger, Miss M. Coole, Miss A. Cordy and Miss P. Howland for technical assistance.

Literature Cited

1. McBurney, L. F. in: "Cellulose and Cellulose Derivatives", Part 1, Ott, E., ed., p. 174 et seq., Interscience Publishers Inc., New York, 1954.
2. Nikitin, N. I., "The Chemistry of Cellulose and Wood", p. 585 et seq., Israel Program for Scientific Translations, Jerusalem, 1966.
3. Rozmarin, G. N., Russian Chemical Reviews, (1965) 34, 854.
4. Segal, L., in: "Cellulose and Cellulose Derivatives", Part V, Bikales, N. M., and Segal, L., eds., p. 736 et seq., Wiley-Interscience, New York 1971.
5. Kilzer, F. J., in: "Cellulose and Cellulose Derivatives", Part V, Bikales, N. M., and Segal, L., eds., p. 1015 et seq., Wiley-Interscience, New York, 1971.
6. Zeronian, S. H. and Menefee, E., Appl. Polymer Symp. (1976) No. 28, 869.
7. Bryant, G. M. and Walter, A. T., Text. Res. J., (1959) 29, 211.
8. Conrad, C. M., Harbrink, P. and Murphy, A. L., Textile Res. J., (1963) 33, 784.
9. Zeronian, S. H., J. Appl. Polym. Sci., (1971) 15, 955.
10. Zeronian, S. H., J. Appl. Polym. Sci., (1965) 9, 313.
11. Zeronian, S. H. and Miller, B. A. E., Textile Chemist and Colorist, (1973) 5, 89.
12. Davidson, G. F., J. Text. Inst., (1941) 32, T109.
13. Segal, L., Creely, J. J., Martin, A. E., Jr., and Conrad, C. M., Text. Res. J., (1959) 29, 786.
14. ASTM Standards, Part 15, American Society for Testing and Materials, Philadelphia, 1970.
15. Henley, D., Arkiv. Kemi., (1961) 18, 327.
16. Segal, L. and Timpa, J. D., Svensk Papperstidn., (1969) 72, 656.
17. Back, E. L. and Didriksson, E. I. E., Svensk Paperstidn., (1969) 72, 687.
18. Goring, D. A. I., Transactions Cambridge Symp. Consolidation of the Paper Web., p. 555, B. P. and B. M. A., London, 1966.
19. Higgins, H. G., J. Polym. Sci., (1958) 28, 645.
20. Valentine, L., Chem. Ind. (London) (1956) 1279.
21. Sharples, A., J. Polym. Sci., (1954) 13, 393.
22. Immergut, E. A. and Ranby, B. G., Ind. Eng. Chem., (1956) 48, 1183.
23. Nevell, T. P. and Zeronian, S. H., Polymer (1962) 3, 187.
24. Atalla, R. H. and Nagel, S. C., Polymer Letters, (1974) 12, 565.
25. Herman, P. H., "Physics and Chemistry of Cellulose Fibres". Elsevier Publishing Co., Inc., New York, 1949.
26. Dmitrieva, O. A., Potapova, N. P. and Sharikov, V. I., Zh. Prikl. Khim., (1964) 37, 1583.
27. Davidson, G. F., J. Text. Inst., (1943) 34, T87.

28. Corbett, W. M., in: "Recent Advances in the Chemistry of Cellulose and Starch". Honeyman, J., ed., p. 126 et seq. Heywood & Co. Ltd., London, 1959.
29. Cabradilla, K. E. and Zeronian, S. H., to be published.

14

Infrared and Raman Spectroscopy of Cellulose

JOHN BLACKWELL

Department of Macromolecular Science, Case Western Reserve University, Cleveland, OH 44106

The technique of polarized infrared spectroscopy was first used to study the structure of cellulose by Tsuboi(1), Marrinan and Mann(2,3), and Liang and Marchessault(4-6), in the late 1950s. At that time the crystal structures of celluloses I and II were unknown, and the spectroscopic investigations were undertaken to determine the orientations of the O-H and CH_2 groups, and hence to predict the hydrogen bonding network between the chains. The polarity of neighboring chains in the structures was unknown, and a variety of possible hydrogen bonding schemes needed to be considered. The infrared results allowed for elimination of some of these possibilities, but it was not possible to select the actual structure for any of the polymorphs.

In the last three years, the structures of cellulose I and II have been determined by x-ray diffraction methods. Figures 1 and 2 show the refined structures of celluloses I and II respectively, as determined in this laboratory(7-10). The methods used in this work are described elsewhere in the proceedings of this meeting by Blackwell, Kolpak and Gardner(11). The structures have also been determined independently by Sarko and coworkers(12,13), and the conclusions for cellulose I have been confirmed by French(14). Cellulose I consists of an array of _parallel_ chains, i.e. all the chains in a particular crystallite have the same sense. In contrast, neighboring chains in cellulose II are _antiparallel_ i.e. have alternating sense. The hydrogen bonding scheme is determined satisfactorily for both forms. In cellulose I all the CH_2OH side-chains have the _tg_ conformation; each residue forms two intramolecular hydrogen bonds (O3-H···O5' and O2'-H···O6), and one intermolecular hydrogen bond (O6-H···O3) to the neighboring chain along the _a_ axis. As a result, the cellulose I structure is an array of hydrogen bonded sheets of chains, which have a relative stagger of approximately a quarter of the fiber repeat, but are otherwise identical. In cellulose II, half of the chains form sheets identical to those in cellulose I. For the chains of opposite

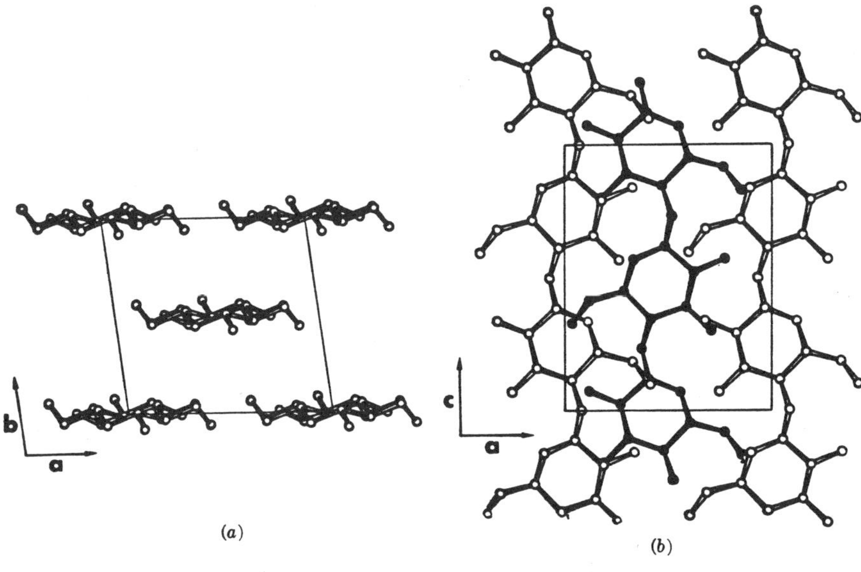

Figure 1. The structure of cellulose I: (a) ab *projection; (b)* ac *projection (8)*

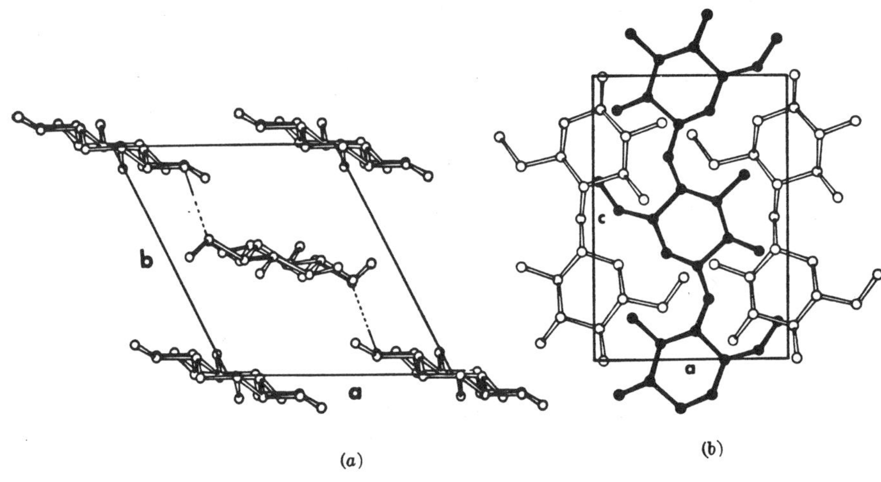

Figure 2. The structure of cellulose II: (a) ab *projection; (b)* ac *projection (10)*

polarity, however, the CH_2OH groups have the gt conformation such that the intramolecular hydrogen bond between 06 and 02'-H is not possible. These two hydroxyl groups both form intermolecular hydrogen bonds: 06-H•••02 to the next chain along the a axis and 02'-H•••02 to the next chain in the 110 plane i.e. on the long diagonal of the ab projection of the unit cell (see Figure 2a). This additional intermolecular hydrogen bond probably contributes to the higher stability of cellulose II.

Now that the structures are determined, the situation with respect to vibrational spectroscopy of cellulose is reversed, in that we are now attempting to explain the observed spectra in terms of the known structures. Even so, the spectra of cellulose are highly complex, and their interpretation is not straight forward. However, such studies are necessary so that we can understand the structural origin of the vibrational spectra of polysaccharides. Cellulose is one of the simplest polysaccharides and detailed structures are available for more than one polymorphic form. If we can interpret the spectra in terms of the different structures, we can then apply these techniques to more complex carbohydrates, for which the structures have not been determined.

In this paper I will review the work we have done in the last few years towards understanding the vibrational spectra of cellulose. In particular, we have studied the spectra of a number of model compounds, including deuterated glucoses, which have enabled us

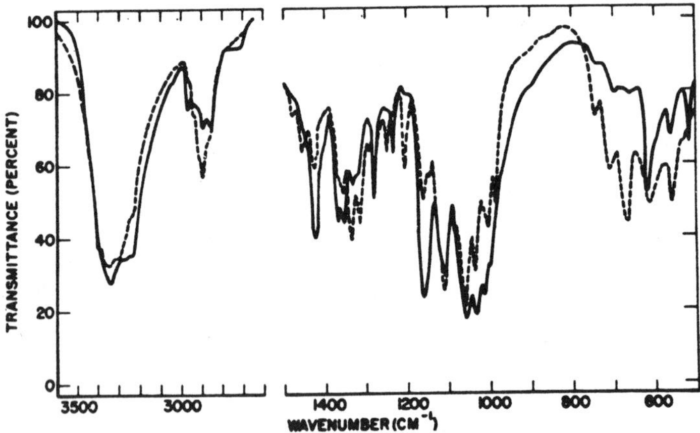

Journal of Applied Physics

Figure 3. Polarized infrared spectrum of oriented fibers of Valonia *cellulose. —— electric vector parallel to the chain axis; – – – electric vector perpendicular to the chain axis* (15).

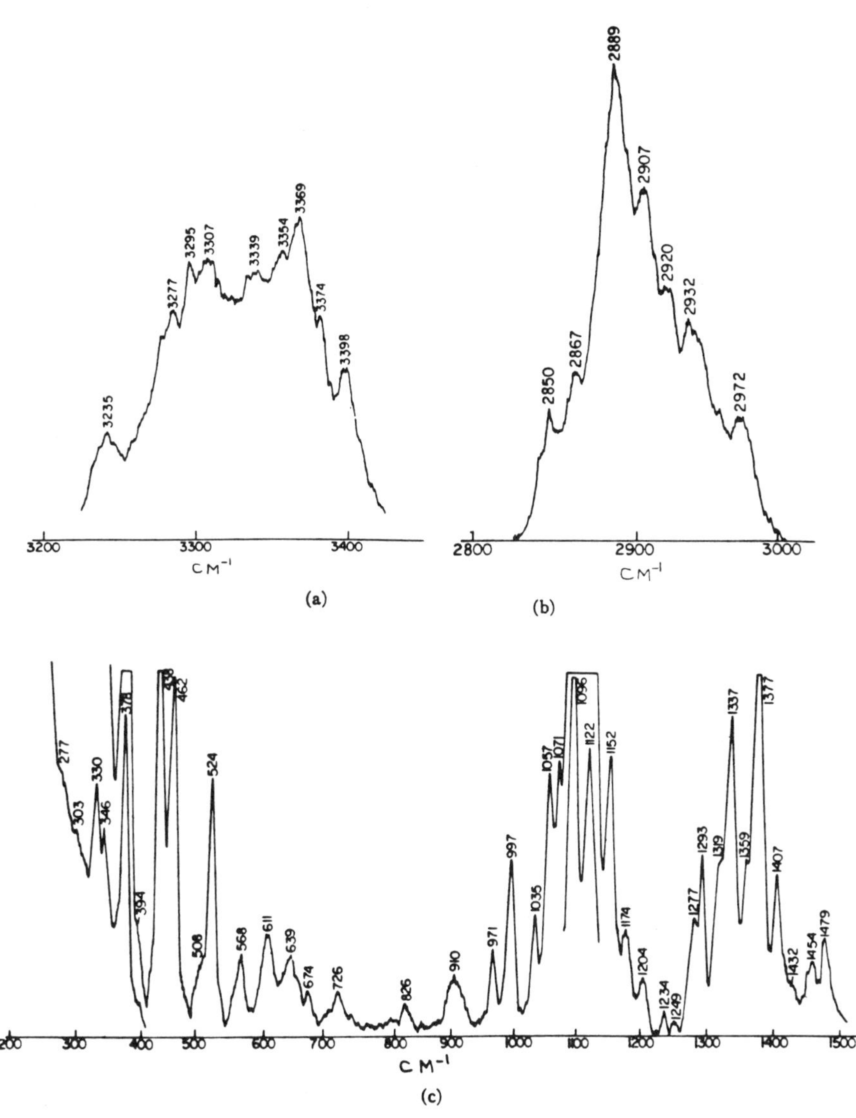

Journal of Applied Physics

Figure 4. Raman spectrum of an unoriented specimen of Valonia *cellulose: (a) O–H stretching region, (b) C–H stretching region, and (c) 1500–200-cm*$^{-1}$ *region (15).*

to build up a list of band assignments, especially for the mixed modes in the region below $1500cm^{-1}$. We have also applied normal coordinate analysis to predict the vibrational modes for an isolated cellulose chain.

Observed Spectra for Cellulose

The polarized infrared spectra of *Valonia* cellulose(I) are shown in Figure 3. These spectra were obtained(15) for thin fibers drawn from the purified cell wall and arranged parallel on a silver chloride window. The spectra were recorded using a Perkin Elmer 521 infrared spectrophotometer, equipped with a gold grid polarizer. The frequencies and dichroisms of the observed bands are given in Table I.

The Raman spectrum of *Valonia* cellulose is shown in Figure 4. These data were obtained(15) for an unoriented section of the cell wall which was soaked in water and allowed to dry on a glass slide. The spectrometer consisted of an argon ion laser, a Spex 1400 double monochrometer and a photomultiplier detector. The observed frequencies are listed on Table I.

Band Assignments

Identification of the infrared and Raman frequencies is necessary before any structural interpretations can be made. The assignments we have made so far are listed in Table I, in which we have built on the previous work described in references 1-6. The O-H and C-H stretching bands can be identified relatively easily in their separate regions of the spectra. Seven frequencies are resolved in the C-H stretching region and match the seven expected for a single glucose: five C-H plus the symmetric and antisymmetric CH_2 stretching modes. The bands at 2932 and $2850cm^{-1}$ were assigned to the antisymmetric and symmetric CH_2 stretching modes respectively by Liang and Marchessault(4). Frequencies close to these values for α-D-glucose disappear when the CH_2 group is converted in CD_2. These bands for cellulose have perpendicular and parallel dichroism respectively, and are consistent with either the *tg* or the *gt* conformation in cellulose I, but rules out the *gg* conformation. The *tg* conformation occurs only rarely in crystalline structures of sugar monomers and oligomers, and the *gt* conformation was thought the most likely CH_2OH orientation in cellulose(16), until *tg* was identified by the x-ray work. These two bands have the same dichroisms in the spectra of cellulose II where there is a mixture of *tg* and *gt*.

Six O-H stretching bands are seen in the infrared spectrum of cellulose I, of which four have parallel and two have perpendicular dichroism. The structure determined by x-ray diffraction has intramolecular hydrogen bonding for the O3-H and O2-H groups, and these hydroxyl bonds are approximately parallel

Table I

VIBRATIONAL SPECTRA OF CELLULOSE I. CALCULATED FREQUENCIES FROM NORMAL COORDINATE ANALYSIS, OBSERVED INFRARED AND RAMAN FREQUENCIES AND INTENSITIES, INFRARED DICHROISMS, AND EXPERIMENTAL BAND ASSIGNMENTS

Calc. Freq. (cm^{-1})		Obs. Freq. (cm^{-1}) Intensities and Dichroisms		Assignments
A Modes	B Modes	I.R.	Raman	
		3408w[a] ⊥	3398w	
		3376w ∥	3374w	
			3369m	
3398	3398	3347vs ∥	3354m	
3398	3398		3339m	
3398	3398	3306w ⊥	3307m	O-H Str.
			3295m	
		3271m ∥	3277m	
		3238m ∥	3235w	
2961	2961	2966w ∥	2972w	C-H Str.
2946	2945	2942w ⊥	2932m	CH_2 Antisym. Str.
2941	2942	2919vw	2920w	
2937	2937	2911vw ⊥	2907m	C-H Str.
2933	2932	2894m ⊥	2889m	
2929	2929	2866w ⊥	2867w	
2868	2868	2853w ∥	2850w	CH_2 Sym. Str.
1485	1485	1482w ⊥	1479m	
1434	1434	1455w ⊥	1454w	C-O-H Def.
1424	1423	1426m ∥	1432vw	CH_2 Def.
		1405w	1407m	C-O-H, C-C-H Def.
1372	1372	1372m ∥	1377s	
1368	1368	1360w ∥		
1355	1355	1357m ∥	1359vw	C-C-H Def.
1331	1333	1334m ⊥	1337s	CH_2 and C-O-H Def.
1327	1326	1315m ⊥	1319vw	
1309	1310			
1299	1297	1297w ⊥	1293m	
1285	1284	1280m ∥	1277w	CH_2-O-H Def.
1282	1282	1270w ⊥		
1246	1246	1249w ⊥	1249vw	C-C-H Def.
1239	1242			
1229	1230	1233w ∥	1234vw	C-O-H Def.
1206	1205	1205w ⊥	1204vw	CH_2 and C-O-H Def.
1182	1187			
1169	1170	1163s ∥	1174w	
1152	1148	1148w ⊥	1152m	
1141	1137	1130w		

Table I (continued)

Calc. Freq.(cm^{-1})		Obs. Freq.(cm^{-1}) Intentsities and Dichroisms			Assignments
A Modes	B Modes	I.R.		Raman	
1113	1122	1112s	⊥	1122vs	
1098	1097	1090vw	⊥	1090vs	C-O-H Def.
1090	1074			1071w	
1055	1060	1060vs	∥	1057m	C-C-H Def.
1043	1037	1035vs	∥	1035w	C-O-H Def.
		1011m	∥		
986	1008	1000w	⊥	997m	
961	964	984w	⊥	971w	
942	956				
893	897	893vw		910w	C1-H, CH_2 and C-O-H Def.
				825vw	
		750w	⊥		
727	669	710w	⊥	726vw	
665	627	668m	⊥	674vw	
601	572	624m	⊥	639vw	
		618w		611w	
568	546	565vw	⊥	568w	
532	526	535vw	⊥	524s	
				508w	
457	448	460w	∥	462vs	
435	427	445w	⊥	438vs	
387	386	394vw			
		378s			
	362	366vw			
347	356	346w			
		330m			
302	299	303w			
291	288	277vw			
250					
240	242				
233	240				
220	237				
182	231				
142	192				
121	159				
79	92				
64	84				
38	59				
31	14				

[a]Key: vs-very strong, m-medium, vw-very weak, w-weak.

to the fiber axis. The O6-H group is oriented approximately perpendicular to the fiber axis. This 2:1 parallel to perpendicular orientation of the hydroxyls is consistent with the 4:2 ratio for the dichroism of the O-H stretching bands. The unit cell contains six nonidentical hydroxyl groups; some coupling of these motions is likely giving rise to splitting such that nine frequencies are resolved in the Raman spectrum. [This will be discussed further below]. The main differences between cellulose I and II occur in the O-H stetching region, where the phase transformation is characterized by the development of two strong bands in the infrared spectrum at 3448 and 3480cm^{-1}. In general the O-H stretching frequencies of cellulose II are at higher wave numbers than those of cellulose I, which has been interpreted as due to weaker hydrogen bonding in form II. This contrasts with the results from x-ray diffraction, where the average hydrogen bond length is 2.80Å in cellulose I and declines to 2.72Å in cellulose II. Although the x-ray refinement cannot resolve individual bond lengths, this average difference may reflect a slightly tighter packing of the chains in cellulose II, and is consistant with the higher stability of this form. The infrared results may be due to the change in polarity of the chains: the antiparallel arrangement may result in cellulose II being a less polar structure, with a result that there is less perturbation of the free O-H stretching frequency due to the effects of hydrogen bonding.

Cellulose is one of the few polysaccharides giving detailed spectra in the O-H and C-H stretching regions, for which a crystalline nonhydrated structure is necessary. Most polysaccharides give poorly resolved spectra in these regions due to poor crystallinity and absorbed water. For this reason, if useful information is to be obtained for these structures, it must come from the region below approximately 1600cm^{-1}. The 1600-800cm^{-1} region for cellulose contains a large number of frequencies. It is significant, however, that the same frequencies are observed for other glucose monomers, oligomers, and polymers. It is likely that these frequencies are due to stretching and bending modes for the common glucose residue, and thus assignments for one compound should be transferable to the group as a whole.

The infrared spectra of crystalline specimens of glucose were compared with those of three C-deuterated D-glucoses, deuterated at C1, C6 (CD_2), and both C1 and C2, which allowed identification of modes containing contributions from C1-H, C2-H and CH_2 deformations(17). In addition, the Raman spectra of D-glucose, maltose, cellobiose, and dextran, in both H_2O and D_2O solutions were compared, from which the C-O-H deformation modes were identified. These assignments were then transfered to cellulose, and are listed in Table I.

Modes at 1426, 1334, 1280, 1205, 1011, and 895cm^{-1} are identified as having extensive contribution from CH_2 deformations.

The $1426cm^{-1}$ band in the infrared spectrum declines in intensity on conversion of cellulose I to cellulose II(18), which is correlated with a change in the conformation of half of the $-CH_2OH$ groups from *tg* to *gt*. In the Raman spectra of amylose, the first three of the above frequencies show changes in frequency and intensity when V-amylose is converted to the B-form(19). The latter change is thought to involve changes in the orientation and hydrogen bonding of the $-CH_2OH$ group.

The similarity of the spectra of the glucose oligomers and polymers in the $1500\text{-}800cm^{-1}$ region is not maintained at lower frequencies. Below $800cm^{-1}$ the spectra are highly characteristic of the particular compound, and the polymorphic forms of cellulose differ substantially in this region. Atalla(20) has studied the Raman spectra of different celluloses, and has shown that the spectra of cellulose III and IV below $800cm^{-1}$ are both composites of those for cellulose I and II. X-ray work indicates that same basic 2_1 chain conformation for all of these structures, and we believe that the spectral differences are probably due to differences in the side chain conformation and particularly the chain packing. Currently we have no way of interpreting these interesting differences in terms of the structures. This region contains predominently skeletal modes, for which it is not possible to predict the effects of packing forces.

Normal Coordinate Analysis

In order to obtain a better understanding of the origin of the frequencies in the deformation region, we have performed normal coordinate analysis for isolated α- and β-D-glucose(21,22) molecules and a single cellulose chain(23). For cellulose, the carbon and oxygen coordinates were taken from the x-ray analysis by Gardner and Blackwell(8). The hydrogens were added at the appropriate bond lengths and angles. The normal modes of vibration were then calculated using a modification of the Wilson GF matrix method(24), as described by Borio and Koenig(25). The chain possesses a 2_1 screw axis, with the result that the normal modes can be factored into two sets, the A and B species, which are respectively symmetric and antisymmetric relative to the screw axis. The force constants were transfered from the valence force fields reported for aliphatic acids(26) and for ethers(27), which we had previously used in our calculations for α- and β-D-glucose(21,22). The calculations were for an isolated chain and did not consider intermolecular forces, except that the force constants for O-H stretching and deformation were for hydrogen bonded molecules.

The calculations for cellulose(23) yield 122 non zero frequencies, which are divided into 61 A and B pairs, as listed in Table I. Negligible splitting occurs between the A and B modes in all but a few cases in the range $1500\text{-}800cm^{-1}$. Hence most of the observed frequencies in this region can be assigned to super-imposed A and B modes. This is probably due to the large size of the glucose residue, such that to a first approxiation the observed frequencies correspond to motions of an isolated monomer unit, i.e. there is negligible contribution from inter-residue coupling through the glycosidic bond. Where splitting does occur, there is appreciable motion of the linkage atoms. This means that for most polysaccharides, calculations of only the Amodes would be adequate when studing the 1500-800cm^{-1} region, which is an appreciable saving of computer time for helices with higher symmetry, as in our own calculations for the 6_1 helix of V-amylose (28).

The calculated frequencies are listed alongside the observed frequencies in Table I, where it can be seen that good agreement exists between them. However, with such a large number of observed and calculated frequencies, some measure of agreement is inevitable, and the match must be judged on other criteria, specifically how well the atomic motions and potential energy distribution for the calculated modes are in agreement with the experimental assignments. On this basis, the predictions are in good agreement, in that the calculated frequencies are not only close to the observed but also correspond to modes which contain the contributions indicated from the deuteration studies. This agreement is not perfect, and may in some cases be fortuitous since the complex modes correspond to motion of a number of groups. Better agreement between the observed and calculated frequencies could probably be achieved by refinement of the force field, but this is not justified at present.

Figure 5 shows three examples of predicted modes, which have been selected since they are general interest in spectroscopic studies of carbohydrates. Figure 5a shows the A mode at 1424cm^{-1} which is assigned to the observed frequency at 1432cm^{-1}. The observed frequency was identified as a CH_2 deformation mode from the deuteration studies, and the atomic displacements show that this is predominantly due to CH_2 symmetric bending with small contributions from C4-C5-H and C5-C4-H in-plane bending. This mode is one of the few almost pure group frequencies in this region; most of the other predicted modes are more complex, consisting of motions of numerous atoms. Figure 5b shows the atomic displacements for the A mode at 1331cm^{-1}. This mode is assigned to the observed frequency at 1334cm^{-1}, which has been identified as a complex mode containing CH_2 and C-O-H deformations. The atomic displacements show this to be a very complex mode which contains extensive contributions from CH_2 and COH motions. The neighboring predicted modes do not contain both contributions. Figure 5c shows the calculated (A) mode at 893cm^{-1} which is assigned the the observed frequency at 895cm^{-1}. There has been interest in the origin of this frequency since it was shown by Barker et al. (29) that an inrared band at ∿895cm^{-1} is characteristic of β-D-glucose

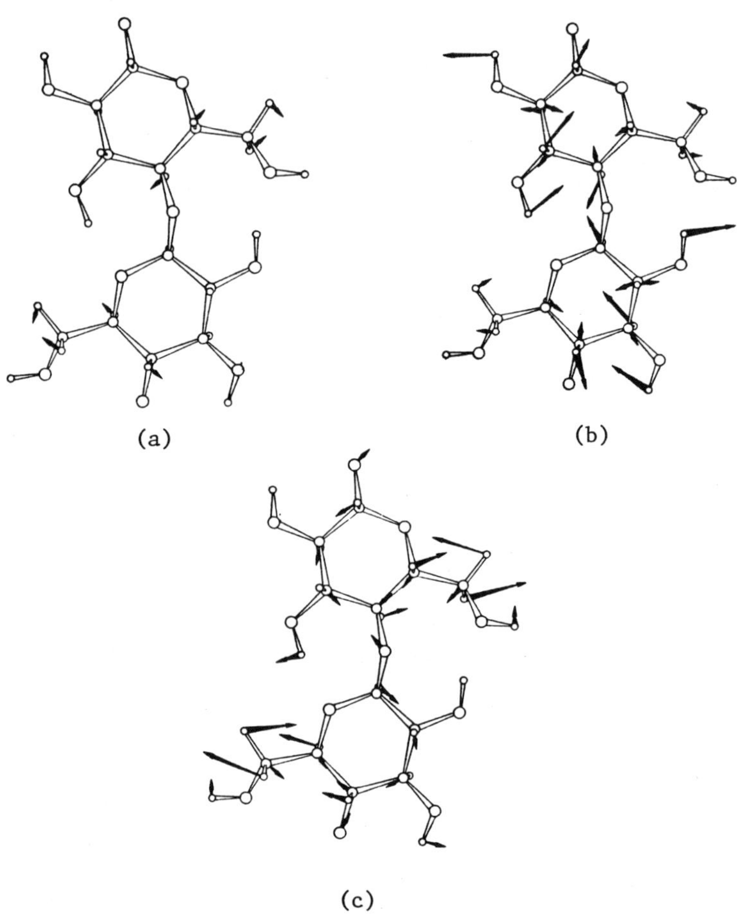

Figure 5. Calculated A modes for cellulose at (a) 1424 cm^{-1}, (b) 1331 cm^{-1}, and (c) 893 cm^{-1} (19)

residues; α-D-glucose residues yield a band at ∿840cm^{-1}. The deuteration studies show that this is also a complex mode containing C-O-H, CH_2 and (significantly) C1-H contributions. The predicted mode contains all these contributions and neighboring modes do not have the C-O-H contribution.

In the C-H stretching region the expected seven modes are predicted and the assignments are confirmed. Three O-H stretching frequencies are predicted, which occur at the same frequency due to the simplicity of the model chosen. However, examination of the atomic displacements shows that only one of these is a pure group vibration corresponding to motion of the O6-H group only. The other two are approximately 75/25 and 25/75 percentage mixtures of the motions of O3-H and O2-H. This information is particularly important since it indicates that even for an isolated chain the O-H stretching modes cannot be assigned to individual groups. In the crystal structure the intermolecular bonding *via* hydrogen bonds is expected to lead to further mixing, resulting in the splitting observed in the infrared and Raman spectra.

In conclusions it can be seen that the vibrational modes of cellulose are highly complex and that much more remains to be done before the spectra can be fully understood. Nevertheless, these spectra contain considerable structural information and further investigations are merited so that we can study polysaccharide structures in solution and in gels, where such information is inaccessible by x-ray methods.

Acknowledgements

The contributions of J.J. Cael and J.L. Koenig to the work described here is gratefully acknowledged. The research was supported by NSF Grant No. GB 34205 and N.I.H. Research Career Development Award No. AM 70642.

Literature Cited

1. Tsuboi, M., J. Polymer Sci. (1957) 25, 159.
2. Marrinan, H.J., and Mann, J., J. Polymer Sci. (1956) 21, 301.
3. Mann, J., and Marrinan, H.J., J. Polymer Sci. (1958) 32, 357.
4. Liang, C.Y., and Marchessault, R.H., J. Polymer Sci (1959) 37, 385.
5. Liang, C.Y., and Marchessault, R.H., J. Polymer Sci. (1959) 39, 369.
6. Marchessault, R.H., and Liang, C.Y., J. Polymer Sci. (1960) 43, 71.
7. Gardner, K.H., and Blackwell, J., Biochem. Biophys. Acta (1974) 343, 232.
8. Gardner, K.H., and Blackwell, J., Biopolymers (1974) 13, 1975.
9. Kolpak, F.J., and Blackwell, J., Macromolecules (1975) 8, 563.
10. Kolpak, F.J., and Blackwell, J., Macromolecules (1976) 9, 273.

11. Blackwell, J., Kolpak, F.J., and Gardner, K.H, Adv. in Chemistry, this issue.
12. Sarko, A., and Muggli, R., Macromolecules (1974) 7, 486.
13. Stipanovick, A.J., and Sarko, A., Macromolecules (1976) 9, 851.
14. French, A.D., and Murphy, V.G., Adv. in Chemistry, this issue.
15. Blackwell, J., Vasko, P.D., and Koenig, J.L., J. Appl. Phys (1970) 41, 4375.
16. Blackwell, J., and Marchessault, R.H., in "Cellulose and Cellulose Derivatives", (ed. N. Bikales and L.E. Segal) Wiley, New York (1971) 5, 1.
17. Vasko, P.D., Blackwell, J., and Koenig, J.L., Carbohydrate Res. (1971) 19, 297.
18. McKenzie, A.W., and Higgins, H.G., Svensk Paperstidn. (1958) 61, 893.
19. Cael, J.H., Gardner, K.H., Koenig, J.L., and Blackwell, J., Carbohydrate Res. (1973) 29, 123.
20. Atalla, R.H., Applied Polymer Symposia (1976) 28, 659.
21. Vasko, P.D., Blackwell, J., and Koenig, J.L., Carbohydrate Res (1972) 23, 407.
22. Cael, J.J., Koenig, J.L., and Blackwell, J., Carbohydrate Res. (1974) 32, 79.
23. Cael, J.J., Koenig, J.L., and Blackwell, J., J. Chem. Phys. (1975) 62, 1145.
24. Wilson, E.B., Decius, J.C., and Cross, P.C., Molecular Vibrations, McGraw Hill, New York (1965).
25. Boerio, F.J., and Koenig, J.L., J. Polymer Sci. A-2 (1971) 9, 1517.
26. Snyder, R.G., and Zerbi, G., Spectrochem. Acta (A) (1967) 23, 391.
27. Brooks, W.V.F., and Haas, C.M., J. Chem. Phys. (1967) 71, 650.
28. Cael, J.J., Koenig, J.L., and Blackwell, J., Biopolymers (1975) 14, 1885.
29. Barker, S.A., Bourne, E.J., Stacey, M., and Wiffin, D.H., J. Chem. Soc (1954) 3468.

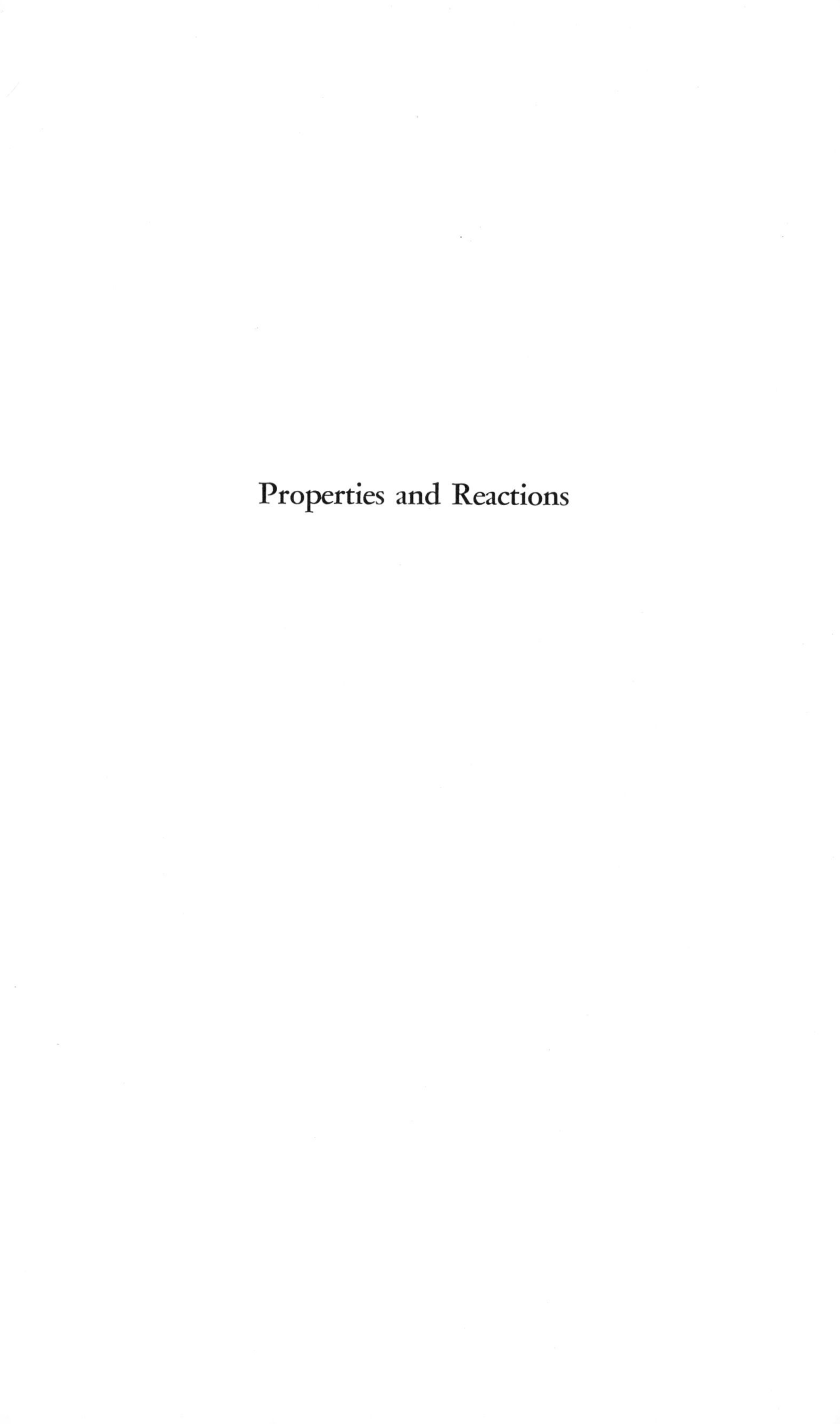

Properties and Reactions

15

Teichoic Acids: Aspects of Structure and Biosynthesis

I. C. HANCOCK and J. BADDILEY

Microbiological Chemistry Research Laboratory, University of Newcastle upon Tyne, Newcastle upon Tyne, UK

The walls of Gram-positive bacteria comprise peptidoglycan and teichoic acid in comparable amounts, together with smaller quantities of protein. In a few cases teichuronic acids rather than teichoic acids are present, and under certain growth conditions the latter may be replaced by the former. Whereas the mechanism of biosynthesis of peptidoglycan from nucleotide-sugar precursors through undecaprenyl phosphate intermediates has been established in considerable detail, relatively little was known hitherto about the route for the synthesis of the other wall components or the mechanism of their mutual attachment to form a wall. Emphasis will be given here to the synthesis of the poly(ribitol phosphate) teichoic acids and their attachment to peptidoglycan.

The assumption that the precursors of glycerol and ribitol teichoic acids are cytidine diphosphate glycerol (CDP-glycerol) and cytidine diphosphate ribitol (CDP-ribitol) respectively (1) was shown to be correct when it was demonstrated that these nucleotides, when incubated with membrane preparations from appropriate organisms, gave polymers of glycerol phosphate and ribitol phosphate (2). The acceptor of the teichoic acid chain is a membrane lipid believed to be identical to (3) or closely similar to the membrane teichoic acid (lipoteichoic acid)(4). This acceptor is called lipoteichoic acid carrier (LTA).

Further progress in understanding the mechanism of assembly of wall polymers has followed the discovery (5) that in a mutant of *Staphylococcus aureus* H which lacks the usual N-acetylglucosaminyl substituents in its poly(ribitol phosphate) wall teichoic acid the attachment of the teichoic acid chain to a muramyl residue in the peptidoglycan is indirect; interposed between the terminal phosphate of the teichoic acid and the 6-hydroxyl group of a muramyl in the peptidoglycan is a linkage unit comprising a trimer of glycerol phosphate (Fig. 1).

```
                                                  /
                                                 /
                                                GlcNAc
                                               /
(Ribitol-P)-Glycerol-P-Glycerol-P-Glycerol-P-MurAC
                                              /
                                             /
                                           GlcNAc
                                            /
                                           /
```

Figure 1. Teichoic acid–peptidoglycan complex

This linkage unit also occurs in the wild S. aureus H (6) and similar linkage units have been found in Bacillus subtilis W23, where the teichoic acid is a poly(ribitol phosphate) with glucosyl substituents, and in Micrococcus 2102, where the wall contains a poly(N-acetylglucosamine 1-phosphate) (unpublished work with A. R. Archibald, J. Coley and E. Tarelli. It seems likely then that linkage units containing glycerol phosphate residues are widespread and might occur in all walls containing teichoic acids. It follows that the biosynthesis of linkage units forms an integral part of the process of wall polymer assembly.

The glycerol phosphate units in the poly(glycerol phosphate) chain of membrane teichoic acid originate from phosphatidyglycerol rather than from CDP-glycerol (7, 8), so the function of CDP-glycerol in bacteria that do not possess a glycerol-containing wall teichoic acid was not clear. In independent studies the present authors (9) and Bracha and Glaser (10) have demonstrated that the origin of the linkage unit is indeed CDP-glycerol. Incorporation of isotope from CDP-(C-14)glycerol into polymeric material occurred readily with membrane preparations, but only if CDP-ribitol and UDP-N-acetyl glucosamine was also present (Fig. 2). This differs markedly from the synthesis of poly(ribitol phosphate)-LTA where the only nucleotide requirement is CDP-ribitol.

	No addition	+UDPGlcNAc	+CDP-ribitol	+UDPGlcNAc +CDP-ribitol
c.p.m. in polymer	72	307	165	9447
% incorporation	0	0.11	0.059	3.45

Figure 2. Incorporation from labelled CDP–glycerol into polymer by S. aureus *H membrane*

It was also shown (9) that toluene-treated whole cells are able to incorporate glycerol phosphate into their wall polymer from externally supplied CDP-glycerol. Further evidence that the linkage unit originates from CDP-glycerol was provided by acid hydrolysis of water-soluble polymeric material synthesised from (P-32)CDP-glycerol by a membrane preparation, when muramic acid (P-32)phosphate was a product (unpublished observation).

The relationship between the synthesis of poly-(ribitol phosphate)-LTA, and of linkage unit in the system described above, can be better understood from an examination of lipids extractable from the membrane by butanol (9). CDP-glycerol contributes glycerol phosphate residues to a butanol-soluble lipid in the presence of UDP-N-acetylglucosamine but in the absence of CDP-ribitol. Although the structure of this lipid has not yet been elucidated, it has properties consistent with that of the undecaprenyl N-acetylglucosaminyl-N-acetylmuramylpeptide pyrophosphate of peptidoglycan synthesis to which a chain of three glycerol phosphate residues has been attached at the 6-position on muramic. The formation of such an intermediate from CDP-glycerol and endogenous lipid intermediates of peptidoglycan synthesis would not require CDP-ribitol but might be stimulated by UDP-N-acetylglucosamine either allosterically or through conversion of the endogenous peptidoglycan lipid intermediate containing the monosaccharide residue (muramyl) to that containing the required disaccharide (glucosaminylmuramyl).

It is suggested (Fig. 3) that this lipid to which the linkage unit has been attached next accepts the poly(ribitol phosphate) chain by transfer from its LTA

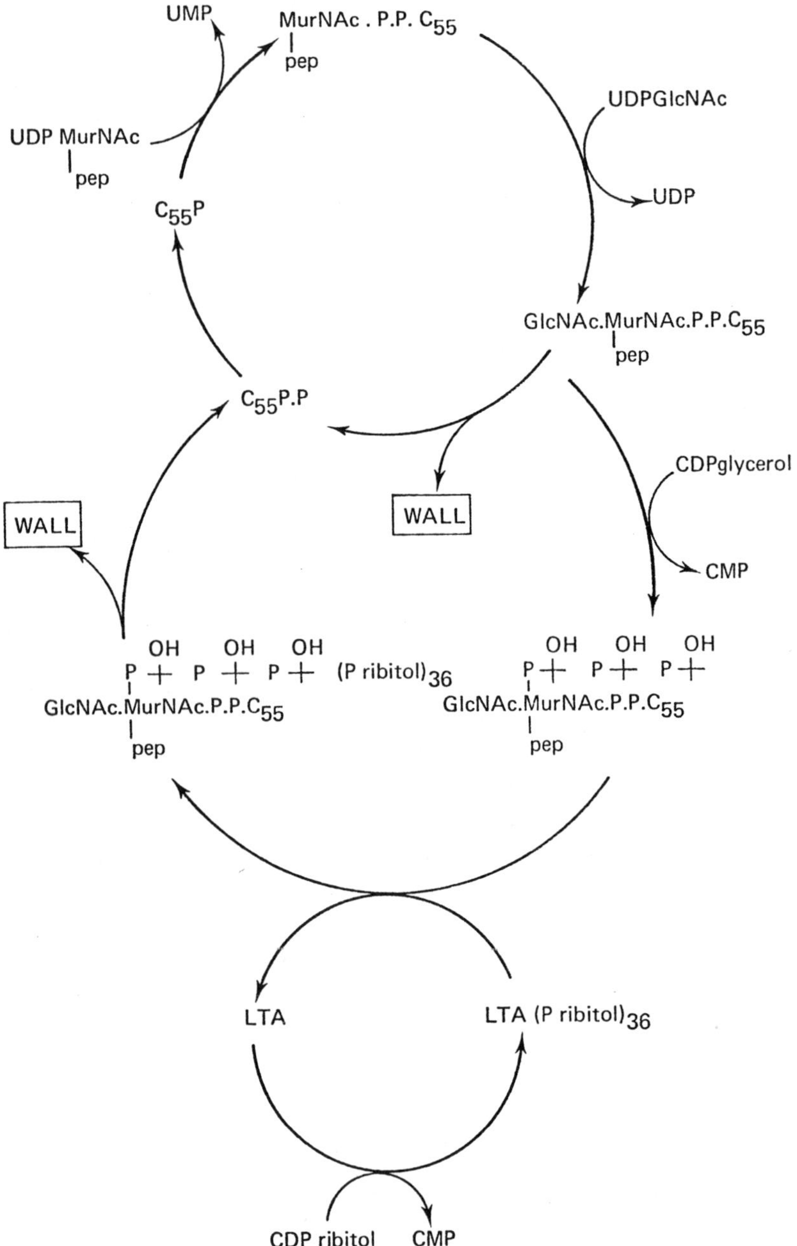

Figure 3. Scheme for the biosynthesis of teichoic acid–linkage unit

complex. The final product is thus an undecaprenyl pyrophosphate derivative containing the complete teichoic acid - linkage unit - disaccharide peptide, which could then be incorporated into growing peptidoglycan chain. It is noteworthy that, although this final lipid possesses an undecaprenyl residue, it would be water-soluble because of the large number of polyol phosphate groups; thus label from CDP-glycerol would be observed in water-soluble polymer only on completion of the synthesis of this complex, i.e. only when all of the nucleotide precursors are present.

The biosynthetic route receives support from studies with preparations from Micrococcus 2102 and *B. subtilis* W23. Not only are the nucleotide requirements consistent with the proposal, but in *B. subtilis* W23 it has been possible to separate the water-soluble polymers into two fractions by chromatography (unpublished work with G. Wiseman). One fraction contains ribitol phosphate residues from CDP-ribitol but no residues from CDP-glycerol, thus corresponding to poly(ribitol phosphate)-LTA. The other macromolecular product contains residues originating from CDP-ribitol and CDP-glycerol, corresponding to teichoic acid-linkage unit-disaccharide undecaprenyl pyrophosphate.

The antibiotic tunicamycin interferes with metabolic routes involving derivatives of undecaprenyl phosphate (11) or dolichyl phosphate (12). In particular it inhibits the synthesis of the N-acetylglucosaminyl-N-acetylmuramylpeptide derivative of undecaprenyl pyrophosphate in peptidoglycan synthesis. Thus it was possible that it might interfere with subsequent steps in the synthesis of teichoic acid shown in Fig.3. In the *S. aureus* system it inhibited (50%) the incorporation of glycerol phosphate into water-soluble polymeric material from CDP-glycerol at 2μg/ml of antibiotic, i.e. below the level required for complete inhibition of growth of the organism. Moreover, in the *B. subtilis* W23 system, it selectively inhibited the synthesis of the polymer which contained glycerol phosphate residues from CDP-glycerol, but did not affect the systhesis of the other polymer.

We conclude that, although further work is required to establish the details of the mechanism of biosynthesis of the teichoic acid-linkage unit-peptidoglycan complexes in the walls of bacteria, the process appears to require the attachment of a linkage unit at the lipid intermediate stage, and this is followed by attachment of the main teichoic acid chain.

Literature Cited

1. Armstrong, J. J., Baddiley, J., Buchanan, J. G., Carss, B., Greenberg, G. R., J. Chem. Soc.(1958) 4344-4354.
2. Burger, M., Glaser, L., J. Biol. Chem.(1964) 239 3168-3177.
3. Fielder, F., Glaser, L., J. Biol. Chem.(1974) 249 2684-2689.
4. Duckworth, M., Archibald, A. R., Baddiley, J., FEBS Letters(1975) 53, 176-179.
5. Heckels, J., Archibald, A. R., Baddiley, J., Biochem. J.(1975) 149, 637-647.
6. Coley, J., Archibald, A. R., Baddiley, J., FEBS Letters(1976) 61, 240-242.
7. Emdur, L. I., Chiu, T. H., Biochem. Biophys.Res. Communc.(1974) 59, 1137-1144.
8. Glaser, L., Lindsay, B., Biochem. Biophys. Res. Communc.(1974) 59, 1131-1136
9. Hancock, I. C., Baddiley, J., J. Bacteriol.(1976) 125, 880-886.
10. Bracha, R., Glaser, L., J. Bacteriol.(1976) 125, 872-879.
11. Bettinger, G. E., Young, F. E., Biochem. Biophys. Res. Communc. (1975) 67, 16-21.
12. Tkacz, J. S., Lampen, J. O., Biochem. Biophys. Res. Communc. (1975), 65, 248-257.

16

Secondary Lignification in Conifer Trees

HERBERT L. HERGERT

ITT Rayonier Inc., 605 Third Ave., New York, NY 10016

During the past one hundred years, organic chemists have devoted an immense amount of effort to determining the structure of lignin, the second most abundant polymer in nature. Conventional approaches to structural elucidation, i.e. identification of degradation products, etc., yielded largely indifferent results until Freudenberg's classic biosynthetic work, first reported in the 1950's, opened the door to an understanding of gross structure and mechanism of formation. Subsequent research has led to general agreement on types of linkages, molecular size, etc., but fine structural details are still being debated. Part of the reason for lack of consensus on fine structure may well be the tendency of lignin chemists to ignore the possibility that lignin structure may vary with the physiological state of the wood from which the lignin was isolated. It is a rare report, for example, that provides information as to whether the wood sample was obtained from heartwood or sapwood. Also missing is data on age of sample, size of tree, growth rate, presence of tension or compression wood, etc.

Most likely the lack of regard for wood physiology in lignin studies continues to be based on a presumption that lignin structure is uniform (as that of cellulose) from the center of the tree to the periphery. Freudenberg postulated (1) the formation of lignin precursors such as coniferin in the cambium. These phenylpropane glucosides subsequently diffuse into differentiating cells adjacent to the cambium where they are hydrolyzed and the aglycone oxidized by enzymes located in the walls of the differentiating xylem. Subsequent polymerization results in the deposition of lignin in and between the cell walls (middle lamella). This hypothesis, based on strong chemical evidence, certainly leads to a belief in lignin homogeneity within the wood of a given tree.

Microscopic and histological work by plant physiologists, however, suggest that "lignification" of some wood cell types is delayed past that of others. Bauch and co-workers (2-5), for example, show that certain types of ray cells and bordered pit membranes remain unlignified in sapwood. Polyphenols or "lignin" are subsequently deposited during heartwood formation. Furthermore, Wardrop's recent work (6) demonstrates that, contrary to Freudenberg's hypothesis, lignification is under individual cell control; hence structural variation from cell to cell is within the realm of possibility. Examples of anticipated differences might be end-wise polymerization where the reaction is rapid (fibers or tracheids) or bulk polymerization (parenchyma cells) where lignification is slow based on the model studies of Lai and Sarkanen (7) on structural variation in dehydrogenation polymers of coniferyl alcohol.

Our interest in this subject came about through studies on Brauns' Native Lignin (BNL) from conifers. In 1939, F. E. Brauns reported (8) the isolation of a soluble lignin product from black spruce wood which he termed "native lignin." It was obtained by extraction of wood with ethanol or acetone followed by reprecipitation into ether and hot water to remove impurities. He considered this material to be identical with nonsolvent-extractable wood lignin. No attention was given, however, to whether the content of the material might be higher in the heartwood or the sapwood of the tree since it was assumed to be unrelated to heartwood formation. Since that time, BNL has been isolated by many other investigators from a number of woody sources, including bark, again without regard to a possible heartwood relationship.

Preliminary work by the author on yields of BNL from sapwood and heartwood of five coniferous species showed a much higher content of BNL in the heartwood. This suggested that BNL might be mainly biosynthesized at the heartwood-sapwood boundary. While Bauch and co-workers' studies suggested deposition of lignin-like materials in rays, pits and cell lumina during the formation of heartwood, they did not establish whether these substances are Brauns' Native Lignin, insoluble polymers identical with cell wall lignin, or some other type of phenolic polymer system. If lignin biosynthesis is not restricted to the first year of tree growth, as generally believed, but might also occur secondarily at a much later time, one might expect to find significantly larger quantities of Brauns' and Klason (72% sulfuric acid insoluble) lignins in the heartwood than sapwood and in the outer bark than in inner bark, respectively.

The objective of the present study was, therefore, to establish the presence of a "secondary" lignification system, if indeed such exists. It was anticipated that this could be achieved by (a) establishing a structural relationship between BNL and solvent-insoluble cell wall lignin, (b) determination of the physical location in wood and bark of BNL (for convenience the present study was limited to conifer trees since they have a much simpler wood anatomy than the hardwoods), and (c) determination of the extent and structure of organic solvent-insoluble secondary lignin, i.e. "lignin" laid down at the sapwood-heartwood boundary and in outer bark.

Experimental

Wood Samples. Trees were cut, cross-sectioned with a chain saw to one inch thickness, chipped into appropriate fractions, air-dried, and reduced in a Wiley mill to pass a 20-mesh sieve. Old-growth (24.8 cm. radius, 127 annual rings) and second growth (11.4 cm., 22 rings) western hemlock (*Tsuga heterophylla*) were obtained in Grays Harbor, Washington. Northern white cedar (*Thuja occidentalis*, 9.5 cm., 90 rings) was obtained near Syracuse, New York; lodgepole pine (*Pinus contorta*, 8.9 cm., 60 rings) from Prince George, B.C., Canada; slash (*P. elliotti*, 9.8 cm., 21 rings, 3.7% heartwood) and longleaf (*P. palustris*, 7.3 cm., 56 rings, 26.5% heartwood) from Nassau County, Florida; baldcypress (*Taxodium distichum*, 14.0 cm., 85 rings) from Waycross, Georgia; and sitka spruce (*Picea sitchensis*, 55.9 cm., 550 rings, heartwood sample selected from rings 429-465 and sapwood, 500-550) from northern Vancouver Island, B.C., Canada.

Wood samples were successively extracted with petroleum ether, ethyl ether, acetone or ethanol, and hot water. Extractive-free wood was analyzed for acetyl, uronic acid, ash, Klason lignin, and sugars by chromatography. Cellulose content was calculated from the measured glucan content (corrected for loss of sugar during hydrolysis) minus that portion of glucose associated with galactoglucomannan. A computer program was developed to facilitate the calculation.

Isolation of Brauns' Native Lignin. Western hemlock sawdust (45% heartwood by weight) was extracted with 10 x 20 vol. portions of acetone. The extract was concentrated and poured into water, the precipitate being filtered, dried, redissolved and precipitated into diethyl ether. This procedure was repeated twice to give a pinkish-tan product in 1.2% yield based on oven-dry weight of unextracted wood. TLC indicated contamination with hydroxymatairesinol, so the product was extracted with chloroform until lignans were shown to be absent (yield loss 45%). Pertinent analytical data are shown in Table I and compared with milled wood lignin prepared from western hemlock sapwood as previously described (10).

Separation of BNL from possible polyphenolic contaminants was achieved by dissolving 25 mg. of the crude product in 10 ml. acetone or acetone-dry dioxane mixture and adding redistilled cyclohexylamine dropwise until incipient precipitation took place. The precipitate was filtered, washed with dilute hydrochloric acid, water, and dried. Infrared spectral comparisons were made before and after treatment. Model compounds were treated with cyclohexylamine; results are presented in Table II.

PMR spectra of TMS derivatives were prepared by suspending 15 mg. of the polymer in an acetone or DMSO solution of HMDS and TMCS. The mixture was dried under vacuum and then dissolved in spectrograde carbon tetrachloride. The solution was filtered through sintered glass and sealed in vials after HMDS, TMCS and TMS were added to retain stability of the solution prior to spectral determination on a Varian 60MHz instrument. Peak areas were integrated and structural assignments made as shown in Table III based on a parallel study of TMS derivatives of phenolic model compounds.

BNL was isolated from serial sections of western hemlock, amabalis fir and other coniferous species by pre-extraction of wood meal with ethyl ether followed by acetone extraction (Sohxlet extractor). The acetone extract was dried, extracted with 4x25 volumes of hot distilled water, dried, and re-extracted with chloroform to remove lignans. The yield of the acetone-soluble, hot water- and chloroform-insoluble fraction was considered to represent the BNL content of a particular sample and is so reported in Tables IV and V.

Results

Comparison of Brauns' solvent-soluble native lignin with milled wood lignin from sapwood showed that the products were very similar (Table I). Acidolysis in dioxane-water (9:1), with or without hydrogen chloride catalyst, yielded typical lignin degradation products, i.e. coniferyl alcohol, coniferyl aldehyde, vanilloyl methyl ketone, vanillin, guaicol, beta-hydroxy coniferyl alcohol, alpha-hydroxy guaicylacetone, beta-hydroxy aceto-

Table I

Comparison of Western Hemlock BNL and Sapwood MWL

	BNL	MWL
Carbon, %	63.9	63.3
Hydrogen, %	5.9	5.9
Methoxyl, %	14.5	15.0
Phenolic OH, %	4.5	2.9
A/A_{1500} at ($cm.^{-1}$)		
1745	.27	-
1715	.23	.20
1655	.25	.37
1595	.63	.52
1135	.82	.95
1085	.68	.75
1030	.85	.95
970	.41	.33
850	.31	.29
810	.33	.26

Table II

Behavior of Model Compounds Upon Treatment with Cyclohexylamine

Solvent	Model Compound	Results
Acetone	Diterpene Resin acids	Nearly quantitative precipitation
	Fatty acids	No precipitate
	Conidendrin	No precipitate
	Hydroxymatairesinol	No precipitate
	Protocatechuic acid	White crystalline precipitate
	4-Methyl Catechol	No precipitate
	Pyrocatechol	No precipitate
	Plicatic acid	Gummy precipitate
	Hemlock BNL	Tan precipitate
	Hemlock Tannin	No precipitate
Ethyl acetate	Plicatic acid	Quantitative white precipitate
	Pyrocatechol	No precipitate
	Cinnamic acid	Crystalline, cream-colored precipitate
	Cinnamaldehyde[a]	No precipitate
	Cedar polyphenols	Partial precipitation
Dioxane-acetone[b]	Hemlock BNL	Tan precipitate
Water	Plicatic acid	No precipitate

[a]Freshly distilled or washed with dilute sodium hydroxide solution to remove cinnamic acid contaminant. [b]9:1 mixture.

Table III

PMR Spectrum of Western Hemlock BNL TMS Derivative

δ Value of Peak	Range	% of Total H	Structural Assignment
6.60	6.25-8.00	31	Aromatic H
6.20	5.50-6.25	5	Beta Vinyl
5.30, 4.97, 4.85	4.10-5.50	14	H on Alpha and beta C-OH
3.70	3.10-4.10	46	Methoxyl
3.10, 2.55	2.25-3.10	4	Alpha and gamma methylene

Table IV

Distribution of Lignin in Old-growth Western Hemlock Wood[a]

Fraction No.	1	2	3	4	5	6	7
Sapwood(s) or Heartwood(s)	S	S	S/H	H	H	H	H
Annual Rings	15	10	13	15	23	21	30
Green Sp.G., g/cc.	0.46	0.46	0.43	0.44	0.42	0.41	0.42
Radius, cm.	1.6	1.6	2.5	3.8	5.7	4.5	5.1
Lignan, %	-	0.10	0.14	0.86	0.93	0.70	0.69
BNL, %	0.11	0.10	0.18	1.02	0.83	0.41	0.30
Klason Lignin, %	30.4	29.5	29.1	29.6	31.2	31.6	34.0
Cellulose, %	38.4	39.9	42.1	41.8	39.1	38.6	37.6
Ratio KL/cell.	.792	.739	.691	.708	.816	.819	.906

[a]Lignan and BNL, unextracted wood basis; Klason Lignin and cellulose, extractive-free basis.

Table V

Distribution of Brauns' Native Lignin (BNL), Klason Lignin (KL) and Cellulose in Various Conifer Wood Fractions

Species	Fraction[a]	BNL,[b] %	KL,[c] %	Cellulose,[c] %	Ratio KL/Cell.
Northern White Cedar	S	0.02	29.9	44.5	0.672
	H	2.86	33.3	41.3	0.806
Lodgepole Pine	S	0.14	25.7	45.0	0.571
	H	0.26	26.8	41.3	0.650
Slash Pine	S	0.12	27.5	46.2	0.595
	H	0.14	29.4	39.2	0.750
Longleaf Pine	S	0.06	26.6	45.0	0.592
	H	0.07	29.6	42.1	0.703
Sitka Spruce	S	0.13	26.0	44.3	0.587
	H	1.25	28.1	41.4	0.679
Baldcypress	S	0.80	33.7	-	-
	H	2.18	34.9	-	-
Western Hemlock[d]	S	0.21	31.6	42.8	0.738
	H	1.38	33.5	33.4	1.003

[a]S = Sapwood, H = Heartwood; [b]Unextracted wood basis; [c]Extractive-free basis; [d]Second-growth, small diameter.

guaicone, and traces of guaiacylglycerol, all of which indicate the substantial presence of β -O linked phenylpropane units. Chemically speaking, BNL must be defined as lignin.

However, there are some minor but important differences. The infrared spectrum shows the presence of a lactonic absorption band at 1745-1755 cm.$^{-1}$ strongly suggestive of the incorporation of a lactonic lignan within the structure of the BNL. A variety of polymer fractionation techniques was employed including absorption on silica gel, celite, etc. but none of them resulted in a polymer without the lactone peak. Wood and bark frequently contain polyflavanoid tannins which co-occur with the BNL fraction. Until now, no truly satisfactory method has been available for complete separation of the softwood bark tannins from their usual co-occuring contaminant of Brauns' native lignin (BNL). Conversely, attempts to separate pure BNL fractions from tannin and lignan contaminants are tedious and usually only partly successful. Quite by accident it was noted that the addition of cyclohexylamine to acetone or related solvent solutions of aromatic carboxylic acids resulted in quantitative precipitation of a crystalline amine salt. This technique is a well-known one in the literature for the separation of fatty and diterpenoid resin acids, but there was no indication that it had utility in the difficult separations previously enumerated.

Addition of cyclohexylamine to a solution of a variety of aromatic compounds resulted in precipitation of those compounds containing a carboxyl group as shown in Table II. The necessary requirement for this reaction appeared to be an appropriate solvent, i.e., one in which the substrate is soluble but the cyclohexylamine salt is not. Since Brauns' native lignin (BNL) and other related solvent-soluble lignins gave a precipitate from this treatment, it must be concluded that these materials contain carboxyl groups. This represents a new finding concerning lignin structure. It also offers an interesting separation technique for purifying polyflavanoid wood and bark extractives which are invariably contaminated with varying quantities of BNL. The tannin fraction can be dissolved in acetone or derivatized to solubilize it in this solvent and then treated with cyclohexylamine to precipitate the BNL contaminant. MWL also gives a precipitate by this treatment, indicative of a small quantity of carboxyl groups, but MWL has such limited solubility in the solvents used for the test that it is of limited utility. Infrared spectra of BNL precipitated with cyclohexylamine and washed with dilute hydrochloric acid to destroy the amine salt still showed that 1750 cm.$^{-1}$ absorption. This was considered to be further evidence of its origin from the BNL polymer and not from a contaminant.

An improved proton magnetic resonance spectrum technique was devised. Conversion of lignin into a TMS derivative soluble in carbon tetrachloride revealed considerable fine structure not readily noticeable in previously reported lignin PMR studies. The different types of protons in a western hemlock heartwood BNL are shown in Table III. Of particular interest is the 5% content of beta vinyl protons, equivalent to one coniferyl alcohol or aldehyde group out of each six C6 - C3 monomeric units in the molecule, and the 4% content of alpha and/or gamma methylene protons which are suggestive of about 10% of the monomers being derived from lactonic lignans such as conidendrin or matairesinol.

Further light was shed on the BNL question by determination of the content across the radius of the tree. Yield from the heartwood was approximately ten times higher than that obtained from the sapwood of an old growth tree (Table IV). The same phenomenon was noted in a second-growth tree (Table V). With the exception of the BNL from the outermost sapwood, the infrared spectra of sapwood BNL was virtually identical with sapwood MWL. The former differed in the presence of a more intense carbonyl peak at 1715 $cm.^{-1}$ ascribable to an impurity or an aromatic ester. The heartwood BNL samples all showed the 1750 $cm.^{-1}$ absorption which is believed, as previously noted, to originate from the incorporation of lactonic lignans in the polymer.

Not only is there a substantial deposition of BNL upon heartwood formation, but there also appears to be an increase in non-solvent-soluble Klason lignin (KL). Since the KL is determined on a summative basis, the extent of increased lignin content in heartwood would not necessarily be revealed by comparative summative analyses unless it was the only constituent increasing. There is no present indication that additional cellulose is biosynthesized at the heartwood-sapwood boundary. Consequently, comparison of the ratio of KL to cellulose in sapwood and heartwood provides the needed information. (It should be noted that this comparison must be undertaken with care. Wood of the same growth pattern, i.e. ratio of summerwood to springwood and annual ring width, must be used. Compression wood must be rigorously excluded since it is known to have an abnormally high lignin content.) Summative analyses of 15 western hemlock trees were available from a separate study in our laboratory. The KL content of heartwood averaged 1.3% (extractive-free wood basis) higher than the sapwood; the ratio of lignin to cellulose was .062 higher. Based on an assumed constant cellulose content, almost 10% additional lignin is laid down at the heartwood-sapwood boundary. This finding suggests that there is a post mortem deposition of "lignin" in both soluble and solvent-insoluble forms in the heartwood. Comparison of the infrared spectra of dioxane-HCl lignins from extractive-free sapwood and heartwood showed the same behavior as the BNL, i.e. the heartwood lignin showed a weak absorptive band at 1750 $cm.^{-1}$ absent in sapwood lignin. Furthermore, chromatography of acidolysis products from the heartwood showed the presence of spots with Rf values and color reactions characteristic of conidendrin and matairesinol along with the usual assortment of lignin-derived phenylpropane derivatives. The additional lignin laid down in the heartwood must, therefore, also contain incorporated lactonic lignans.

In an attempt to establish the site of BNL deposition, western hemlock heartwood was separated into summerwood and springwood fractions. The content of BNL was higher in springwood (1.89%) than in summerwood (1.76%), generally following the distribution of the Klason lignin (32.51 vs. 29.63%). The ratio was not exactly the same in both fractions, so Sarkanen's hypothesis (11) cannot be correct that BNL represents the lowest molecular weight fraction of wood lignin and this accounts for its organic solvent solubility. The higher ratio of BNL to Klason lignin in springwood suggests, in our opinion, that it is formed in pit torii, fiber lumen surfaces and parenchyma cells. Each of these has a higher ratio to total cell wall volume in springwood than in summerwood.

To determine whether the secondary deposition of lignin is unique to western hemlock or is general to conifers, similar studies were carried out on unrelated species. The content of BNL was measured in a small diameter, 29 year old western red cedar. The BNL content of the heartwood was approximately ten times higher than that of the sapwood (2.7 vs. 0.2%). The infrared spectrum of the heartwood BNL differed from that of sapwood in that it contained a strong lactone band at 1760 $cm.^{-1}$ and a series of peaks (1198, 1118, 1090 and 790 $cm.^{-1}$) characteristic of plicatin and closely related lignans. All of these peaks were retained after the cyclohexylamine purification procedure. This is strongly indicative that the heartwood BNL is derived in part from plicatin and related compounds. Approximately 0.6% additional Klason lignin is deposited in the heartwood, based on the higher Klason lignin-cellulose ratio in the heartwood. Acidolysis lignin from western red cedar heartwood also shows a lactone peak. This suggests that the insoluble lignin deposited in the heartwood has a similar structure to that of the BNL. Similar results were noted in the closely related northern white cedar. The content of BNL was approximately two hundred times higher in the heartwood than the sapwood (Table V). Infrared spectra of the sapwood BNL and BNL from the heartwood showed important differences. The sapwood BNL showed a significantly more intense carbonyl peak at 1710 $cm.^{-1}$ than normally expected in a typical wood lignin. This same behavior was also noted in the spruce wood sapwood BNL (see below). The heartwood BNL, on the other hand, showed a number of absorption peaks (especially the intense pair of bands at 1118 and 1090 $cm.^{-1}$) strongly indicative that it is derived at least in part from plicatin, plicatic acid and related compounds. Since the plicatin series of compounds are characteristic of the Thuja genus, the BNL of western red cedar and northern white cedar is also characteristic of the genus and differs in structure from that of other coniferous genera.

The Klason lignin content of white cedar heartwood is 3.4% higher, on an extractive-free basis, than that of the sapwood. This suggests that further amounts of polymer, in addition to the BNL, are deposited in the heartwood during its formation. Here again, infrared spectrum of heartwood dioxane-HCl lignin shows the same set of weak lignan-related absorption bands superimposed on the basic lignin spectrum.

A large diameter, old growth amabalis fir was separated into six sections from the center of the tree to the cambium. The heartwood contained 4-5 times more BNL than the sapwood (0.29 vs. 0.06%). The heartwood BNL infrared spectra showed a lactone peak at 1755 $cm.^{-1}$ and a series of bands in the 1200-800 $cm.^{-1}$ region similar to matairesinol and hydroxymatairesinol which are present as extractives in amabalis fir. These bands were not present in the sapwood BNL.

Two different approaches were used to measure the variation of BNL content in baldcypress. In the first of these, 14 serial sections from the pith to the bark were extracted with ether to remove diterpenes, wax and low molecular weight phenols (sugiresinol, etc.) and then extracted with 95% ethanol. The ultraviolet absorption of the extracts was measured, the absorbence being calculated back to the wood basis. The guaiacyl absorption band at 275 nm., characteristic of BNL is 8-10 times higher in the heartwood than in the sapwood. Direct isolation studies show a three times higher content of BNL in heartwood than sapwood in baldcypress and in the closely related species pondcypress. Klason lignin contents are also higher in the heartwood than sapwood, but only modestly so.

The BNL and Klason lignin content in a small diameter lodgepole pine were higher in the heartwood than in the sapwood (Table V). Using the ratios of Klason lignin to cellulose, however, shows that the lignin laid down in the heartwood is 3.5% on a wood basis (25.7 x 650/571 = 29.2%), i.e. a 13.6% increase of heartwood lignins compared to sapwood lignins. The probable explanation for this increase is that parenchyma cells of many pine species are largely unlignified in the sapwood while in the heartwood they may have a 40-50% lignin content based on ultraviolet spectra of microtome sections. This is also confirmed by a much higher K number of parenchyma screenings from pine heartwood chemical pulp compared to the screened fibers (K number is a measure of lignin content). The infrared spectra of sapwood and heartwood BNL from pine show little differences from each other or from milled wood lignin from the same species. Most pine species do not contain significant quantities of lignans in the heartwood. Evidently the pine genus does not incorporate lignans into the BNL or insoluble lignin laid down in the heartwood, pinoresinol being a possible exception. Only small diameter pine trees were available for this study so it is possible that a large diameter tree with a substantial heartwood content would give different results.

Small diameter plantation growth longleaf and slash pine trees showed essentially the same phenomena as the lodgepole pine sample. The BNL content is quite low and only slightly higher in the heartwood than in the sapwood. A substantial quantity of insoluble lignin is laid down in the heartwood, 7.1% in the case of the sample of slash pine and 4.9% in longleaf pine. This result may be too high because of the difficulty in matching the density of the heartwood with the sapwood. The center of the southern pine tree generally contains a zone of low density wood, referred to as juvenile wood, which has a higher lignin to cellulose ratio than wood located at a distance of more than 10-12 annual rings from the pith. The apparent increase in heartwood lignin content may be, in part, due to this change. This question was checked by examining summative analyses of a separate set of eight southern pine tree sections. Heartwood KL content averaged almost 1% higher than sapwood. Comparing lignin/cellulose ratios, this represents a net increase of 13% lignin upon heartwood formation.

Thus far, little has been said about bark. Analyses for lignin in bark are made considerably more difficult by the high content of solvent-soluble and insoluble polyflavanoids. Inner and outer bark of longleaf pine were separated by hand, extracted with a solvent sequence similar to that used on wood and the fractions subsequently purified to give the results shown in Table VI. No BNL could be detected in the living inner bark, but it was obtained in 1.1% yield from outer bark. The Klason lignin was similarly low in inner bark but relatively high in the outer bark. Assuming that no cellulose is formed during the conversion to outer bark, about a seven fold increase in Klason lignin takes place. The reason for this seems to be related to the cell types in bark. Most of the cellular material in pine is parenchymatic. Upon death of these cells (outer bark formation) substantial deposition of lignin, xylan and galactoarabinan formation takes place. This phenomenon is probably not much different than in wood except in the latter, parenchyma cells are less than ten percent of the total volume of the wood, whereas in bark they are the predominant cell type. Preliminary chemical analyses of the pine bark wood lignin fractions suggest a somewhat higher para-hydroxyphenyl to guaiacyl ratio than in wood.

Table VI

Distribution of Lignin and Polyphenols in Longleaf Pine Bark

Fraction	Inner Bark	Outer Bark
BNL, %	None	1.1
"Phenolic Acid", % [a]	16.0	32.1
Klason Lignin, %[a,b]	7.8	27.4
Cellulose, %	38.0	19.8
Ratio K. Lignin/Cellulose	0.20	1.38

[a]Extractive-free basis. [b]Phenolic acid-free

Conclusions

Based on the work summarized in the preceding paragraphs, the following conclusions were reached:

1. Brauns' Native Lignin in conifers is mainly synthesized during heartwood formation.
2. Based on this work, which supplements the microscopic studies of Bauch and co-workers (2-5), the BNL appears to be mainly located in parenchyma cells, pit torii and the fiber lumen.
3. Brauns' Native Lignin is accompanied by an organic solvent-insoluble counterpart determinable as Klason lignin. It, too, is mainly formed at the sapwood-heartwood boundary. We propose the term, secondary lignification, for the deposition of these soluble and insoluble polymers.
4. The amount of the insoluble, secondary lignin is larger

than indicated by direct summative analyses. It can be calculated by comparison of lignin/cellulose ratios of heartwood and sapwood samples of identical growth pattern.

5. The structure of secondary lignin (with the possible exception of pine) differs from primary lignin in its higher content of copolymerized lignans which are species dependent.
6. The same phenomena appear to be in operation in bark on a much elevated level since bark has a much higher parenchyma content than wood.
7. Since heartwood contains these secondary lignin polymers with copolymerized lignans, it should not be used for primary lignin structural studies. Sarkanen and co-workers (12) have previously recognized this problem in their work but it seems to have been ignored by most other lignin structure investigators.

Abstract

Serial analyses of conifer wood and bark show that Brauns' Native Lignin (BNL), a solvent-soluble polymer formerly considered to be most representative of whole wood lignin, is mainly biosynthesized during heartwood and outer bark formation. A corresponding insoluble fraction (0.2-5% in heartwood and 10-30% in outer bark) co-occurs with BNL. While these secondary lignins are structurally similar to primary wood lignin, they show measurable variations because of the incorporation of species-dependent phenols such as matairesinol (hemlock), plicatic acid (cedar), sugiresinol (redwood), etc. into the polymer. The presence of these materials in wood and bark has implications for lignin structural studies (the use of heartwood will give misleading results) and in pulping or wood-treating (the polymers appear to be deposited in such a way as to inhibit liquid transfer).

Literature Cited

1. Freudenberg, Karl and Neish, A. C., Constitution and Biosynthesis of Lignin, Springer-Verlag, Berlin, 1968.
2. Bauch, J., Liese, W. and Scholz, F., Holzforschung (1968), 22, 144-153.
3. Bauch, Josef, and Berndt, Heide, Wood Science and Tech. (1973), 7, 6-19.

4. Bauch, J., Schweers, W., and Berndt, H., Holzforschung (1974), 28, 86-91.
5. Parameswaran, N., and Bauch, J., Wood Science and Tech. (1975), 9, 165-173.
6. Wardrop, A. B., Applied Polymer Symposia (1976), No. 28, 1041-1063.
7. Lai, Yuan-Zong, and Sarkanen, K. V., Cellulose Chem. and Tech. (1975), 9, 239-245.
8. Brauns, F. E., J. Am. Chem. Soc. (1939), 61, 2120.
9. For references see Hergert, H. L., "Infrared Spectra," pp. 267-297, in Lignins, Occurrence, Formation, Structure and Reactions, ed. by K. V. Sarkanen and C. H. Ludwig, 1971, Wiley-Interscience, New York.
10. Hergert, H. L., J. Org. Chem. (1960), 25, 405-413.
11. Lai, Y. Z., and Sarkanen, K. V., "Isolation and Structural Studies," pp. 179-180 in Lignins, Occurrence, Formation, Structure and Reactions, op. cit.
12. Sarkanen, K. V., Chang, Hou-min, and Allan, G. G., Tappi (1967), 50, 583-587, 587-590.

Contribution No. 165 from the research laboratories of ITT Rayonier Inc., 605 Third Avenue, New York, New York, 10016.

17

Electron Donor Properties of Tertiary Amines in Cellulose Anion Exchangers

DOROTHY M. PERRIER and RUTH R. BENERITO

Southern Regional Research Center, New Orleans, LA 70179

The preparation and properties of diethylaminoethyl (DEAE) celluloses in which the tertiary amine groups are in the free amine form have been reported (1,2). Use of such groups within the cellulose matrix for the preparation of mono-quaternary ammonium ions and di-quaternary ammonium ions by reactions with alkyl halides and dihaloalkanes, respectively, has been reported (3). Recently, we have studied the use of cellulose anion exchangers in the preparation of exchangers having oxidation-reduction properties, as well as anion exchange properties. Initially, the source of the reducing or oxidizing power resided in the anion, stabilized by the quaternary ammonium groups of the cellulose matrix. Cellulose exchangers in which the tertiary amine groups of DEAE cottons acted as ligands to Ni (II) and Cu (II) cations were also prepared and were found to be paramagnetic (4). This is a report of the possible use of the free electrons on the tertiary amine groups of DEAE cottons as the donor in formation of compounds with acceptor molecules. The 2,3,5,6-tetrachloro-p-benzoquinone was selected as one of the acceptor molecules because its reactions with triethylamine have been studied extensively (5,6). The other donor selected was tetracyanoethylene. Recently, the electron paramagnetic resonance of the adduct formed between TCNE and triethylamine has been reported (7).

Experimental

Reagents. The 2,3,5,6-tetrachloro-p-benzoquinone (chloranil), tetracyanoethylene (TCNE), and chloroform were reagent grade chemicals obtained from Eastman Organic Chemicals.* Absolute anhydrous methanol, reagent grade, was obtained from Matheson Coleman & Bell Manufacturing Chemists, tertiary butanol, Baker analyzed reagent, from J. T. Baker Chemical Co.; and anhydrous $CuCl_2$ from the British Drug Houses Ltd.

Fabrics. The nonaqueous preparations of sodium cellulosates in fabric form and their subsequent reactions at room temperature with β-chloroethyldiethylamine in t-butanol to form the free amine forms of diethylaminoethyl (DEAE) cottons of varrying nitrogen contents have been reported (1,2). For a DEAE cotton of approximately 2% N, cotton sheeting (4.4 oz/yd^2) was reacted with a 1.25 M sodium methoxide solution, and the resultant sodium cellulosate was reacted for 24 hr with a solution (4% by weight) of β-chloroethyldiethylamine in t-butanol.

The adducts were formed by reacting DEAE fabric with a chloroform solution of chloranil or TCNE in a nitrogen atmosphere. The yellow green chloranil adduct and the reddish brown TCNE adduct were washed several times in $CHCl_3$ before being analyzed.

Chemical Analyses. Nitrogen contents of all modified cottons were determined by the Kjeldahl method and are reported as total milliequivalent per gram of fabric (meq N/g fabric).

Chlorine and metal ion analyses of adduct products were determined by Galbraith Laboratories.

Infrared absorption spectra of fabrics were obtained on a Perkin Elmer 137 by the KBr disc method.

Reflectance spectra in the uv-visible region were obtained on the Varian Model 17 Spectrophotometer, with diffuse reflectance baseline set at 90% transmittance for MgO in the 250-700 nm range.

Techniques of electron emission spectroscopy for chemical analyses (ESCA) were applied to the DEAE cottons before and after their treatments with the electron acceptors. ESCA spectra were obtained on a Varian IEE spectrometer, Model VIEE-15, that had a Mg K_α X-ray source. Spectra were obtained on samples of fabric. All binding energies were given against a C_{1s} line of 285.6 electron volts (eV) as reference. These spectra, analyzed to yield relative amounts of nitrogen of different binding energies, apply to analyses of surfaces only. A DuPont 310 curve analyzer was used to resolve overlapping peaks. Peak positions are reported with a precision of ± .5 eV.

The electron spin resonance spectra (ESR) of the fabrics were determined with a Varian 4502-15 spectrometer system equipped with a variable temperature accessory and a dual sample cavity.

Results and Discussion

DEAE Cotton-Chloranil Adduct. Typical elemental analyses of DEAE cottons, before and after reactions of the amine groups with chloranil or with $CuCl_2$, are given in Table I. The characteristic binding energies of the N_{1s}, Cl_{2p}, and Cu_{2p} electrons for the reagents and adducts are also given. Included in the table are corresponding data to show effects of adding $CuCl_2$ to the DEAE cotton-chloranil adduct and of adding chloranil to the DEAE cotton-$CuCl_2$ adduct.

TABLE I. Binding Energies (eV) of Electrons Characteristic of Elements in Cellulose Anion Exchangers

	N_{1s}(eV)			Cl_{2p}(eV)		Cu_{2p}(eV)				N	Cl	Cl^-	Cu^{+2}
										Weight %			
1[a]	--	400.1[40][b]	398.1[60]	--	197.6	--	--	--	--	2.02	.88	.86	--
2	401.6[56]	400.3[28]	398.3[16]	200.2	197.9	--	--	--	--	2.02	6.28	4.76	--
3	401.7[39]	400.9[27]	399.3[34]	--	198.2	--	953.2	--	933.4	2.10	6.24	5.16	.47
4	401.8[70]	399.3[20]	397.9[10]	--	198.2	963.0	955.1	944.6	935.0	1.86	6.56	6.49	4.11
5	401.4[73]	400.2[9]	398.9[18]	200.7	198.0	958.0	952.1	944.7	933.5	--	--	--	--
6	--	--	--	200.9	--	--	--	--	--	--	--	--	--
7	--	--	--	--	199.4	964.0	955.7	945.3	936.0	--	--	--	--

a - 1 is diethylaminoethyl cellulose (DEAE) in free amine form
2 is adduct of DEAE and chloranil
3 is adduct of 2 + $CuCl_2$
4 is DEAE + $CuCl_2$
5 is adduct of 4 and chloranil
6 chloranil
7 $CuCl_2$

b - Bracketed number is percentage of total N of each binding energy

Wet elemental analyses of DEAE Cottons before and after reaction with chloranil showed that 4 meqs Cl/g were added for every 3 meq N/g. The ratio of meqs $C\bar{l}$ ions to meqs of organic chlorine per gram fabric was 3/1 when the reaction was carried out in nonpolar solvents.

ESCA spectra of the chloranil adducts were used to estimate relative amounts of nitrogens in the DEAE cotton that had become more positively charged as a result of reaction with chloranil. The binding energies, $E_{B.E.}$, of the N_{1s} electrons observed in the ESCA spectra of nonaqueously prepared DEAE-cottons have been discussed in earlier reports (3,8). In free amine groups, the $E_{B.E.}$ of the N_{1s} electron is approximately 399 eV. Quaternization of these free amine groups with methyl iodide to form $\overset{+}{N}R_4$ groups gives a spectrum equivalent to $E_{B.E.}$ of approximately 402.5 eV for the N_{1s}, or of at least 3 eV higher than that for the free amine nitrogen. These differences can be easily differentiated in the ESCA spectra. However, the $E_{B.E.}$ of the N_{1s} electron in some $\overset{+}{N}R_3H$ groups can differ from that of the $\overset{+}{N}R_4$ groups by only 1 eV. Therefore, at times it is difficult to differentiate between true quaternary nitrogens and those in the hydrosalts of tertiary amine.

The N_{1s} peak with $E_{B.E.}$ of 398.3 eV, characteristic of relatively negative nitrogen in free tertiary amine groups of DEAE cotton, was still present after reaction with chloranil, but in a lesser amount than in the original DEAE cotton. In addition, N_{1s} peaks with $E_{B.E.}$ values of 400.6 and 401.6 eV were present. Relative areas of these peaks were used to estimate that 16% of the tertiary amine groups of the original DEAE fabric were unreacted. Based on ESCA and wet analyses, it was concluded that of the 84% reacted, three amine nitrogens reacted with one chloranil molecule to yield a product of mole ratios of $3N/3\bar{Cl}/1$ organic chlorine. The following is one possible structure (5) for the product:

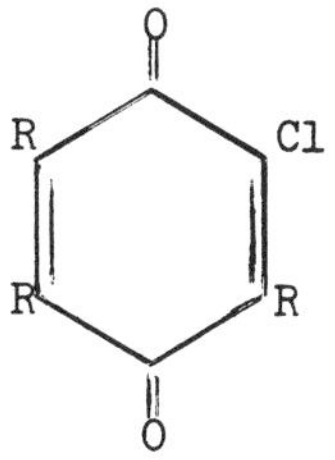

(Structure: 1,4-benzoquinone ring bearing O at positions 1 and 4, Cl and three R substituents) $3C\bar{l}$ where R is $(CellOC_2H_4-\overset{+}{N}(C_2H_5)(H)-CH=CH-)$

Ng and Hercules (9) used ESCA spectra to support structures for various donor-chloranil adducts. Absence of ESCA shifts in the spectra of the adducts formed by weak donors such as hexamethylbenzene (HMB) with chloranil, indicated nonbonding in the ground state. A shift in the $E_{B.E.}$ of the N_{1s} electron of tetramethylphenylenediamine (TMPD) by +2 eV when it reacted with chloranil was attributed to formation of D^{+2} cations to the extent of 14%.

The shifts in the $E_{B.E.}$ of the N_{1s} electrons of DEAE cottons on complexing with chloranil were from +2 to +3.5 eV. There was a decrease in the $E_{B.E.}$ of $Cl_{2p_{1/2}}$ **electrons** of -0.7 eV and an increase in the intensity of the peak at approximately 198 eV characteristic of Cl^- ions. Thus, some $Cl_{2p_{1/2}}$ electrons experience a shift of only -0.7 eV (the organic chloro atoms), and those going to ions (Cl^-) a shift of -3.0 eV.

ESCA spectra of the C_{1s} and O_{1s} electrons were also studied, but the shifts in the $E_{B.E.}$ of these elements were not significant. Even though the $E_{B.E.}$ of the C_{1s} electron in chloranil is at 286.3 eV, its appearance at 284.6 eV in the DEAE cotton-chloranil complex is equivalent to that in DEAE cottons. Similarly, $E_{B.E.}$ of the O_{1s} electrons of 532.1 eV in chloranil, 531.9 in the DEAE cotton-chloranil complex and 532.0 in DEAE cotton were equivalent.

Wet elemental analyses showed that DEAE cottons coordinate 0.5 mm of Cu (II) for every meq N per g of fabric. The ratios of meqs amine nitrogen/mm Cu II/meqs Cl^- per g fabric were 2/1/3. In this complex, the N_{1s} spectra showed three types of nitrogens, as was the case in the DEAE cotton-chloranil complex. As in the spectra of the other complex, a small percentage of unreacted amine nitrogens was present as indicated by the $E_{B.E.}$ of lowest value, approximately 398 eV. In the copper complex, only 10% of the tertiary amines were unreacted. In donations of electrons to either Cu (II) ions or to chloranil, two types of more electropositive nitrogens are formed. The highest $E_{B.E.}$, that approximating 402 eV, is characteristic of nitrogens in quaternary ammonium ions. That peak of intermediate $E_{B.E.}$ is similar to the peak of a nitrogen in a hydrosalt of a tertiary amine group, or it could be the result of a back donation in the formation of donor-acceptor bonds as when adducts are formed. The ratio of nitrogens having the highest $E_{B.E.}$ to nitrogens with intermediate $E_{B.E.}$ was 2/1 for the chloranil complex and 3.5/1 for the $CuCl_2$ complex.

When anhydrous $CuCl_2$ is reacted with nonaqueous DEAE cottons, the complex formed gives ESCA spectra for the $CU_{2p_{1/2}}$ and $CU_{2p_{3/2}}$ electrons similar to those observed with $CuCl_2$ in that each $E_{B.E.}$ has a satellite at a higher binding energy than the main peak. Addition of chloranil to the DEAE cotton-$CuCl_2$ complexes lowers the $E_{B.E.}$ of the $Cu_{2p_{3/2}}$ electrons from 935.0 to 933.5 eV without affecting its satellite, and lowers the $E_{B.E.}$ of the $Cu_{2p_{1/2}}$ electrons from 955.1 to 952.1 and its satellite from 963.0 to 958.0 eV. These changes indicate complexing of Cu (II) with added chloranil. Addition of $CuCl_2$ to the already complexed chloranil of the DEAE cotton-chloranil complex results in the lower values of 953.2 and 933.4 eV for the $E_{B.E.}$ values for the $Cu_{2p_{1/2}}$ and $Cu_{2p_{3/2}}$ electrons, but no satellites are present.

The DEAE cotton-chloranil product was paramagnetic. A typical ESR spectrum obtained at room temperature is shown in Figure 1.

The singlet is not characteristic of a free electron on a nitrogen, which would generate a triplet. The g value of 2.0059 was found in the study. Thus far, very little research has been carried out on single crystals formed by charge transfer for D^+A^- reactions in the solid state. With chloranil as the acceptor of electrons from tetramethylphenylene-diamine (TMPD) rather than from a triethylamine group, it was found that the hyperfine splitting of the ESR signal changed with time. The absence of a triplet characteristic of a radical cation in such complexes has been explained in several ways (10).

In the pure crystals of TMPD-chloranil, the D^+ ion was absent and it was presumed that both the D^{++} ions and A^- ions were present, but in small concentrations in a predominantly molecular lattice. The components of the g-term gave an average g value of 2.0052 similar to that obtained in this study for the DEAE cotton-chloranil complex.

Comparisons of IR spectra of DEAE cottons before and after reaction with chloranil were made to gain evidence of reaction. New and strong bands at 6.3 μ characteristic of N^+R_4 groups, and those bands characteristic of conjugated C=C groups at 6.2 and 6.6μ were evident in the spectra of the fabrics containing the DEAE cotton-chloranil adducts. In addition, there were also weak bands at 5.95μ, characteristic of C=O, and bands at 7.4μ and 10.1μ characteristic of C-Cl groups.

Visible reflectance spectra of the cottons containing the DEAE cotton-chloranil adducts were compared with those of the DEAE cottons and of known products of chloranil. The DEAE cottons absorb only in the 280-300 nm range. Chloranil absorbs at 285 nm; its adduct with triethylamine absorbs initially in the 320-380 nm region, but at longer wave lengths (380, 435, 787 nm), on standing. Its sodium salt absorbs at 425 and 452 nm. Absorption spectra of aliphatic amines and chloranil have been thoroughly investigated (5,11). It has been shown that the semiquinone anion, $(C_6Cl_4O_2^-)$ is responsible for absorption at 426 and 452 nm. After long standing, the yellowish solution, as well as the solid green adduct formed between triethylamine and chloranil, absorb at 380, 435, 770, 393, 440, and 889 nm. It is not surprising therefore that our DEAE cotton-chloranil complex formed on the fabric absorbed throughout the 300-750 nm region. The fabric adduct is dark green in nonpolar solvents and a lighter green in polar solvents. Foster also reported that the solid adduct formed between triethylamine and chloranil showed a strong ESR single line absorption.

Examination of redox powers of the fabrics containing the chloranil adduct showed that the product is a good reducing reagent. Nitroblue tetrazolium in the presence of $\overline{CN}$ can be reduced by the adduct but not by the original chloranil or DEAE cotton used separately.

The DEAE cotton-chloranil reaction must have resulted in some crosslinking between cellulose chains, because the products were insoluble in cupriethylenediamine.

DEAE Cotton-TCNE Adduct. Nitrogen analyses of DEAE cottons before and after reaction with TCNE by the Kjedahl method showed that two meqs N/g of cotton were added for each meq N/g of cotton in the original DEAE cotton. A DEAE cotton (1.44 meq N/g) contained 4.29 meq N/g after TCNE treatment, for example. This indicates reaction of 2 amino nitrogen groups of DEAE cotton with one TCNE to yield the following product:

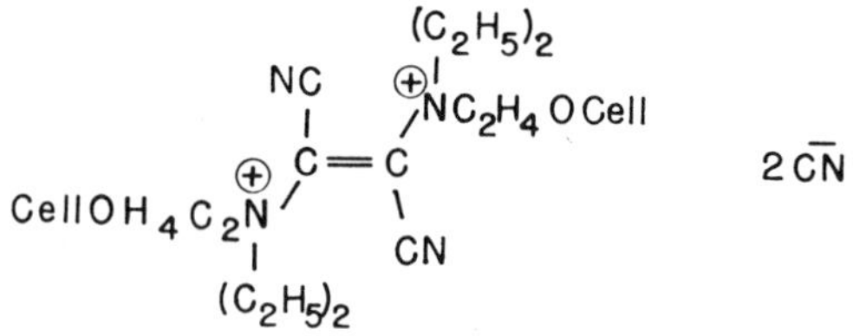

$$\mathrm{CellOH_4C_2\overset{\oplus}{N}(C_2H_5)_2{-}C(CN){=}C(CN){-}\overset{\oplus}{N}(C_2H_5)_2C_2H_4OCell} \qquad 2\,\overline{CN}$$

Data in Table II show the $E_{B.E.}$ values for the C_{1s}, O_{1s}, and N_{1s} electrons from all fabrics and of the $Cl_{2p_{1/2}}$ and $Cl_{2p_{3/2}}$ and of the $Cu_{2p_{1/2}}$ and $Cu_{2p_{3/2}}$ electrons, when these elements were present. As previously reported, DEAE fabrics contain about 30% of the nitrogens in a somewhat more positive form than that in the free amino groups. Probably in the preparation of DEAE cottons, some free amine nitrogens react with excess β-chloroethyldiethylamine or an organic solvent or a proton. The $E_{B.E.}$ values for elements in TCNE are also included.

The ESCA spectra of the N_{1s} electrons again showed three different $E_{B.E.}$ values. The ratio of relative amounts of highest to intermediate peaks for this complex was 1/1. The intermediate peak in this case is a combination of amino nitrogens that became more electropositive during reaction and TCNE nitrogens that became more negative. Initially it was assumed that the lowest $E_{B.E.}$ of 397.7 eV was due to the $\overline{CN}$ in combination with free amino nitrogens. However, reported values for $E_{B.E.}$ of the $\overline{CN}$ are questionable (12). Therefore, the $\overline{CN}$ ions of the DEAE cotton-TCNE adduct were exchanged for $\overline{Cl}$ by use of excess NaCl solution. Analyses of the N_{1s} spectra for $E_{B.E.}$ values of 401.3, 399.1 and

TABLE II. Binding Energies (eV) of Electrons in Elements of Treated Fabrics

Fabrics[a]	C_{1s}	O_{1s}	Cl_{2p}	N_{1s}			Cu_{2p}	
DEAE	284.6(2.7)[b]	532.0(2.2)	197.6(2)	401 [40][c]	398.1[60]	--	--	--
DEAE-TCNE	285.3(3.0)	532.2(2.3)	--	401.3[26]	399.1[26]	397.7[46]	--	--
DEAE-TCNE-$CuCl_2$	285.0(2.7)	532.3(2.2)	197.9(3.5)	401.6[34]	399.0[35]	397.8[31]	952.4	932.9
DEAE-$CuCl_2$	285.3(2.7)	532.7(2.3)	198.2(4.1)	401.8[70]	399.3[20]	397.9[10]	963.0 944.6	955.1 935.0
DEAE-$CuCl_2$-TCNE	285.3(3.0)	532.7(2.1)	198.1(2.7)	401.7[61]	398.8[28]	398.1[11]	--	935.9
TCNE	289.3_{sh} 285.3	--	--	400.2	--	--	--	--

[a] DEAE is diethylaminoethyl cotton in the free amine form
DEAE-TCNE is DEAE fabric complexed with tetracyanoethylene
DEAE-TCNE-$CuCl_2$ is DEAE complexed first with TCNE and then treated with $CuCl_2$
DEAE-$CuCl_2$ is the complex formed with $CuCl_2$
DEAE-$CuCl_2$-TCNE is the $CuCl_2$ complex subsequently treated with TCNE

[b] (#) is width at half height

[c] [#] is relative percentage of total N

397.7 eV for a DEAE cotton-TCNE complex, before and after exchange with $\overline{Cl}$ ions, give relative peak areas of 1/1/2 before and 1/1/1 after extraction. Use of NaI rather than NaCl to exchange $\overline{CN}$ ions from another piece of the same DEAE cotton-TCNE adduct resulted in oxidation of some $\overline{I}$ to free I_2. Nevertheless, ratios of peak areas were changed to 1/1/0.5. It was shown that the $E_{B.E.}$ of the $\overline{CN}$ was that of the lowest of the three values observed for N_{1s} spectra.

On the assumption that only the unreacted ($:NR_3$) groups were responsible for the peak of lowest $E_{B.E.}$ after exchange with excess NaI, it was calculated that the number of nitrogens quaternized (characterized by highest $E_{B.E.}$) was twice that left as unreacted tertiary amines (characterized by lowest $E_{B.E.}$). This information, together with wet analyses of DEAE cotton and the DEAE cotton-TCNE complex, showed that for every amino N in the DEAE cotton that reacted, one nitrogen from the TCNE remained in the bound or organo form ($E_{B.E.}$ = 399.1), and 1.5 nitrogens from the TCNE were in the $\overline{CN}$ form. Thus, the ratio of amine nitrogens reacted/bound CN/$\overline{CN}$ was calculated to be 2/2/3. These data show that 70% of the amino nitrogens of DEAE cotton were quaternized by TCNE and that most were in the highest electropositive state, because the ratio of nitrogens in highest to intermediate $E_{B.E.}$ was 10/1 compared to 3.5/1 for the DEAE cotton-$CuCl_2$ complex, and 2/1 for the DEAE cotton-chloranil complex.

The increase of +3 eV in the $E_{B.E.}$ of the N_{1s} electrons of DEAE cottons on reaction with TCNE, and the decreases of -1.1 and -2.5 eV in the $E_{B.E.}$ of the N_{1s} electrons of TCNE, are indicative of a strong reaction between the donor and acceptor in which amino nitrogens become quaternized and the cyano groups bound to the ethylenic carbons are less electropositive than in the original TCNE.

Addition of anhydrous $CuCl_2$ to the DEAE cotton-TCNE adduct resulted in a small addition of Cu (II) to the fabric, but the $E_{B.E.}$ values for the $Cu_{2p_{1/2}}$ and $Cu_{2p_{3/2}}$ were at 952.4 and 932.9 eV, respectively. These ESCA shifts were -2.7 and -2.0 eV from those observed with the DEAE cotton-$CuCl_2$ complex. In addition, neither satellite was present when the Cu (II) was added to the DEAE cotton-TCNE complex. Also, when TCNE was added to the DEAE cotton-$CuCl_2$ complex, a small addition of nitrogen was obtained and only the main $Cu_{2p_{3/2}}$ peak of $E_{B.E.}$ at 935.9 eV was observed.

The DEAE cotton-TCNE adduct was also paramagnetic, giving a singlet in its ESR spectra. In Figure 2 is a typical spectrum obtained at room temperature. The g-value was 2.0032. This ESR singlet is not characteristic of an electron on the N of the cation or of a free electron on an anion formed by a charge transfer complex. Stamires and Turkevich (7) were able to obtain an 11-line spectrum by reacting TCNE with triethylamine in chloroform. The 11 lines with a separation of about 1.6 gauss and an intensity ratio of 1/4/10/16/19/16/10/4/1, were attributed to the anion radical formed by the capture of one electron by TCNE. The g-value of the center of the complex was close to that of a free

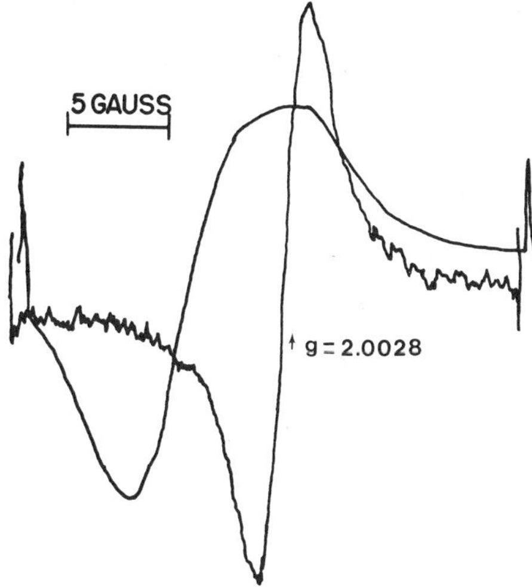

Figure 1. ESR signal of a DEAE cotton after its reaction with 2,3,5,6-tetrachloro-p-benzoquinone (chloranil) in chloroform

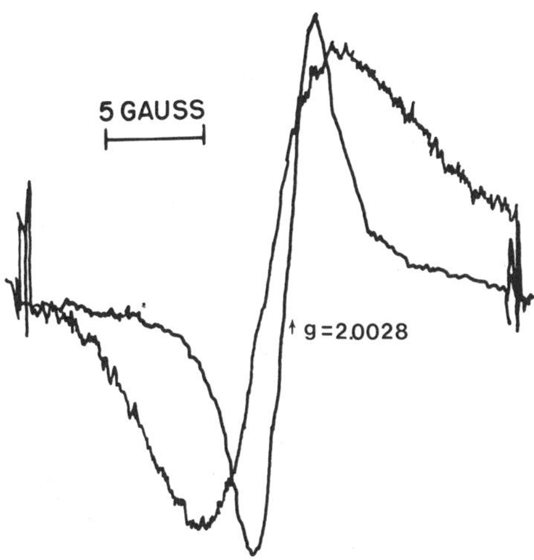

Figure 2. ESR signal of a DEAE cotton after its reaction with tetracyanoethylene (TCNE) in chloroform

electron. The g-value of the singlet spectrum obtained in our adduct was less than that obtained with the DEAE cotton-chloranil adduct. The IR spectra of DEAE cotton-TCNE adducts on the fabric give strong absorption bands at 4.6μ, characteristic of C≡N, and at 5.95 and 6.1 μ characteristic of C=N groups. The strong bands at 6.2 and 6.7μ were indicative of conjugated double bonds. These marked differences in spectra after DEAE cotton had been treated with TCNE indicated strong reaction between donor and acceptor. The redox powers of TCNE were also different after it was reacted with DEAE cotton. TCNE, a good oxidizing reagent alone, was also an effective oxidizing reagent when complexed with DEAE cotton. After reaction with DEAE cotton, the resultant fabric was a peroxide generator, it fluoresced, and it produced chemiluminescence.

Summary

Interactions between tertiary amine groups of DEAE cottons as an n type of donor, D, and TCNE or p-chloranil as the π type of acceptor, A, were large enough to produce products with absorptions in the uv-visible range that were not characteristic of the individual components, and to produce electron transfers in the ground state resulting in paramagnetic properties of the D-A complex within the cellulose matrix. ESR data indicated absence of unpaired electrons located on nitrogen atoms of the amine groups and suggest transformation of outer (π) D-A complexes to inner (σ) $D^{+}A^{-}$ or $D^{++}A^{--}$ type complexes.

ESCA spectra were used in conjunction with wet analyses to support proposed structures for the products.

Acknowledgement

The authors express their appreciation to Mr. Oscar Hinojosa for ESR spectra.

Literature Cited

1. Berni, Ralph J., Soignet, Donald M., and Benerito, Ruth R., Textile Res. J., (1970), 40, 999.
2. Soignet, Donald M., Berni, Ralph J., and Benerito, Ruth R., J. Appl. Polymer Sci., (1971), 15, 155.
3. Perrier, Dorothy M. and Benerito, Ruth R., J. Appl. Polymer Sci., (1975), 19, 3211.
4. Perrier, Dorothy M. and Benerito, Ruth R., J. Appl. Polymer Sci., (1976), 20, 000.
5. Foster, R., Rec. Trav. Chim., (1964), 83, 711.
6. Mulliken, Robert S., J. A. C. S., (1952), 74, 811.
7. Stamires, Dennis N. and Turkevich, John, J.A.C.S., (1963), 85, 2557.
8. Soignet, Donald M., Berni, Ralph J., and Benerito, Ruth R., Anal. Chem., (1974), 46, 941.

9. Ng, Kung T. and Hercules, David M., J.A.C.S., (1974), 97, 4168.
10. Foster, R., "Organic Charge-Transfer Complexes," p. 269, Academic Press, New York, 1969.
11. Shah, Salish B. and Murthy, A. S. N., J. Phys Chem., (1975), 70, 322.
12. Siegbahn, Kai, Nordling, Carl, Faklman, Anders, Nordberg, Ragnar, Hamrin, Kjell, Hedman, Jan, Johansson, Guirilla, Bergmark, Torsten, Karlson, Sven-Erik, Lindgren, Ingvar, and Lindberg, Bernt, "ESCA Atomic, Molecular and Solid State Structures Studied by Means of Electron Spectroscopy," Almquist and Wiksells AB, Stockholm, 1967.

*Mention of companies or commercial products does not imply recommendation by the U. S. Department of Agriculture over others not mentioned.

18

Pyrolysis of Cellulose after Chlorination or Addition of Zinc Chloride

YUAN Z. LAI, FRED SHAFIZADEH, and CRAIG R. McINTYRE

Wood Chemistry Laboratory, University of Montana, Missoula, MO 59812

We have been concerned with the mechanism of pyrolytic reactions involved in the initiation, propagation and suppression of cellulosic fires (1-6). Since chlorine derivatives and salts are extensively used as flame retardants, pyrolysis of cellulose after chlorination or addition of zinc chloride has been investigated in order to determine the reactions and principles involved in suppressing the production of flammable volatiles which fuel the flaming combustion.

Preparation of Samples

Samples of 6-chloro-6-deoxycellulose with varying degrees of chlorination were prepared by treating microcrystalline cellulose in DMF with methanesulfonyl chloride at 63°C (7). The products and their properties are listed in Table 1.

Samples of cellulose containing varying amounts of $ZnCl_2$ were prepared by adding the calculated proportion of catalyst in methanol and evaporating the solvent under vacuum.

Discussion

Thermal Reactions of Chlorinated Celluloses. The thermogram of pure cellulose is shown in Figure 1. In this figure, the DTA, TG and DTG curves reflect the sequence of physical transformations and chemical reactions as the sample is heated at a constant rate.

The DTA curve contains a broad decomposition endotherm in the range of 300-375°C centered at 340°C, that is accompanied by a rapid weight loss leaving 6.5% of carbonaceous residue at 400°C.

TABLE 1

Preparation of 6-Chloro-6-deoxycellulose

Sample	Reaction Conditions		Product		
	Temp (°C)	Time (hr)	Color	Chlorine (%)	D.S.[a]
A	63	1.5	White	0.40	0.018
B	63	2.5	White	0.50	0.023
C	63	3.7	White	1.16	0.053
D	63	7.1	White	1.96	0.091
E	63	19.6	Tan	4.12	0.192
F	63	48.6	Tan	5.72	0.270

[a]Degree of substitution calculated from the chlorine content.

The DTG curve, indicating the rate of weight loss, shows a sharp peak at 351°C.

Chlorinated samples gave entirely different thermograms. They decomposed about 100°C lower and gave more charred residue, as illustrated in Figure 2, for a sample (B) with a degree of substitution (D.S.) of 0.023. This sample shows a broad endotherm in the range of 200-360°C with overlapping peaks near 225° and 280°C, and a residue of 30% at 400°C, with the maximum rate of weight loss occurring at 300°C.

The nature of chemical reactions and weight loss occurring within the range of 150-300°C were investigated by monitoring the volatile products formed and analysis of the residues.

Hydrogen chloride, detected as AgCl, started to evolve at ∿ 200°C, indicating that the first endothermic peak involves dehydrohalogenation of the molecule. More than 80% of the combined chlorine disappeared on heating the sample to 250°C. The endotherm at 225°C, corresponding to the dehydrohalogenation, became more prominent with a higher degree of chlorination. For example, sample F, with D.S. of 0.27, showed only a broad endothermic peak at 220°C, as illustrated in Figure 3.

ESR scanning of the pyrolysis residues within the range of 150-300°C indicated that the initial decomposition reactions are heterolytic, but subsequent reactions, which take place above 230°C are accompanied by the formation of stable free radicals.

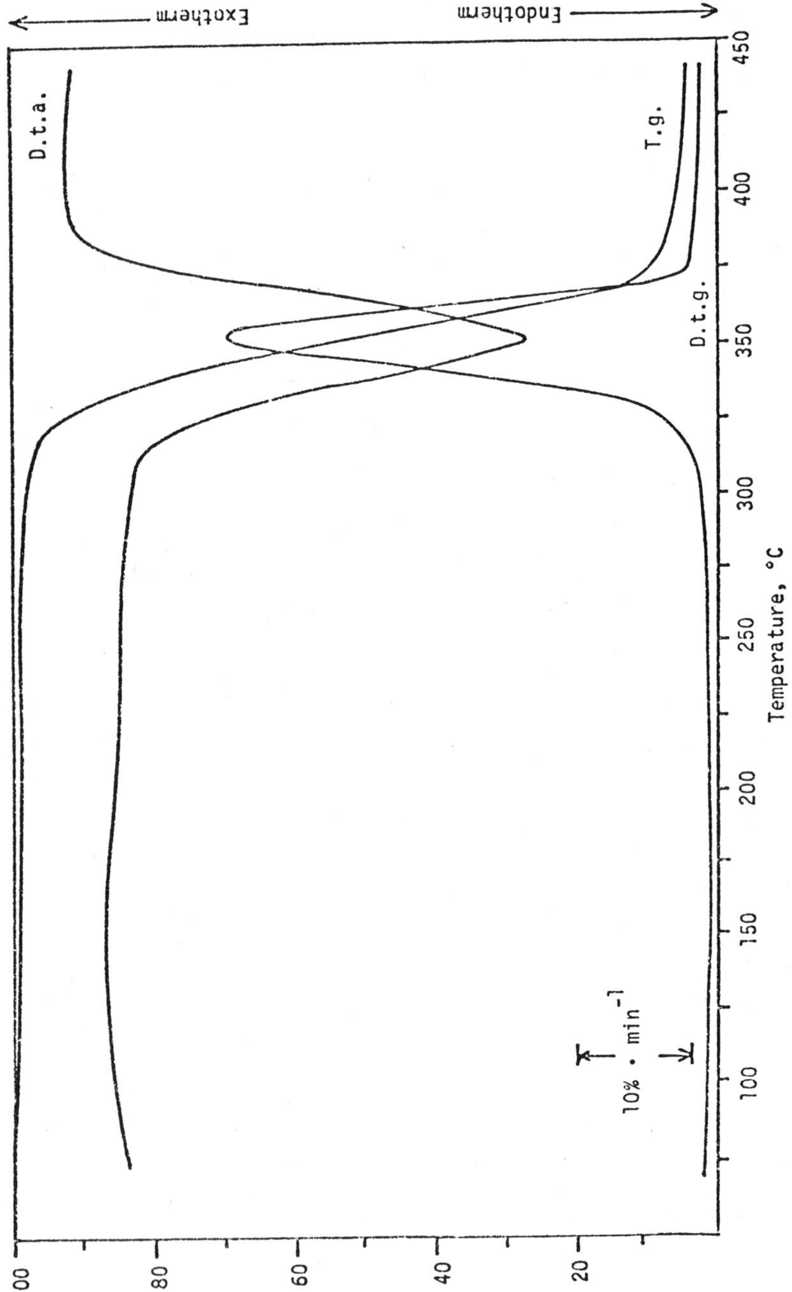

Figure 1. Thermal analysis curves of microcrystalline cellulose

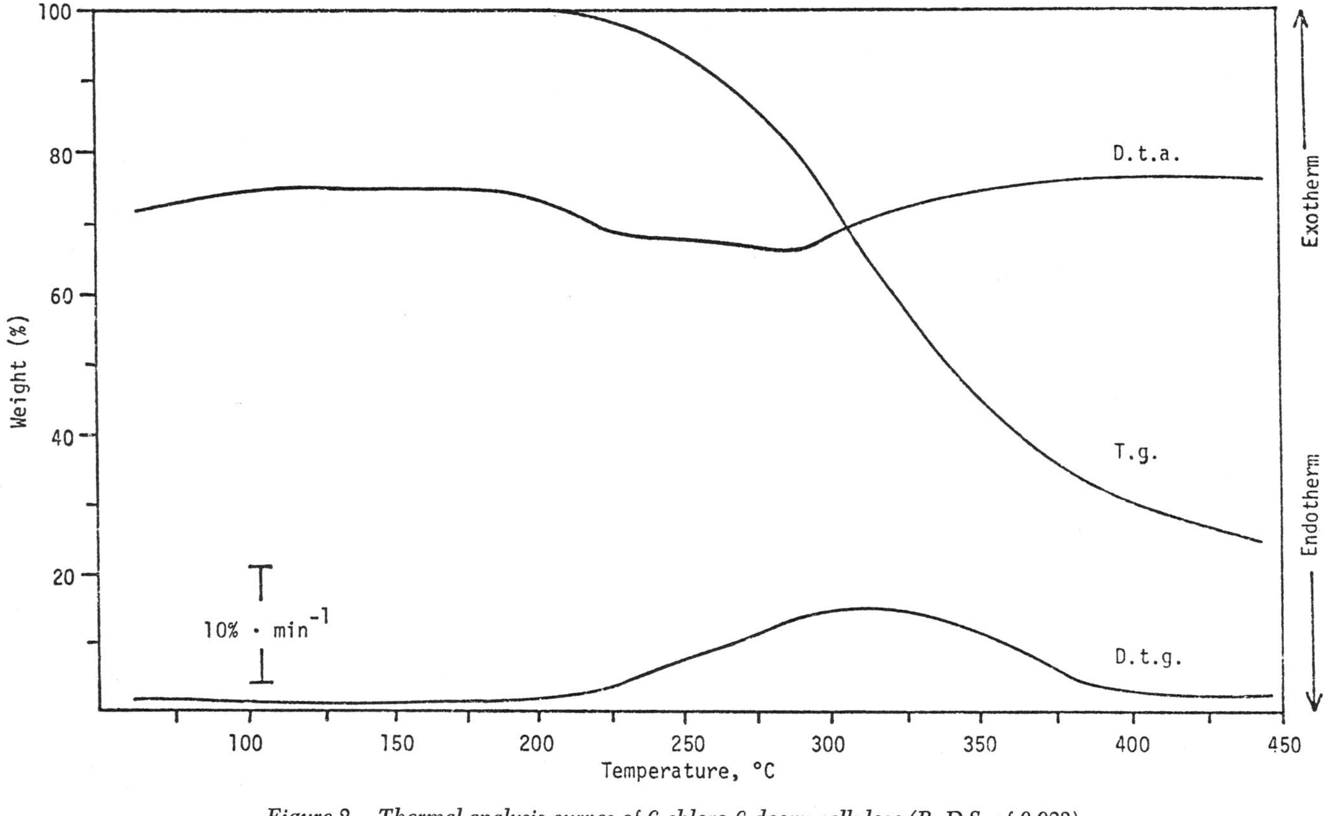

Figure 2. Thermal analysis curves of 6-chloro-6-deoxy cellulose (B, D.S. of 0.023)

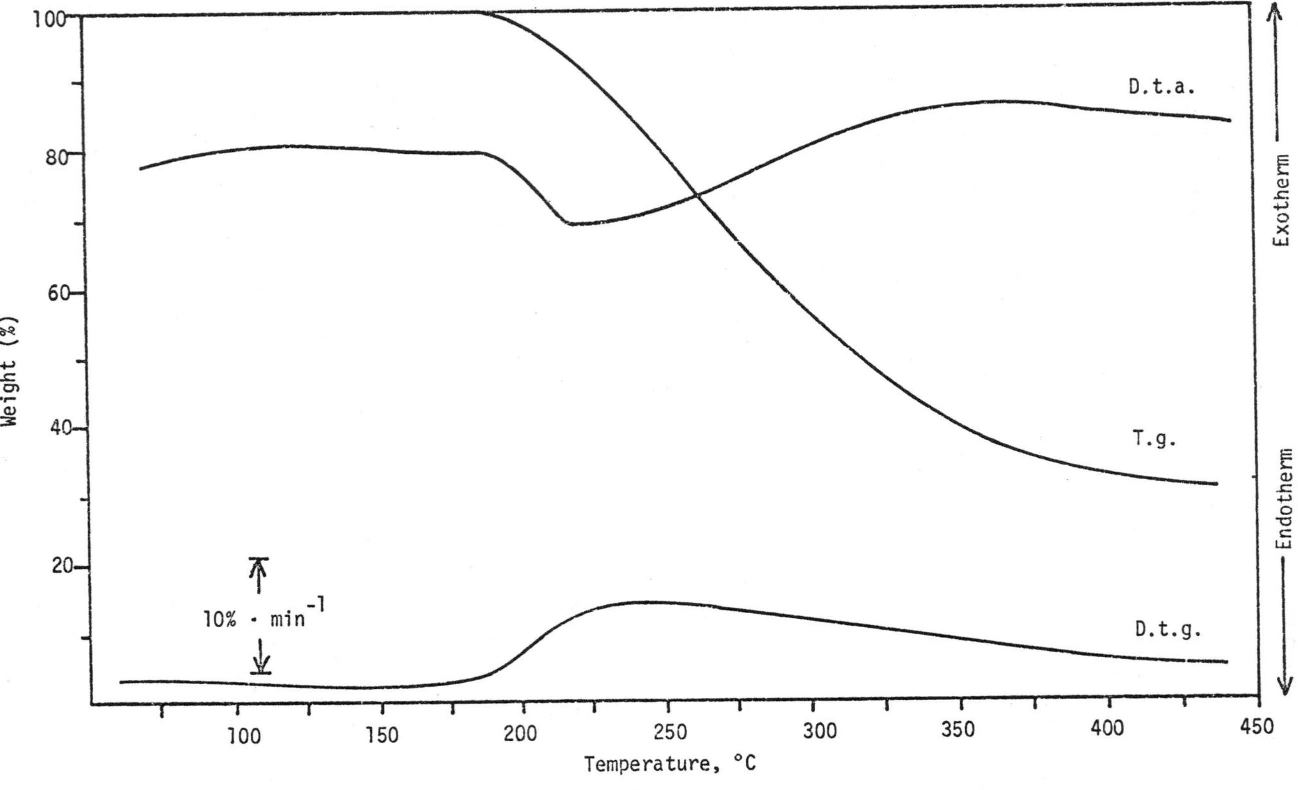

Figure 3. Thermal analysis curves of 6-chloro-6-deoxy cellulose (F, D.S. of 0.270)

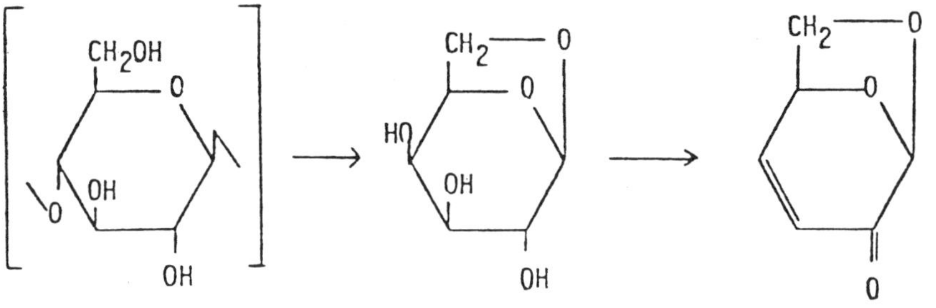

Scheme 1. Production of levoglucosenone

The scanning mass spectroscopy data for the volatiles formed at various temperatures are summarized in Figure 4. These data show the evolution of water (m/e 17), carbon dioxide (m/e 44), 2-furaldehyde (m/e 95 and 96), 5-methyl-2-furaldehyde (m/e 109 and 110), β-angelicalactone (m/e 98) and levoglucosenone (1,6-anhydro-3,4-dideoxy-β-D-glycero-hex-3-enopyranos-2-ulose) (m/e 98 and 126) occurred at the early stages of heating.

Formation of these ions is consistent with the chemical analysis of volatile products shown in Table 2. The results show that chlorinated samples, in contrast to pure cellulose, give a substantial amount of levoglucosenone as shown in Scheme 1.

Table 3 and Figure 5 summarize the thermal analysis features of chlorinated cellulose with various degrees of chlorination. They show that charring increases with the degree of substitution up to a D.S. of 0.023. Additional chlorination after that has little or no effect. Thus, only a small quantity of combined halogen is necessary to achieve the maximum charring effect. Also, pre-heating the sample causes a sharp decrease in the formation of levoglucosenone and 2-furaldehyde. For example, sample B on pre-heating to 250°C lost ∿ 80% of its original chlorine content and gave only 3.2% of levoglucosenone on further heating as compared to 12.5% levoglucosenone obtained from the unheated sample. This investigation shows that hydrogen chloride, liberated in the early stages of heating, could catalyze a series of dehydration reactions. Further condensation and carbonization of the dehydration products, such as levoglucosenone, reduces the production of combustible volatiles. The condensation and carbonization reactions are catalyzed more effectively by Lewis acids, such as $ZnCl_2$, than by Arrhenius acids (6).

Thermal reactions of cellulose after addition of zinc chloride, shown in Figure 6, reflect the charring reactions. The thermal analysis curves of a sample of cellulose containing 5% of $ZnCl_2$ shows a broad decomposition endotherm in the range of 200-358°C, which peaks at 333°C with a shoulder near 275°C. The decomposition endotherm is accompanied by ∿ 64% of weight loss, leaving 36% of charred residue at 400°C.

The initial decomposition reactions occurring within the

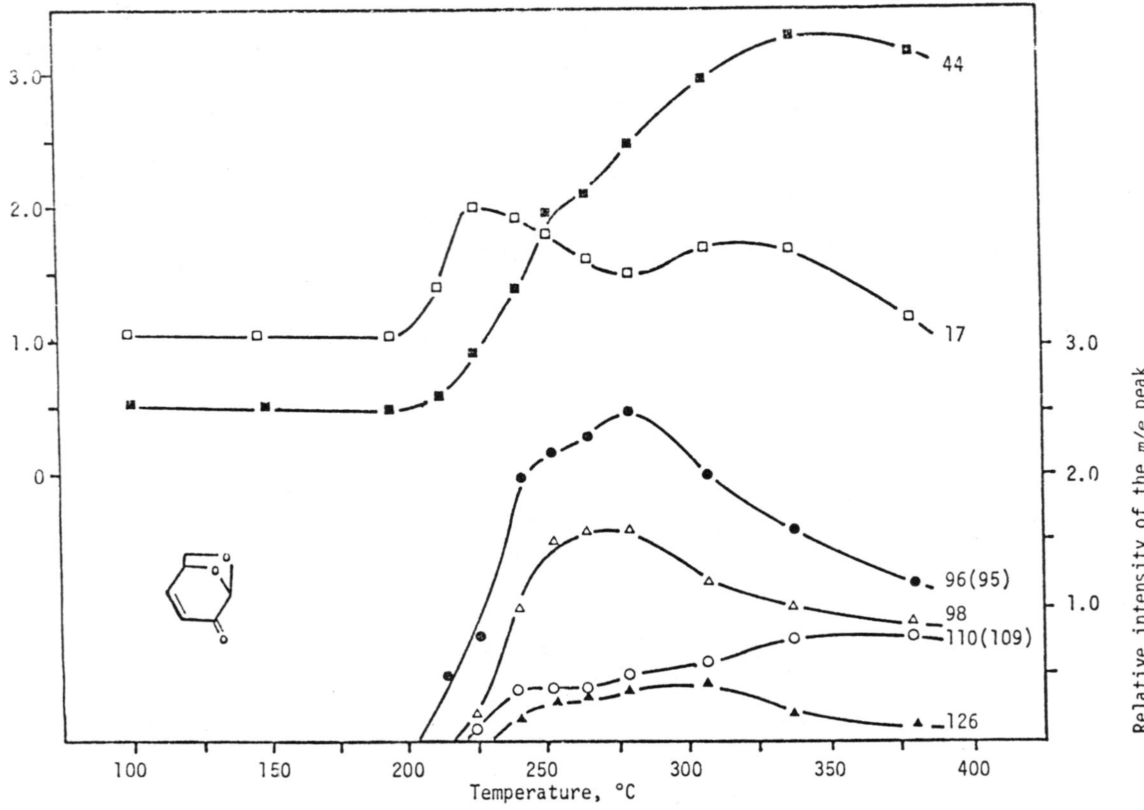

Figure 4. Mass spectrometry of the pyrolysis products of 6-chloro-6-deoxy cellulose (F, D.S. of 0.270)

TABLE 2

Volatile Products of 6-Chloro-6-deoxycellulose and its Partially Degraded Residue Formed on Rapid Heating to 500°C

Substrate			Products Formed on Rapid Heating to 500°C (%)				
D.S.	Preheating °C	Weight Loss (%)	2-Furaldehyde	5-Methyl-2-furaldehyde	β-Angelica-lactone	Levoglucosenone	Levulinic Acid
0.000	--	--	1.2	0.5	0.2	1.4	T
0.018	--	--	1.8	0.7	0.4	12.8	T
0.023	--	--	1.0	0.8	0.7	12.5	T
	200	2.1	0.2			3.3	
	220	2 8	0.3			3.0	
	240	5.1	0.3			3.1	
	250	7.1	0.4			3.2	
	260	10.2	0.3			1.3	
	280	17.8	0.3			1.3	
	300	29.5	0.3			0.6	
0.053	--	--	1.0	0.9	0.6	13.4	T
0.091	--	--	1.1	0.6	0.4	10.6	0.2
0.192	--	--	1.3	0.6	0.4	11.4	1.2
0.270	--	--	1.1	0.6	0.4	8.3	1.2

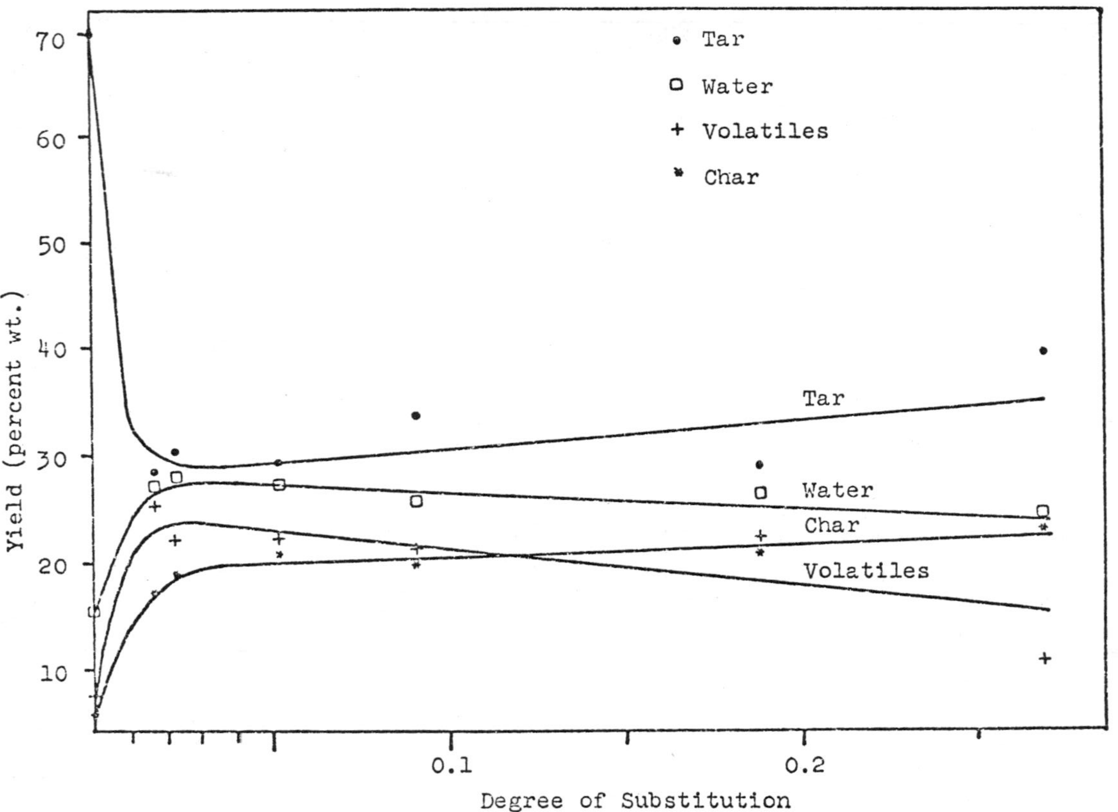

Figure 5. Pyrolysis products of 6-chloro-6-deoxy-cellulose at 500°

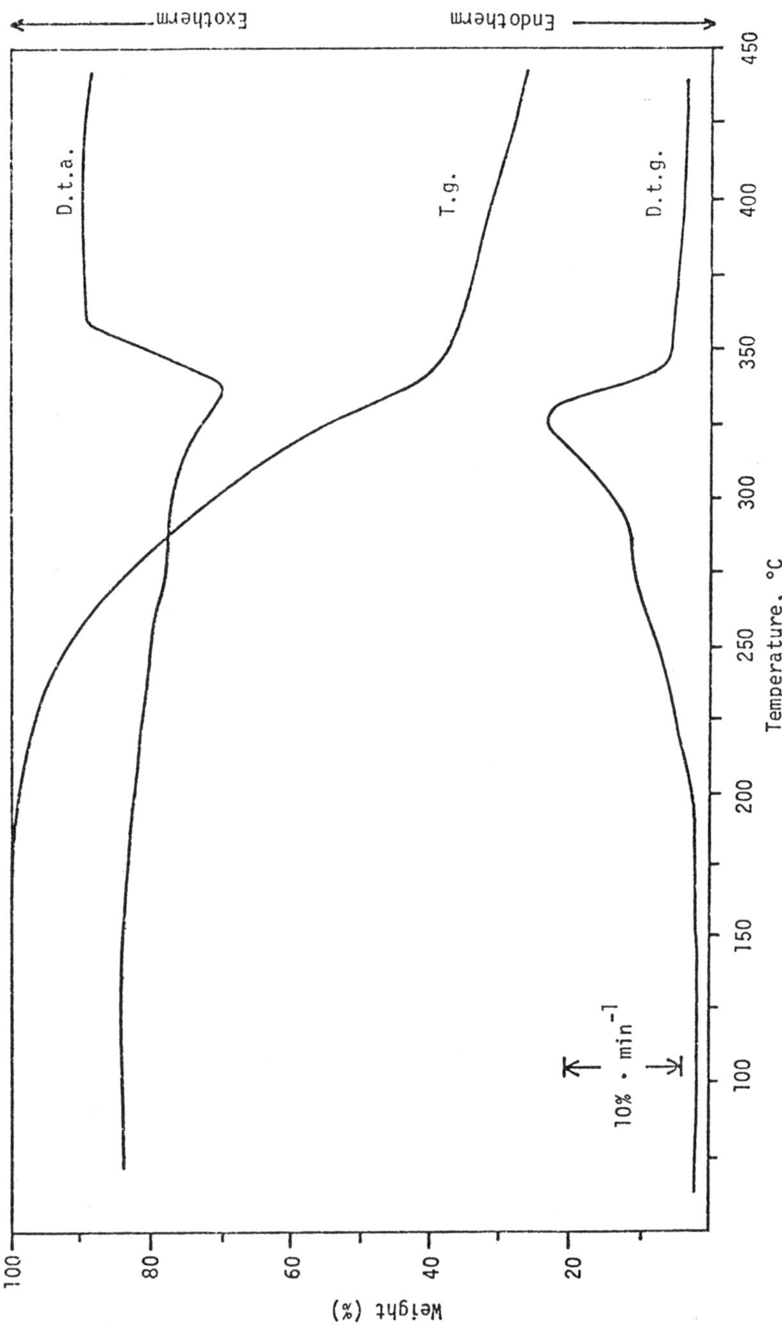

Figure 6. Thermal analysis curves of cellulose in the presence of 5% zinc chloride

TABLE 3

Thermal Analysis Features of 6-Chloro-6-deoxycellulose

Sample	Degree of Substitution	D.t.a. peaks Range °C	D.t.a. peaks Peak °C	D.t.g. peaks Dec. °C	T.g. Residue at 400°C (%)
		6-Chloro-6-deoxy cellulose			
A	0.018	210-360	225[a],285	300[a],335[a]	12.0
B	0.023	200-360	225[a],280[a]	300,325[a]	30.0
C	0.053	200-350	225,280	285	29.2
D	0.091	200-345	225,280	230[a],280	32.0
E	0.192	205-340	225,275	215,260	31.5
F	0.270	190-340	220	240	31.5
		Cellulose			
		300-375	340	351	6.5

[a]A shoulder

range of 200-330°C were investigated by physical and chemical analysis of the residues obtained at different temperatures. The analytical results are summarized in Table 4. The IR data showed that formation of carbonyl (∿ 1710 cm^{-1}) and unsaturated (∿ 1610 cm^{-1}) groups resulting from dehydration reactions takes place in the early stages of heating. The fine structure of cellulose, indicated by the fingerprint region (1500-800 cm^{-1}), was practically destroyed on heating to 310°C with 32% of weight loss. The water-soluble fraction, after borohydride reduction and acid hydrolysis, was found to consist mainly of unhydrolyzable condensation products and smaller quantities of D-glucose and 3,6-anhydro-D-glucose.

ESR spectroscopy of the heated sample containing 5% $ZnCl_2$ showed that the radical formation associated with the charring reactions takes place at a relatively low temperature of 230°C, corresponding to 4% weight loss. Therefore, it was concluded that $ZnCl_2$ catalyzes the nonglycosidic condensation and charring reactions at the very early stages of heating. This accounts for the low yield of monomeric products and large proportion of residues discussed above.

TABLE 4

Analysis of Cellulose Samples Containing 5% Zinc Chloride Pyrolyzed to Different Temperatures

Properties	Temperatures, °C				
	235°	270°	280°	310°	330°
Weight Loss (%)	5.5	14.1	16.91	32.36	45.67
Residue	94.5	85.9	83.09	67.64	54.33
Color	Gray	Brown	Brown	Deep Brown	Black
IR spectra					
Carbonyl group ($1700\text{-}1720\ cm^{-1}$)	Very Weak	Weak	Moderate	Moderate	Strong
Unsaturation groups ($1610\ cm^{-1}$)	--	Weak	Weak	Moderate	Strong
Fingerprint region for cellulose ($1500\text{-}800\ cm^{-1}$)	Strong	Moderate	Moderate	Weak	Very Weak
Water-soluble material (%)	9.2	9.8	8.6	12.4	5.8
After reduction and acid hydrolysis					
3,6-anhydro-D-glucose (%)	(1.4)[a]	(3.5)	(3.7)	(2.2)	(1.9)
D-glucose (%)	(11.2)	(11.2)	(13.3)	(11.0)	(T)

[a]Number in parentheses are percentages of the water-soluble fraction.

TABLE 5

Thermal Analysis Features of Cellulose Containing Zinc Chloride

Additive	D.t.a. peaks Range °C	D.t.a. peaks Peak °C	D.t.g. peaks °C	T.g. Residue at 400°C (%)
Neat	300-375	340	351	6.5
1%	275-365	345	340	12.2
3%	200-360	333	275[a],327	30.5
5%	200-358	275[a],333	275[a],327	31.0
7%	175-340	225[a],310[a]	265,310	40.0
10%	175-330	210[a],260[a],305[a]	265,310	41.3

[a]A shoulder

The thermal analysis features of cellulose containing varying amounts of $ZnCl_2$ are summarized in Table 5. These data show that higher proportions of catalyst lower the decomposition temperature even further and result in a very broad endotherm with no distinct peaks, apparently due to the overlapping of a variety of concurrent and consecutive reactions. Also, the extent of charring increases as the proportion of the additive is increased and reaches the maximum value of 40% at ∿ 7% addition.

<u>Isothermal Pyrolysis</u>. Isothermal pyrolysis at 500°C provided a cross-section of the products formed by the sequence of pyrolytic reactions. These products, listed in Table 6, consisted of tar, char, water, carbon dioxide, hydrogen chloride, and a variety of volatile organic compounds, namely 2-furaldehyde, 5-methyl-2-furaldehyde, and levoglucosenone, derived mainly from dehydration reactions.

G.L.C. analysis of the tar fraction, before and after acid hydrolysis, gave the results shown in Table 7. These data indicate that the tar fraction, in general, consisted mainly of unhydrolyzable condensation products, and small amounts of levoglucosan, its furanose isomer, and 3,6-anhydro-D-glucose as the monomeric products. However, the tar fractions from $ZnCl_2$-treated samples, both before and after acid hydrolysis, contain

TABLE 6

Pyrolysis Products of 6-Chloro-6-deoxy cellulose and Cellulose-Zinc Chloride Mixture at 500°C

Product	Yield (%)						
	6-Chloro-6-deoxy cellulose			Cellulose with Zinc Chloride			
	B (0.02)	D (0.09)	F (0.27)	Neat	1%	5%	10%
Acetic acid	T	T	T	T	T	T	T
α-Angelicalactone	T	T	T	T	T	T	T
2-Furaldehyde	0.6	0.7	1.0	0.4	1.4	1.9	2.0
5-Methyl-2-furaldehyde	0.5	0.6	1.0	T	0.4	0.3	0.2
2-Furfuryl alcohol	T	0.1	0.3	T	T	T	T
β-Angelicalactone	T	T	T	T	T	T	T
Levoglucosenone	0.4	0.6	0.3	0.3	0.3	0.1	<1.0
Total acid (HCl, Carboxylic acid)	3.4	4.6	9.7	2.5	2.7		
Carbon dioxide	2.7	2.9	3.5	6.0		3.0	3.9
Water	28.4	26.3	24.0	12.0	19.6	23.1	29.0
Char	22.7	28.4	29.2	8.0	19.6	30.8	42.4
Tar	31.2	25.6	22.6	65.8	48.6	30.5	14.7

TABLE 7

GLC Analysis of Tar Fraction

Product	Yield (%)						
	6-Chloro-6-deoxy-cellulose			Cellulose with Zinc Chloride			
	B (0.02)	D (0.09)	F (0.27)	Neat	1%	5%	10%
Before acid hydrolysis							
3,6-Anhydro-D-glucose	0.5	0.4	1.4	T	3.3	5.9	6.7
1,6-Anhydro-β-D-glucopyranose	5.2	4.0	1.5	46.2	16.8	9.1	4.2
1,6-Anhydro-β-D-glucopyranose	0.4	0.4	0.1	3.6	1.2	0.3	T
After acid hydrolysis							
3,6-Anhydro-D-glucose	8.6	8.2	17.6	1.5	14.4	23.3	37.9
D-Glucose	8.9	8.0	2.1	77.3	30.0	18.2	10.1

higher quantities of monomeric products than those obtained from chlorinated samples.

The difference in the catalytic effects between $ZnCl_2$ and HCl may also be seen from Table 8 which illustrates the major volatile products formed on isothermal pyrolysis at 300° and 400°C. $ZnCl_2$-treated samples gave a slightly higher yield of 2-furaldehyde, but a substantially lower yield of levoglucosenone, which implies that $ZnCl_2$ is more effective in promoting further transformation of levoglucosenone and other dehydration products to nonvolatile materials.

From this experiment, it was concluded that the hydrogen chloride produced from pyrolysis of 6-chloro-6-deoxycelluloses, and the added $ZnCl_2$ catalyze the transglycosylation, dehydration and condensation reactions shown by model compounds. $ZnCl_2$, however, is more efficient in promoting the condensation and carbonization reaction.

TABLE 8

Major Volatile Pyrolysis Products of 6-Chloro-6-deoxy cellulose and Cellulose-Zinc Chloride Mixture formed at 300 and 400°C

Product	Temperature °C	Yield (%)											
		6-Chloro-6-deoxy cellulose						Cellulose with Zinc Chloride					
		A	B	C	D	E	F	Neat	1%	3%	5%	7%	10%
2-Furaldehyde	400	0.9	0.8	0.6	0.5	0.4	0.4	0.4	1.6	2.1	1.8	2.0	2.1
	300	0.3	0.3	0.4	0.4	0.4	0.5	0.4	0.7	1.9	0.8	1.4	1.3
Levoglucosenone	400	7.9	5.3	5.1	4.7	4.8	4.5	0.7	2.1	1.8	2.1	2.2	1.4
	300	7.8	7.7	8.1	9.2	8.5	7.2		1.6	2.2	1.7	2.3	2.2

Acknowledgement

The authors thank Dr. Clayton Huggett and the National Fire Prevention and Control Administration for their interest in the chemistry of cellulosic fires and for supporting this study.

Abstract

The role of halogen derivatives used in flameproofing of cellulosic materials was investigated by pyrolysis of cellulose in the presence of zinc chloride and after various degrees of chlorination at position 6 of the D-glucose units. Both treatments lowered the decomposition temperature and increased the production of char and water at the expense of the combustible volatiles. The volatile fraction contained mainly water and some carbon dioxide, hydrogen chloride, levoglucosenone, 2-furaldehyde, 5-methyl-2-furaldehyde and traces of other dehydration products. The tar fraction consisted mainly of unhydrolyzable condensation products and small amounts of levoglucosan ,its furanose isomer and 3,6-anhydro-D-glucose as the monomeric products. The isothermal and scanning analysis of these products at different temperatures indicated an initial dehydrohalogenation of chlorinated cellulose samples. The hydrogen chloride produced in this manner and the added zinc chloride catalyzed the transglycosylation, dehydration and condensation reactions shown by model compounds. Further heating of the cross-linked or nonglycosic condensation products led to elimination of substituents and carbonization, which could be monitored by ESR spectroscopy of the stable free radicals in the residue.

Literature Cited

1. Shafizadeh, F., Lai, Y.Z. and Nelson, C.R., J. Appl. Polym. Sci., (1976), 20, 139.
2. Lai, Y.Z. and Shafizadeh, F., Carbohyd. Res., (1974), 38, 177.
3. Shafizadeh, F., Appl. Polym. Symp., (1975), 28, 153.
4. Shafizadeh, F., J. Polymer. Sci., Part C, (1971), 36, 21.
5. Shafizadeh, F., and Fu, Y.L., Carbohyd. Res., (1973), 29, 113.
6. Shafizadeh, F., Chin, P.S., Carbohyd. Res., (1976), 46, 149.
7. Horton, D., Luetzow, A.E., and Theander, O., Carbohyd. Res., (1973), 26, 1.

19

A Speculative Picture of the Delignification Process

D. A. I. GORING

Pulp and Paper Research Institute of Canada, Department of Chemistry, McGill University, Montreal, Quebec, Canada H3C 3G1

It is currently accepted that lignin in the cell wall of wood is a crosslinked three-dimensional polymer which forms an amorphous matrix in which the cellulosic material is embedded. During delignification the network is broken down by the action of chemicals and the lignin is dissolved as macromolecules of approximately spherical shape in a wide range of sizes (1). In the present report a closer look is taken at the network concept of lignin in the light of certain recent findings concerning the ultrastructural arrangement of lignin in the wood cell wall (2).

In 1938 Bailey (3) published photomicrographs which clearly showed that the cell wall of wood was lamellar in nature. More recently, by an elegant application of dimensional analysis, Stone, Scallan and Ahlgren (4) showed that the lignin in the cell wall is arranged in a laminar fashion. From a careful electron microscopic study, Kerr and Goring (2) concluded that the lamellae in the cell wall were about twice as thick as the 3.5 nm dimension of the cellulose protofibril. However, like Heyn (5), they found no evidence for lamellae continuous around any significant fraction of the fibre circumference. Rather, the cellulose microfibrils were pictured as ribbon-like structures consisting of 2-4 protofibrils bonded on their radial faces with their tangential surfaces coplanar and parallel to the middle lamella. The lignin was then visualized as being sandwiched between the cellulose ribbons. This "interrupted" lamellar structure (Figure 1) is rather similar to that proposed by Scallan and Green (6) to explain the changes in fibre dimensions which occur during chemical pulping.

It should be noted that in Figure 1 the lignin is also lamellar in structure. In fact, to describe the arrangement of lignin shown in Figure 1 as a three-dimensional network is only approximately correct. A more accurate description would be that the lignin is a two-dimensional network, arranged in the form of a membrane, folded between the microfibrils of cellulose and parallel to the fibre axis. Since the ratio of the volume of cellulose to that of lignin in the secondary wall of a spruce

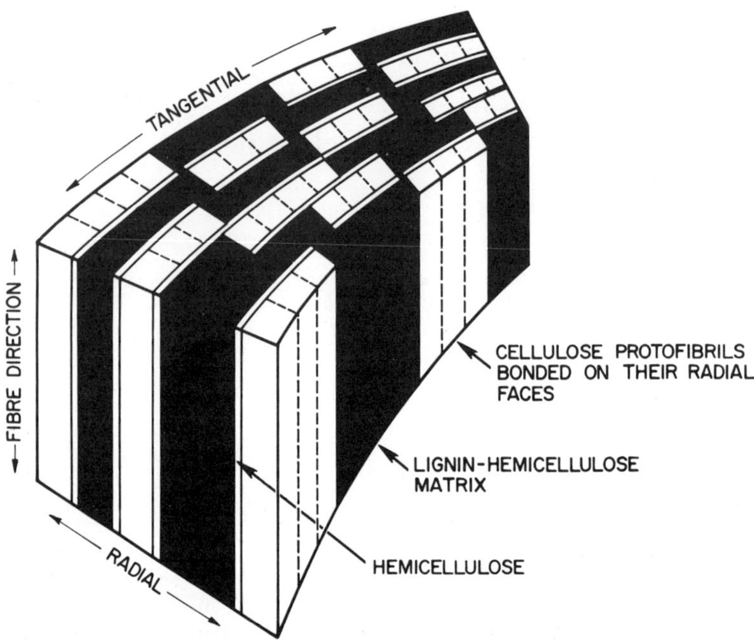

Figure 1. Schematic of the proposed interrupted lamella model for the ultrastructural arrangement of lignin and carbohydrate in the wood cell wall

tracheid (2) is about 45:26 and since the cellulose protofibril is about 3.5 nm in width, the thickness of the lignin substance in the membrane will be about 2 nm.

Let us now try to visualize what will happen to the lignin structure in Figure 1 during chemical pulping. The membrane will be broken up into fragments which will vary widely in size. These fragments will then diffuse through the cell wall to dissolve in the pulping liquor. Most of the smaller fragments will be roughly spherical in shape. However the larger ones will be pieces of the membrane, irregular in shape but with a constant thickness of 2 nm. To a first approximation these larger pieces will be disc-like in shape. In solution a disc-like macromolecule will tend to fold in a random fashion and therefore to approximate the configuration of a sphere. The breakup and solution of the lignin membrane in chemical pulping is depicted graphically in Figure 2.

Some support for the concept described above comes from the data of Luner and Kempf (7) on the thickness of lignin monolayers. These authors made monomolecular films of various fractions of lignin on a water surface and measured the film thickness by means of a Langmuir trough. If the lignin macromolecules

TABLE I: Film Thickness for Lignin Fractions of Various Molecular Weight (from Luner and Kempf (7))

Lignin Sample	Mol. Wt. (wt. av.)	Film Thickness (nm) Calculated for sphere	Experimental
Kraft Demethylated (pine)	2,900	1.8	1.8
Kraft (hardwood)	2,900	1.9	1.7
Kraft (pine)	3,500	2.0	1.7
Dioxane (aspen)	6,500	2.5	1.8
Bjorkman (spruce)	7,100	2.5	1.3
Dioxane II (spruce)	7,400	2.6	1.5
Dioxane IV (spruce)	85,000	5.8	1.7

were rigid spheres then the film thickness would be the diameter of the sphere and would be expected to increase with increase in molecular weight as shown in Figure 3. If, on the other hand, the macromolecules were disc-like fragments of a membrane, they would be expected to lie flat on the surface of the water and to give a film thickness which would remain approximately constant with

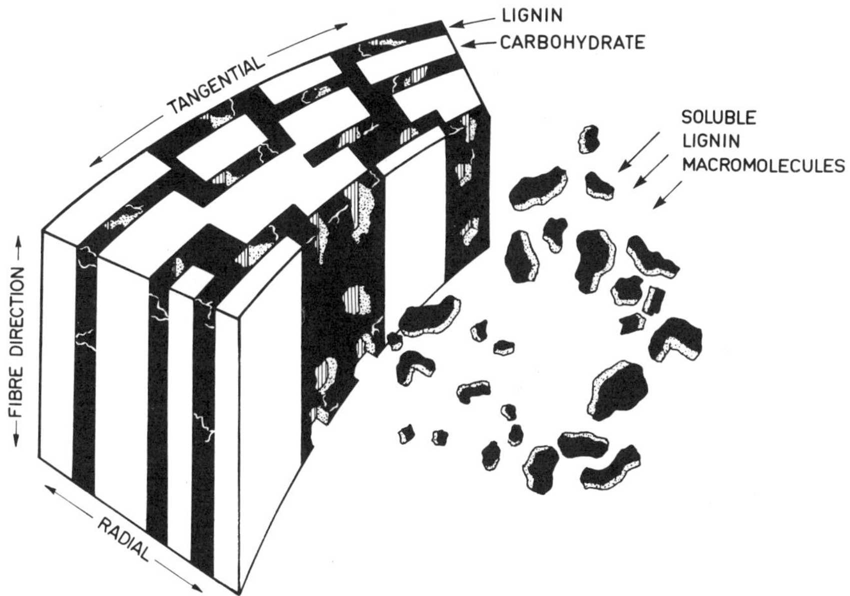

Figure 2. Schematic of the breakdown and solution of lignin during chemical pulping

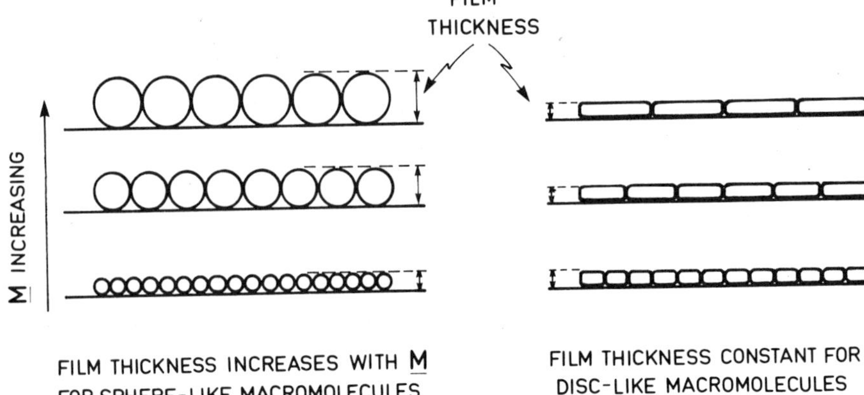

Figure 3. Schematic of the dependence of film thickness on molecular weight (M) for sphere-like and disc-like macromolecules

increase in molecular weight. In Table I, the data of Luner and Kempf are reproduced together with the film thickness calculated on the basis of a spherical configuration for the macromolecule. It is quite evident that the experimentally determined values of film thickness are almost constant and therefore correspond to those expected for a flat disc-like macromolecule rather than for rigid spheres. Furthermore, the mean film thickness for the seven fractions studied is 1.6 nm which is near the value of 2 nm predicted for the thickness of the lignin membrane in the cell wall.

It should be noted that the present concept in no way contradicts the conclusions drawn from earlier work on the solution properties of lignin macromolecules (1). For several different types of lignin the hydrodynamic behaviour in solution was consistent with that expected for a macromolecule having a configuration between that of an Einstein sphere and a random coil. The flexible disc-like configuration for soluble lignin macromolecules arising from the membrane concept of lignin in the cell wall fits this requirement very well.

Literature Cited,

1. Goring, D.A.I., "Polymer Properties of Lignin and Lignin Derivatives", in "Lignins", Ed. Sarkanen, K.V. and Ludwig, C.H., John Wiley, New York, 1971.
2. Kerr, A.J. and Goring, D.A.I., Cellulose Chem. Technol., (1975) 9, 563.
3. Bailey, I.W., Ind. Eng. Chem., (1938) 30, 40.
4. Stone, J.E., Scallan, A.M. and Ahlgren, P.A., Tappi, (1971) 54, 1527.

5. Heyn, A.N.J., J. Ultrastructure Res., (1969) 26, 52.
6. Scallan, A.M. and Green, H.V., Wood and Fiber, in press.
7. Luner, P. and Kempf, U., Tappi, (1970) 53, 2069.
8. Procter, A.R., Yean, W.Q. and Goring, D.A.I., Pulp Paper Mag. Can., (1967) 68, T445.
9. Kerr, A.J. and Goring, D.A.I., Svensk Papperstidn., (1976) 79, 20.

20

Non-aqueous Solvents of Cellulose

BURKART PHILIPP, HARRY SCHLEICHER, and WOLFGANG WAGENKNECHT

Akademie der Wissenschaften der DDR, Institut für Polymerenchemie in Teltow-Seehof, German Democratic Republic

Non-aqueous solvents for cellulose received a great deal of interest during the last decade, as they promised a deeper insight into dissolution processes of cellulose in relation to cellulose structure, as well as alternative procedures for regenerate cellulose fiber spinning. Thus, there is rather ample experimental evidence on a variety of binary and ternary solvent systems for cellulose, as well as on the permanent substitution reactions proceeding during dissolution in some of these systems. Reaction mechanisms have been proposed for several of these processes, but mostly are not proven, and we are still at the beginning of a systematization and generalization in understanding the action of non-aqueous solvent systems on cellulose. Furthermore, experimental facts are rather scarce with regard to the influence of cellulose structure on extend and rate of dissolution and with regard to chain degradation during dissolution.

With this paper, we summarize and interprete results of our own and of others centering on the following problems:

- Is there a generalizable "first step" of the dissolution process to be considered on the concept of electron pair donator-acceptor (EDA) interaction?
- What are the premises for a permanent substitution during dissolution?
- What is the role of cellulose physical structure in connection with non-aqueous solvent systems?

Finally, some general conclusions are drawn, covering non-aqueous and aqueous solvent systems as well.

A survey of non-aqueous solvent systems for cellulose is presented in Table I, systems found in our

TABLE I

Non-aqueous Solvent Systems for Cellulose

Systems			
One-component Systems			trifluoracetic acid (1) ethylpyridinium chloride (2)
Two-component Systems	N_2O_4-polar organic liquid (3-9)	SO_2-amine (10)	CH_3NH_2 - DMSO (12)
	NOCl-polar organic liquid (4,8)	(SO_3-DMF or DMSO) (11)	chlorale-polar organic liquid (8,13)
			paraformaldehyde-DMSO (14)
	$NOHSO_4$-polar organic liquid (9)		NH_3-inorganic salt like NaSCN (15)
Three-component Systems		SO_2-amine-polar organic liquid (8,9 16,17)	NH_3-Na-salt-polar organic liquid like DMSO, ethanolamine (18)
		$SOCl_2$-amine-polar organic liquid (9)	
		SO_2Cl_2-amine-polar organic liquid	

work being marked by framing. Obviously, one-component systems are rather an exception, as only very few single liquids comply with both the demands on a cellulose solvent, i.e. the splitting of H-bonds combined with adduct formation and the solvation of the adducts.

Using a commercial sulphite dissolving pulp, and in some cases also linters or rayon as a starting material, our experimental work has been centered on

- the comparison of the "nitrosylic" compounds N_2O_4, $NOCl$, $NOHSO_4$ in binary systems with a polar organic liquid
- the systematization of 3-component-systems consisting of SO_2, an amine and a polar liquid, extending this principle to the whole series of oxides and oxychlorides of sulfur, i.e. SO_2, $SOCl_2$, SO_2Cl_2, SO_3
- the investigation of amine or ammonia containing systems without an acid anhydride or chloride, like DMSO - CH_3NH_2.

For a discussion of our experimental results obtained with these solvent systems, we centered on the three questions of

- solvent action and solvent stability in relation to solvent composition
- permanent substitution at the cellulose chain during dissolution
- role of cellulose physical structure in the processes of dissolution and reprecipitation.

Solvent Action and Solvent Stability in Relation to Solvent Composition

First of all, two facts may be stated, valid for all solvent systems discussed here:

- A rather big excess of the "active agent" interacting with the cellulose chain, at least a ratio of 3 moles per mol glucose-unit is necessary for complete dissolution.
- Organic liquids being used as a component of the "active agent" and/or as a solvent for this "active agent" can be protic as well as aprotic ones, but have to be of rather high polarity. This high polarity may promote dissolution due to charge separation at the primary adduct formed or due to swelling the cellulose structure. Liquids mainly to be considered here are formamide, dimethylformamide (DMF), ethanolamine and dimethylsulfoxide (DMSO), most of them representing rather strong swelling agents for cellulose by themselves. In some cases

the choice among these liquids is further limited by reaction with the "active agent" affecting the stability of the solvent system.

The variability of the components within one class of sovents is depending on the kind of "active agent". The range of concentration suitable for dissolution of cellulose is determined by the class of solvent, i.e. the kind of "active agent" as well as by the special components chosen, and sometimes by cellulose physical structure, too.

Thus, the binary system CH_3NH_2/DMSO represents a rather special one, quite on the borderline of solvent systems at all: There is no variability of components with regard to amine or polar liquid, a mixture of DMSO and $C_2H_5NH_2$ showing no solvent action. Cellulose can be dissolved completely only in a small concentration range between 10 and 20 % CH_3NH_2, depending on its physical structure, and high DP native cellulose was not completely dissolved at all.

With binary systems consisting of a nitrosylcation-forming compound like N_2O_4, NOCl or $NOHSO_4$ as an "active agent" and a polar organic liquid, this second component may be varied rather widely, as well as the concentration of the nitrosylic compound, without loss of solvent power.

According to published data (8), in some solvents, i.e. DMF, a higher excess of "active agent" is needed in systems with NOCl than in N_2O_4-containing ones, and the variability of the polar liquid is somewhat smaller in the former case. With $NOHSO_4$ the choice of the polar liquid is rather limited due to the limited solubility of this saltlike compound, but several liquids already mentioned i.e. DMSO, DMF, dimethylacetamide proved to be suitable, the rate of dissolutions being higher than in comparable systems with N_2O_4.

In contrast to these "nitrosylic systems", analogous binary mixtures of a polar organic liquid and an oxide or oxychloride of sulfur cannot be classified as "solvents of cellulose", SO_2 and $SOCl_2$ showing no solvent activity sat all, SO_2Cl_2 in formamide or DMF dissolving cellulose after a long time with severe degradation, and SO_3 in DMF sulfating cellulose to a high DS under dissolution and excessive chain degradation.

On the other hand, binary systems of SO_2 and a secondary or tertiary aliphatic amine are able to dissolve cellulose as shown by Hata (10), while NH_3 and primary amines form solid adducts with SO_2, which

TABLE II

Influence of Polar Liquid Component on the Composition of Cellulose Solvent Systems with SO_2 (Sample: Hydrolized Cotton)

Amine	Formamide Molar Ratio		DMSO Molar Ratio		Acetonitrile Molar Ratio	
	Amine:SO_2	Amine/SO_2: Glucose Unit	Amine:SO_2	Amine/SO_2: Glucose Unit	Amine:SO_2	Amine/SO_2: Glucose Unit
Ammonia	1-11[b]	4	1-5[b]	3	no dissolution	
Methyl-amine	0,4- 6	6	0,5-2[b]	6	" "	
Ethyl-amine	not investigated		0,5-1,4	5	" "	
Propyl-amine	0,2- 9	5	0,2-4	3	" "	
Diethyl-amine	0,1-17	3	0,1-7	3	0,25-0,78	14
Triethyl-amine	0,1-1,9[a]	3	0,1-7	3	no dissolution	
Ethylene-diamine	1-7,5[b]	6	no dissolution		" "	

a) beginning phase separation of amine
b) with increasing amount of amine increasing crystalline precipitate

are insoluble in an excess of one of the components. The same happens in systems of $SOCl_2$ or SO_2Cl_2 and any kind of aliphatic amine, thus rendering all these mixtures ineffective as potential solvents for cellulose. Addition of a polar organic liquid as a third component often effects a dissolution of these solid adducts and thus leads to the rather ample spectrum of three-component solvent systems investigated mainly by Nakao (8, 17) and by us (9, 16).

With regard to type of amine and to polar liquid variability is by far the broadest with SO_2 as an acceptor component in complex formation. As shown by the data in Table II formamide proved to be most suitable as a polar liquid component, closely followed by DMSO. In combination with these two liquids, all kinds of amines and even NH_3 can be applied in composing cellulose solvents, and the effective range of component concentrations is rather ample, with several amines an excess of either SO_2 or amine being equally suitable. With other liquids like acetonitrile the choice of amine as well as the suitable concentration range is rather limited. Considering the order of amines secondary amines are obviously most general applicable, followed by tertiary and primary amines and at last by NH_3. The same order holds true with respects of the stability of the solvent systems. Even with the most favorable compositions, the molar ratio of the amine-SO_2-complex to glucose unit has to exceed value of 3, the limitary ratio varying widely with type of amine and polar liquid, and also with the liquid to cellulose ratio (Table III). Obviously, an equilibrium exists in binding of the "active complex" to the cellulose and to the liquid mixture, this equilibrium of course being determined not only by the complex: cellulose ratio but also by the complex concentration in the solvent system.

TABLE III

Influence of Liquid : Cellulose Ratio on Limiting Molar Ratio of SO_2/Amine : Glucose Unit in the System SO_2-Diethylamine-DMSO

Liquid to Cellulose Ratio	Molar Ratio SO_2/Amine: Glucose Unit
100	5
25	3

In solvent systems containing $SOCl_2$ or SO_2Cl_2 as an acceptor in complex formation, formamide or DMSO proved to be outstanding as a polar liquid component, too. But in contrast to SO_2-containing systems, the limited solubility of the amine-acceptor-adducts led to rather small usable concentration ranges of the components. Systems of satisfactory solvent action and stability could be obtained here only by using a secondary or tertiary amine in excess to the oxychloride, and, as shown in Table IV for SO_2Cl_2 containing systems, a rather large molar ratio of acceptor-amine-complex to glucose unit is required for complete dissolution of cellulose. In systems containing SO_3, the role of amines is different, as the addition of an amine to a solution of SO_3 in DMF or DMSO inhibits sulfation as well as dissolution of the cellulose sample, if the molar ratio of amine to SO_3 exceeds a value of 1, indicating an inactivation of SO_3 by binding to the amine.

TABLE IV

Limiting Molar Ratios of SO_2Cl_2 to Glucose Unit and Amine to SO_2Cl_2 for Dissolving of Cellulose

System	Molar Ratio SO_2Cl_2 to Glucose Unit	Molar Ratio Amine to SO_2Cl_2
Propylamine/SO_2Cl_2/Formamide	8	7
Diethylamine/SO_2Cl_2/Formamide	8	4
Triethylamine/SO_2Cl_2/Formamide	4	4
Diethylamine/SO_2Cl_2/DMSO	6	4
Triethylamine/SO_2Cl_2/DMSO	6	3,5

Permanent Substitution at the Cellulose Chain and Chain Degradation in Relation to Solvent Composition

In the binary system CH_3NH_2/DMSO dissolution was accompanied by only rather small chain degradation and no permanent substitution was observed, of course, after regeneration of the cellulose by precipitation in water.

In solvent systems containing an acid anhydride, acid chloride, or $NOHSO_4$, respectively, an esterification of cellulose OH-groups is principially possible, and in most cases can be realized experimentally, the degree of substitution differing widely in dependence of the "active agent" and the polar organic liquid of the solvent systems. Substitution to a rather high DS is generally combined with severe chain degradation, thus excluding the action of these solvent systems as a route to high DP cellulose esters. Some analytic data of cellulose regenerated from systems containing SO_2, $SOCl_2$, SO_2Cl_2, or SO_3 resp. are summarized in Table V. With SO_2-containing systems, no substitution at all was observed. With $SOCl_2$ and SO_2Cl_2 in combination with an amine and a polar liquid small amounts of sulfur as well as chlorine were fixed at the cellulose chain. As already known by work of Schweiger (11), a mixture of SO_3 and DMF can be used to prepare high-DS cellulose sulfates.

TABLE V
Permanent Substitution of Cellulose after Dissolution in 3-Component Systems Containing SO_2, $SOCl_2$, SO_2Cl_2 and SO_3 resp. (Precipitation into Water a), Diethylether b))

System	DS Sulfur	DS Chlorine
SO_2/Diethylamine/Formamide a)	no Substitution	
$SOCl_2$/Diethylamine/Formamide a)	0,073	0,056
SO_2Cl_2/Diethylamine/Formamide a)	0,03	0,015
SO_3/Dimethylformamide b)	1,26	-

With nitrosylic compounds (N_2O_4, $NOHSO_4$) in binary systems with a polar liquid, substitution reactions are rather complex. With all of these compounds, primarily a nitrite ester of cellulose is formed according to

$$\text{Cell-OH} + NO^+ \longrightarrow \text{Cell-O-NO} + H^+$$

The DS of nitrite ester groups in the sample mainly depends on the nitrosylic compound and the mode of precipitation of the sample (Table VI). Parallel to nitrite ester formation, the anionic part of the nitrosylic compound, resp. the corresponding mineral acid, may react with the cellulose, too, forming the appropriate ester. Thus, by the action of $NOHSO_4$ in DMF, or DMSO resp., we finally arrived at a mixed nitrite-sulphate-ester of cellulose (9), with an equimolar substitution of N and S being achieved in a wide range of molar ratio $NOHSO_4$: glucose unit. As shown in Figure 1, the DS steady increases after exceeding a molar ratio of 1:1, and tends to reach a limiting value of 1,5 for each kind of the ester groups complying with a total substitution of all hydroxyl groups. The influence of polar liquid component on the DS reached at fixed conditions of reaction is demonstrated in Table VII, showing a rather loose correlation with donor number (19) and swelling value (20) of the liquids. Chain degradation varied to a large extend with the nitrosylic compound used, being rather small with NOCl and N_2O_4 in the absence of water (compare 4), but arriving at DP-values of 100 to 200 with $NOHSO_4$, starting from a commercial sulphite dissolving pulp.

TABLE VI

Permanent Substitution of Cellulose after Dissolution in 2-Component Systems Containing N_2O_4, NOCl, and $NOHSO_4$ resp.
(Precipitation into Acetone a), Diethylether b))

System	DS Nitrogen	DS Chlorine	DS Sulfur
NOCl/Dimethylformamide a)	0,11	0,06	-
N_2O_4/Dimethylformamide a)	0,23	-	-
$NOHSO_4$/Dimethylformamide b)	1,07	-	1,10

Role of Cellulose Structure in Connection with Non-aqueous Solvent Systems

The influence of cellulose structure on the course of dissolution in aqueous and non-aqueous solvents has been treated in our previous publication (9). Thus, only a short summary will be given here, supplemented by some remarks on cellulose structures regenerated from non-aqueous solvents. Within the suitable range of composition all solvent systems containing a nitrosylic compound or an oxide or oxychloride of sulfur dissolve native cellulose even of high DP rather quickly, within minutes at room temperature, and without residue, comparable to the action of cadoxene, for example. But in comparison to this and other aqueous solvent systems, three points of difference are remarkable:

TABLE VII
Influence of Polar Liquid Component on the Substitution of Cellulose by $NOHSO_4$
(Molar Ratio $NOHSO_4$ to Glucose Unit = 1)

Polar Liquid	$DN^{a}_{SbCl_5}$	LRV[b]	DS-Sulfur
Dimethylsulfoxide	29,8	72	0,71
Dimethylformamide	26,6	25	0,71
Dimethylacetamide	27,8	20	0,42
Dioxane[c]	28,2	15	0,08
Acetonitrile[c]	14,1	8	0,06
Diethylether[c]	19,2	5	0,05
Chloroform[c]	-	6	0,04
Acetone[c]	17,0	5	0,02

a) DN_{SbCl_5} = Donor Number with $SbCl_5$

b) LRV = Liquid Retention Value = Swelling value in the polar liquid concerned (%)

c) $NOHSO_4$ not dissolved completely

1. The range of composition suitable for complete dissolution is generally smaller with non-aqueous solvents.
2. Instead of the "Kugelbauch"-swelling usually observed as an intermediate stage in aqueous solvents, we find a course of dissolution proceeding from the fibre surface, leading to spindle-like fragments as intermediate structures Figure[2, 3].
3. Dissolution of cellulose II, i.e. of rayon or alkali cellulose regenerated by acid washing and drying, proceeded rather slowly in several non-aqueous systems as compared to native cellulose samples of much higher DP, and in some cases no dissolution at all has been observed with regenerated samples even after several days.

As a propable cause we assume a hindrance of reagent transport into the cellulose fibre due to the large volume of the "active agent" in connection with the limited penetrability of the pore system of regenerated cellulose. The enhanced dissolution rate with

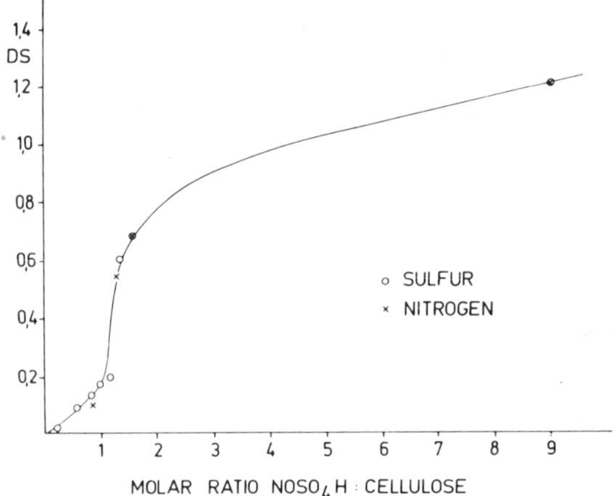

Figure 1. Substitution of cellulose by $NOHSO_4$ in DMF

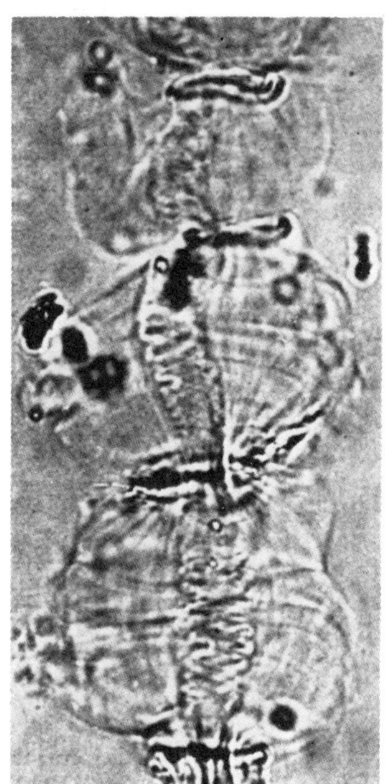

Figure 2. "Kugelbauch"—swelling of cellulose in cuen

cellulose II samples pre-swollen in DMSO seems to confirm this statement. However, some contradiction arises here with respect to the morphological changes observed during dissolution, and possibly also the different lattice structure of cellulose I and II is of some importance here.

Quite a different response to cellulose structure is shown by the binary system DMSO-CH_3NH_2, this "borderline solvent system" dissolving only cellulose samples of rather low state of order and/or low DP, irrespective of lattice type. Thus, rayon staple and hydrolytically degraded wood pulp are dissolved without residue, while linter must be reprecipitated to achieve a dissolution.

With regard to cellulose physical structure obtained by precipitation from non-aqueous solution into an aqueous medium, a rather peculiar behavior of the DMSO-CH_3NH_2-system was confirmed again: The precipitate consisted only partially of cellulose II, showing simultaneously an other lattice type classified by us as cellulose I, while all the other precipitates obtained from non-aqueous and aqueous solutions under the same conditions were composed of cellulose II only. With non-aqueous solvent systems we arrived at precipitates generally showing a trend to a somewhat lower state of order - as measured by X-ray-diffraction - as compared to samples regenerated from aqueous solution, while no significant and generalizable differences could be stated with regard to several accessibility criteria of samples precipitated from aqueous and non-aqueous solutions (21). Obviously, accessibility is determined here predominantly by the procedure of precipitation and not by the type of solvent.

Discussion of the Mechanism of Interaction between Cellulose and Non-aqueous Solvents

Previous considerations on the dissolution mechanism of cellulose in non-aqueous solvents usually were centered on the peculiar solvent system investigated. Starting from the assumption of electron-donator-acceptor (EDA) complex formation in these systems first mentioned by Nakao (8), we arrived at a generalizable qualitative model for the first step of interaction between cellulose and solvent, which permits an interpretation of most of the experimental data acquired for the different solvent systems and which is not in contradiction with the remaining ones.

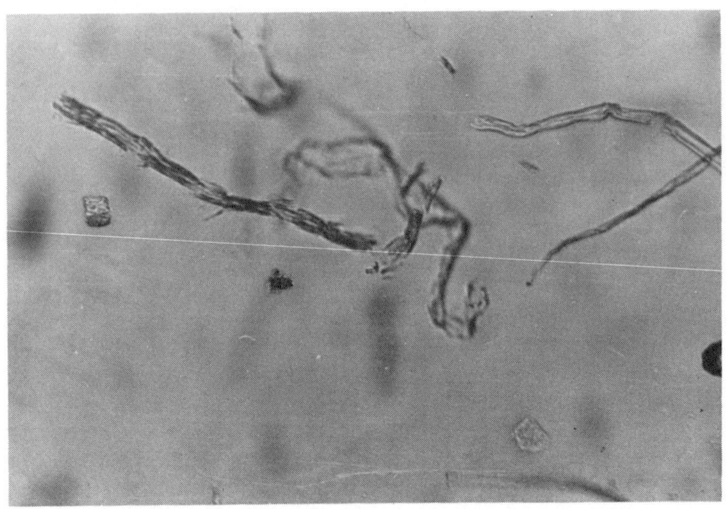

Figure 3. Spindle-like structures of cellulose during dissolution in N_2O_4–DMF

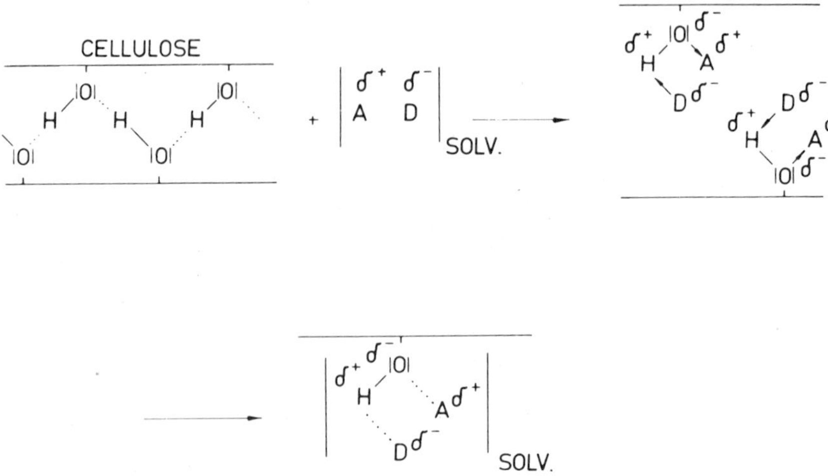

Figure 4. Scheme of EDA-interaction as a first step of dissolution

Principal assumptions of this model are

- the participation of the O-atom and the H-atom of a cellulosic OH-group in EDA-interaction, with the O-atom acting as an n-electron-pair donator and the H-atom as an σ-electron-pair acceptor
- the presence of a donor and an acceptor centre in the "active agent" of the solvent system, both centres being in a steric position suitable for interaction with O and H of the hydroxyl group
- the existence of a definite optimal range of EDA-interaction strength, leading - in connection with the steric position of the donor and acceptor centre and the action of the polar organic liquid - to the optimal amount of charge separation of the OH-groups for dissolving the complexed cellulose chain.

Figure 4 demonstrates in a strongly schematized form these considerations. This first step of interaction, decisive for getting the cellulose into solution, may be followed by a permanent esterification of OH-groups, as shown with $NOHSO_4$ for example, but this esterification is no prerequisite for dissolution, as shown by the very good solvent action of SO_2-containing systems. By means of this EDA-concept with the premises given above, a reasonable mechanism of interaction with cellulose may be plotted down for all non-aqueous solvent systems known today.

Between the binary system DMSO-CH_3NH_2 and cellulosic OH-groups several modes of EDA-interaction are to be considered, as shown in Figure 5. In all possible cases the interaction is rather weak, thus limiting the solvent action of this system and explaining the specifity of components by the necessity of a special steric arrangement of the different charge centres. As a most probable mechanism we discuss a primary interaction between the hydroxylic O-atoms as a donor and the acceptor centre of DMSO, which itself is complexed by the amine, the amine additionally binding the H-atom of the cellulosic hydroxyl group.

Common feature of all "nitrosylic systems" is the interaction of the strong acceptor NO^+, present as a free ion or as an ion dipole, with the hydroxylic O-atom as a donor [Figure 6], followed by complete charge separation in the NO^+X^--dipol and the formation of the nitrite ester of cellulose. Depending on the anionic part of the NOX-molecule resp. on the acid formed by it capturing the H-atom of the cellulosic hydroxyl, a second kind of ester groups can be introduced into the cellulose chain either by direct esterification of free OH-groups, or by transesterification via nitrite ester groups [Figure 7]. Ex-

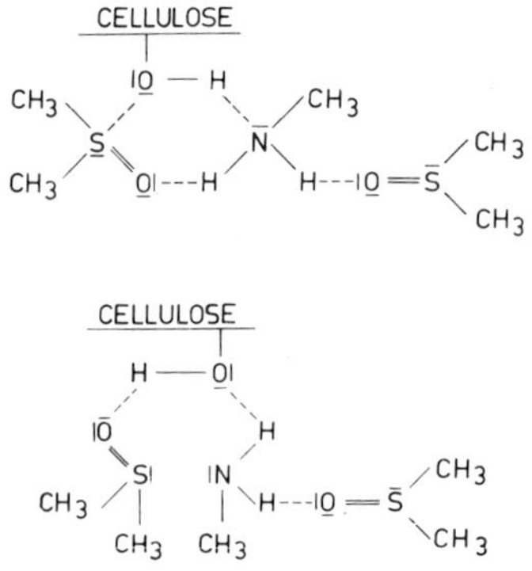

Figure 5. EDA-interaction of cellulose with CH_3NH_2–DMSO

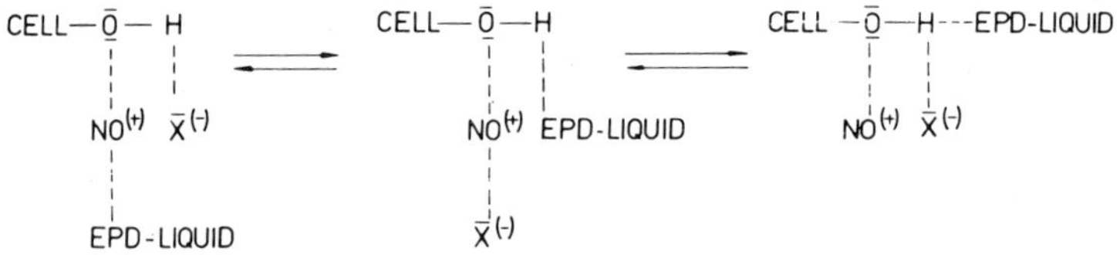

Figure 6. EDA-interaction of cellulose with nitrosylic compounds in strong polar liquids

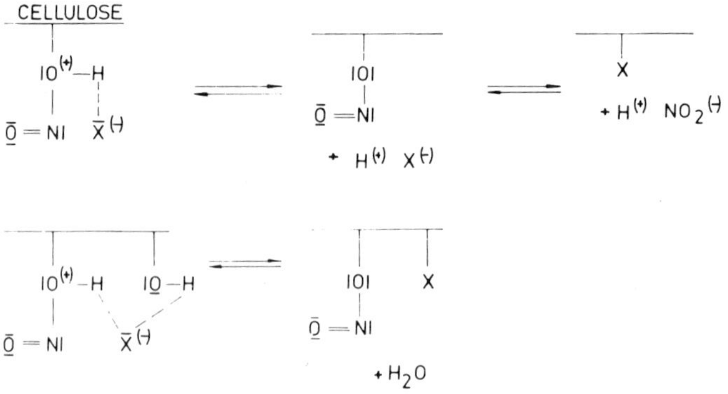

Figure 7. Substitution of cellulose by reaction with nitrosylic compounds

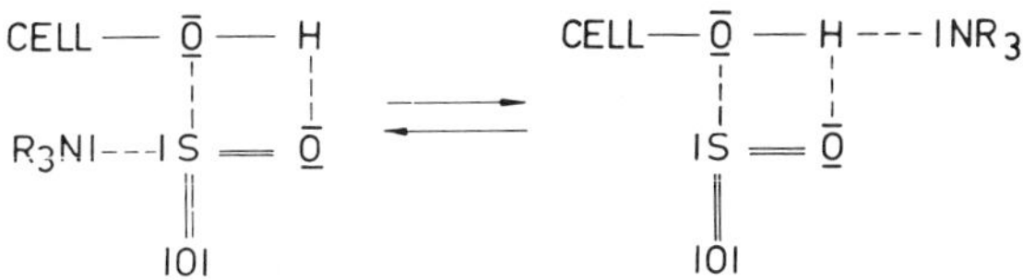

Figure 8. EDA-interaction of cellulose with ternary system SO_2–amine–polar liquid

perimental evidence is in favor of the first reaction path with $NOHSO_4$, while with NOCl a transesterification is more probable.

In systems containing SO_2, $SOCl_2$ or SO_2Cl_2, the acceptor strength of the S-atom is alone not sufficient for a interaction leading to dissolution, but must be supported and increased by the complexing action of an amine, the amine additonally binding the H-atom of the OH-groups. As shown in Figure 8 for SO_2 as an acceptor, this complex dissolves in the polar liquid present as a third component, the interaction with this liquid further promoting polarization of the S-O-bonds of SO_2 and the subsequent interaction with the cellulosic OH-groups.

Applying $SOCl_2$ or SO_2Cl_2 as an acceptor component, full complexation requires an excess of amine, i.e. a minimum molar ratio of amine to $SOCl_2$ or SO_2Cl_2, resp., of 3:1, probably due to a stronger intrinsic dissociation of the acceptor molecule [Figure 9].

General conclusions covering nitrosylic systems as well as oxides or oxychlorides of sulfur containing ones, must be drawn with caution, of course. However, the following statements may be summarized from our experiments and from data published by others:

1. If "solvent power" is defined as the number of moles of "active agent" (active component or active complex)per mole glucose unit needed for dissolving high DP-cellulose, and if DMF and DMSO

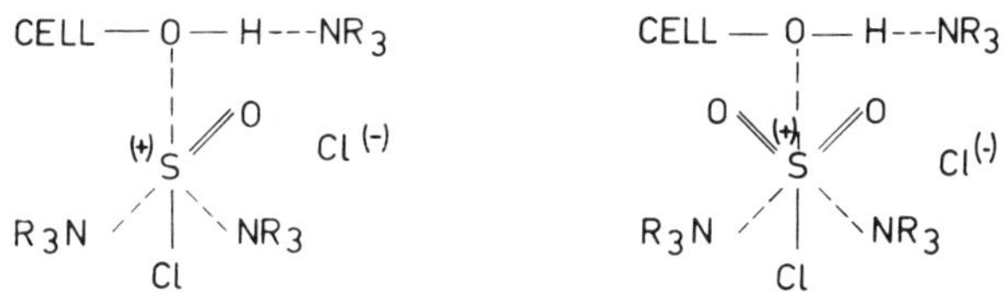

Figure 9. EDA-interaction of cellulose with ternary system $SOCl_2$ or SO_2Cl_2–amine–polar liquid

(with NOX-systems) or formamide and DMSO (with oxides and oxychlorides of sulfur) are considered as polar liquid component, an order of acceptor components with respect to solvent power can be given according to

$$N_2O_4 \geq SO_2 > NOHSO_4 > SO_2Cl_2 > NOCl > SOCl_2.$$

2. The polar liquid component should comply with the rather different requirements of dissolving the NOX-compound or the amine-SO_2 ($SOCl_2$, SO_2Cl_2)-complex, of further polarizing these compounds via EDA-interaction, but not reacting with them in a permanent, irreversible way, and of solvating the EDA-adduct formed with cellulose.
   Formamide and DMSO proved to be outstanding in coming up to these demands.
3. If an amine is to be added to the system, it should have a rather high donor strength combined with a minimal reactivity with regard to the other components of the system, thus tertiary and secondary amines generally being more suitable than primary ones or NH_3.

Final Remarks and Conclusions

Generally speaking, we are still at the beginning of a mechanistic understanding of cellulose dissolution processes in nonaqueous and aqueous systems as well. The EDA-concept used here may be a guideline in this systematization and in searching for new non-aqueous solvent systems if we are aware of its limitations, too. Limitations to be kept in mind are

- the still rather qualitative character of the models used here, due to lack of reliable quantitative data especially or acceptor strength of the compounds concerned. Recent work published by Gutmann (19) might be helpful here, but much experimental work has still to be done, by spectroscopy or conductometry, for example, to quantify these interactions in our special two- or three-component systems;
- the impossibility to give any regard to cellulose morphology in connection with molar volume of solvent components;
- the applicability of the concept just to the first step of cellulose-solvent-interaction.

On the other hand, this EDA-concept may be extended to aqueous solvent systems, too, and thus serve a synoptic discussion of cellulose solution processes in general.

As demonstrated by the series of solvent systems

SO_2-NH_3-polar organic solvent

NH_3-inorganic salt-polar organic liquid

liquid NH_3-inorganic salt

H_2O-inorganic salt

there is no definite gap between aqueous and non-aqueous solvents for cellulose. Applying the EDA-concept to the first step of polymer-solvent-interaction, we can consider as an n-electron pair donator organic amines, OH-groups, and NH_3 as well as anions of inorganic acids, and as an σ-electron pair acceptor tetraalkylammonium-ions and inorganic cations as well as SO_2, $SOCl_2$, SO_2Cl_2, or the C-atom of a carbonyl group.

Figure 10 summarized these considerations in a strongly schematized manner, showing n-σ-electron pair donator acceptor interaction as a unifying aspect of aqueous and non-aqueous cellulose solvents.

Abstract

Several classes of non-aqueous solvent systems for cellulose, i.e. binary systems of a nitrosylic compound and a polar liquid, ternary systems of an amine, an oxide or oxychloride of sulfur and a polar liquid, and the rather special system DMSO-CH_3NH_2 are treated with respect to solvent action in dependence of solvent composition, permanent substitution of cellulose in the dissolution process, and role of cellulose structure. A generalizable qualitative model, based on EDA-interaction is presented for the first step of interaction between solvent system and cellulosic hydroxyl groups.

Literature Cited

1. Valdsaar, H., Dunlap, R., Amer. Chem. Soc., Meeting Atlantic City (1952).
2. Husemann, E., Siefert, E., Makromol. Chemie (1969) 128, 288.
3. Fowler, W.F., Unruh, C.C., McGee, P.A., Kenyon, W. V., J. Am. Chem. Soc. (1947) 69, 1636.
4. Schweiger, R.G., Tappi (1974) 57, 86.
5. Venkateswaran, A., Clermont, L.P., J. Appl. Polymer Sci. (1974) 18, 133.

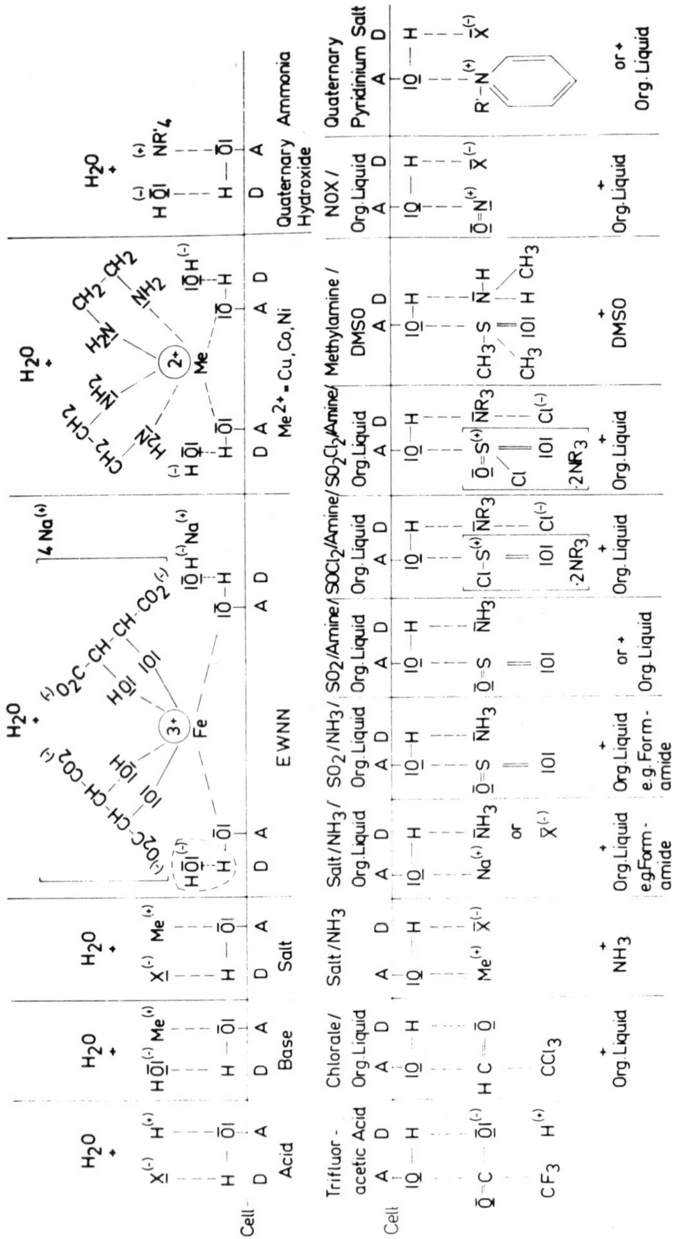

D = ELECTRON PAIR DONATOR
A = ELECTRON PAIR ACCEPTOR
R = ALKYL GROUP OR H
R' = ALKYL OR BENZYLIC GROUP

Figure 10. Comparison of EDA-interaction of cellulose with aqueous and non-aqueous solvent systems

6. Grinshpan, D.D., Kaputskii, F.N., Ermolenko, I.N., Gavrilov, M.Z., Klepcha, V.S., Dokl. Akad. Nauk BSSR (1974) 18, 828.
7. Pasteka, M., Mislovicova, D., Cellulose Chem. Technol. (1974) 8, 107.
8. Nakao, O., Sen-i To Kogyo (1971) 4, 128.
9. Philipp, B., Schleicher, H., Wagenknecht, W., Cellulose Chem. Technol. (1975) 9, 265.
10. Hata, K., Yokota, K., Sen-i Gakkaishi (1966) 22, 96.
11. Schweiger, R.G., Carbohyd. Res. (1973) 21, 219.
12. Koura, A., Schleicher, H., Philipp, B., Faserforsch. u. Textiltechnik (1972) 23, 128.
13. Meyer, K.H., Studer, M., van der Wyk, A.J.A., Mh. Chemie (1950) 81, 151.
14. Johnson, D.C., Nicholson, M.D., Haigh, F.C., Dimethylsulfoxide-paraformaldehyde, a non-degrading solvent for cellulose. 8th Cellulose Conference. Syracuse, New York. May 19-23, (1975).
15. Scherer, P.C., J. Am. Chem. Soc. (1931) 53, 4009.
16. Philipp, B., Schleicher, H., Laskowski, I., Faserforsch. u. Textiltechnik (1972) 23, 60.
17. Yamazaki, S., Nakao, O., Sen-i Gakkaishi (1974) 30, T 234.
18. Schleicher, H., Linow, K.-J., Schubert, K., Faserforsch. u. Textiltechnik (1972) 23, 335.
19. Mayer, U., Gutmann, V., Gregor, W., Mh. Chemie (1975) 106, 1235.
20. Philipp, B., Schleicher, H., Wagenknecht, W., J. Polymer Sci., (1973), Symposium No. 42, 1531.
21. Schleicher, H., Laskowski, I., Philipp, B., Faserforsch. u. Textiltechnik (1975) 26, 313.

21

Effects of Solvents on Graft Copolymerization of Styrene with γ-Irradiated Cellulose

YOSHIO NAKAMURA and MACHIKO SHIMADA

Faculty of Technology, Gunma University, Kiryu, Gunma 376, Japan

Considerable works have been done on radiation-induced graft copolymerization of vinyl monomers onto cellulose during past years. As many of them aimed the formation of graft copolymers with unique properties, few studies have been made on solvents effects for graft copolymerization. Sakurada et at (1 -5) have shown that the addition of small amount of water to reaction system markedly enhanced radiation-induced graft copolymerization of vinyl monomers onto cellulose. Since then, many swelling agents to cellulose were used to aid diffusion of monomers into cellulosic fibers. Dilli and Garnett (6-13) reported that methanol is particularly attractive for the observation of the Trommsdorff effect in radiation-induced grafting of vinyl monomers onto cellulose. From the reports mentioned above, graft copolymerization of vinyl monomers onto irradiated cellulose is considered to be affected by the diffusion of monomers into fibers, the swelling of trunk polymer, and the Trommsdorff effect of solvent on graft polymer radicals. In this paper, these effects of solvents on graft copolymerization of styrene onto irradiated cellulose will be discussed by investigating systematically weight increase after graft copolymerization, extent of grafting of graft polystyrene bonded chemically with irradiated cellulose, amount of cellulose reacted with styrene, graft efficiency, molecular weight of graft polymer, number of reaction sites of graft copolymerization, and decay of cellulose radicals in reaction system.

Experimental

Materials. Scoured cotton cellulose of Egyptian variety was purified by extracting with hot benzene-ethanol mixture (1/1 volume ratio) for 24 hr and washing with distilled water prior to air-drying.

Styrene was purified by passing the monomer through a column filled up with activated almina to remove inhibitors of polymerization. Methanol was distilled before use. Other reagents were reagent grade, and were used without further purification.

Irradiation. After purified cellulose was dried under vacuum at 50°C for 20 hr, it was irradiated under nitrogen atmosphere for 1 hr by Co-60 γ-rays with an exposure rate of 1.0×10^6 R/hr, for grafting reaction.

Degree of Swelling. Diameters of cellulosic fibers and polystyrene tablet were measured with a microscope of 800 magnifications before and after immersion in various solvents for 24 hr at 30°C and averaged. Degree of swelling (D.S.) was calculated as follows:

$$D.S. = Da/Db \qquad (a)$$

where Da stands for diameter after immersion in solvent and Db means diameter before immersion in solvent.

Graft Copolymerization. Graft copolymerization by styrene onto irradiated cellulose was carried out under nitrogen atmosphere at 30°C. The liquor ratio was 100. After graft copolymerization, the samples were washed with methanol followed by washing with distilled water and air-drying. The samples were extracted with hot benzene to remove homopolymer. The apparent extent of grafting (A.E.G.) was determined according to (b):

$$A.E.G. = (Wg-Wo)/Wo \qquad (b)$$

where Wg describes weight of graft copolymerized cellulose after the benzene extraction, and Wo means weight of cellulose before graft copolymerization.

Fractionation. In order to separate the apparent graft samples into three parts, namely, ungrafted cellulose, inside homopolymer, and chemically bonded graft copolymer, the samples were nitrated, or acetylated. the procedure is shown in Figure 1. (i) Nitration. The grafted samples were nitrated according to nondegradative conditions reported by Alexander and Mitchel (14). To remove ungrafted cellulose nitrates completely, the nitrated samples were extracted with hot acetone. After extraction, extracted cellulose nitrates were weighed and the nitrogen content was measured. The degree of nitration was 2.4 per glucose unit. The amount of reacted cellulose (A.R.C.) was calculated as follows:

$$A.R.C. = (1-Wn/1.67)/Wo \qquad (c)$$

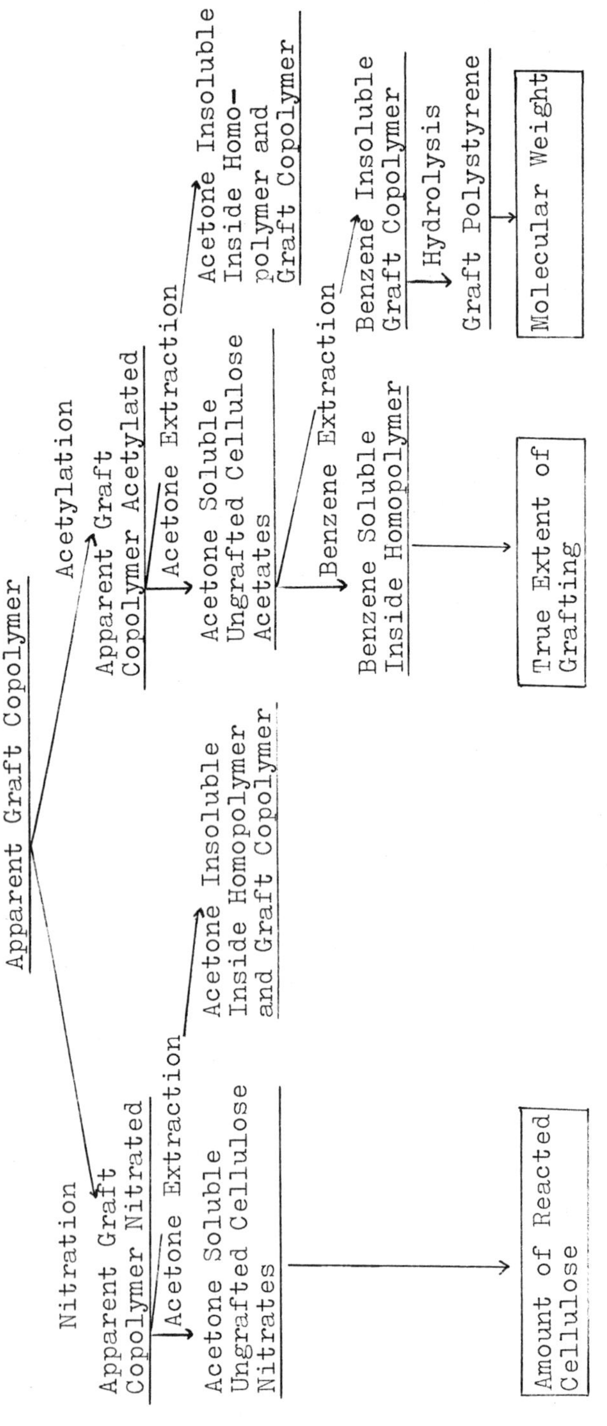

Figure 1. Procedure of fractionation of apparent graft copolymer

where Wn means weight of extracted cellulose nitrates. (ii) Acetylation. As styrene was also nitrated as cellulose, acetylation was carried out to separate inside homopolymer and true graft copolymer from apparent graft copolymer. The apparent grafted samples were acetylated in the mixture of acetic anhydride, glacial acetic acid, and zinc chloride for 48 hr at 60°C (15). The acetylated samples were extracted completely with hot acetone to remove ungrafted cellulose acetates. After insoluble part in acetone was weighed, it was extracted by hot benzene to remove homopolymer which was not extracted by the first benzene extraction because of entanglement. The true extent of grafting (T.E.G.) was desided as follows:

$$\text{T.E.G.} = (Wg - Wo - Wh)/Wo \qquad (d)$$

where Wh stands for weight of inside homopolymer.

Measurement of Molecular Weight. After the acetylated true graft copolymer was dissolved in methylene chloride, the same volume of acetone was added to the solution. The concentrated hydrochloric acid was added till the concentration of hydrochloric acid was 3 N, and the solution was pored into methanol and precipitated (16). After the isolated graft polystyrene was purified, it was dissolved in benzene and viscosity average molecular weight was measured according to (e):

$$[\eta] = 2.4 \times 10^{-4} M^{0.65} \qquad (e)$$

Reaction Sites. Number of reaction sites (N.R.S) contributed to grafting reaction in irradiated 100 g cellulosic fibers was desided as follows:

$$\text{N.R.S.} = (Wt/Mt)/(100/Mc) \qquad (f)$$

where Wt means true extent of grafting expressed in %, Mt stands for viscosity average molecular weight of true graft polystyrene, and Mc describes viscosity average molecular weight of irradiated cellulose.

ESR Measurement. Cellulosic fibers were led into spectrosil ESR tube with glass ampul connected with breakable joint so that the fiber axes were perpendicular to the magnetic field, and irradiated with Co-60 γ-rays for 1 hr at an exposure rate of 1.0×10^6 R/hr at room temperature. After reaction solution was pored into the glass ampul and degassed by repeating freezing and melting, it was led into the ESR tube and connected with irradiated cellulosic fibers at 30°C for ESR measurement of decay of cellulose radicals. ESR spectra were taken with a JES-ME ESR spectrometer with 100 kHz modulation at -196°C.

Results and Discussion

Swelling of Cotton Cellulose and Polystyrene in Various Solvents. As mentioned before, it is generally considered that graft copolymerization of styrene onto γ-irradiated cotton cellulose is dependent on the degree of swelling of trunk polymer by solvent, the rate of diffusion of monomer to the trunk polymer and the Trommsdorff effect of solvents on growing graft polymer radicals. To study these points in detail, some solvents which have various effects on trunk and graft polymers were selected and used as reaction solution. As one of the characters of these solvents, degree of swelling of cotton cellulose and polystyrene tablet in the solvents was measured and tabulated in Table I. Cellulose is remarkably swollen by water and ethylene glycol as compared with other solvents. Polystyrene is dissolved by N,N-dimethylformamide.

Table I Degree of swelling of cotton cellulose and polystyrene tablet

solvent	degree of swelling	
	cotton cellulose	polystyrene
water	1.23	1.01
methanol	1.16	0.92
dimethyl-formamide	0.99	soluble
ethylene glycol	1.28	0.94

Determination of Reaction Systems. When γ-irradiated cellulose was immersed in styrene monomer at 30°C for 1 hr under nitrogen atmosphere, no weight increase was observed. As it is well known that grafting reaction is enhanced by addition of diluents to monomer, methanol was added to styrene in various percentages. Grafting reaction took place by addition of methanol in any proportion to styrene monomer. In order to carry out grafting reaction for cotton cellulose in homogeneous solution, the content of styrene in reaction system was kept at 2 %, and the content of other solvents was changed. In Figure 2, apparent extent of grafting carried out in styrene-methanol-water reaction system at 30°C for 1 hr is shown. The concectration of styrene was 2 %, and the composition of water and methanol was changed. Apparent extent of grafting increased with increase of amount of water. When the amount of water exceeded 34 %, the reaction systems were separated into two phases. So the water content was kept below 34 %. For detailed study on the effects of swelling of trunk polymer by water, two reaction systems were selected, whose water contents were 20 % (reaction system A), and 34 % (reaction system B), respectively. On the other hand, various amount of ethylene glycol was added in place of water to the reaction system B composed of 2 % styrene, 64 % methanol, and 34 % water. The composition of water and ethylene glycol was changed. Apparent extent of grafting in the reaction system composed of styrene, methanol, water and ethylene glycol is shown in Figure 3. The reaction was carried out for 1 hr under nitrogen atmosphere at 30°C. In spite of the fact that ethylene glycol has the same ability with water to swell cellulose as shown in Table I, the apparent extent of grafting decreased with increase of amount of ethylene glycol. From these results, it is guessed that the solvents with capacity to swell trunk polymer is not the only one factor to accelerate grafting reaction. To study this phenomenum in detail, the reaction system composed of 2 % styrene, 64 % methanol, 15 % water, and 19 % ethylene glycol (reaction system C) was selected. The Trommsdorff effect of solvent for graft polymer radicals is also thought to be one of the factors to control grafting reaction. Dimethylformamide dissolves polystyrene and does not swell cellulose as shown in Table I, so dimethylformamide was added to the reaction system composed of 2 % styrene, 34 % water, and 64 % methanol in place of methanol in various percentages. Apparent extent of grafting in styrene-water-methanol-dimethylformamide system at 30°C for 1 hr is shown in Figure 4. The concentration of styrene and water was 2 %, and 34 %, respectively. The composition of methanol and dimethylformamide was changed. The apparent extent of grafting decreased as the content of dimethylformamide increased. For the detailed study on the Trommsdorff effect of solvents for graft polystyrene radicals, the reaction system composed of 2 % styrene, 34 % water, 58 % methanol, and 6 % dimethylformamide (reaction system D) was selected. As mentioned above, four reaction systems were selected to study the effects of solvents on graft copolymerization of styrene onto γ-irradiated cotton cellulose, and tabulated in Table II.

Apparent Extent of Grafting. In Figure 5, apparent extent of grafting is plotted against reaction time. Apparent extent of grafting increased as reaction time proceeded in all cases. Viewed from water

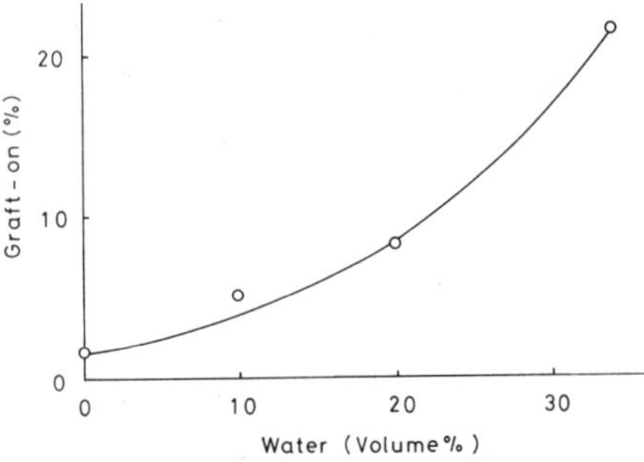

Figure 2. Apparent extent of grafting carried out in styrene–water–methanol reaction system at 30°C for 1 hr. Content of styrene was 2%, and composition of water and methanol was changed.

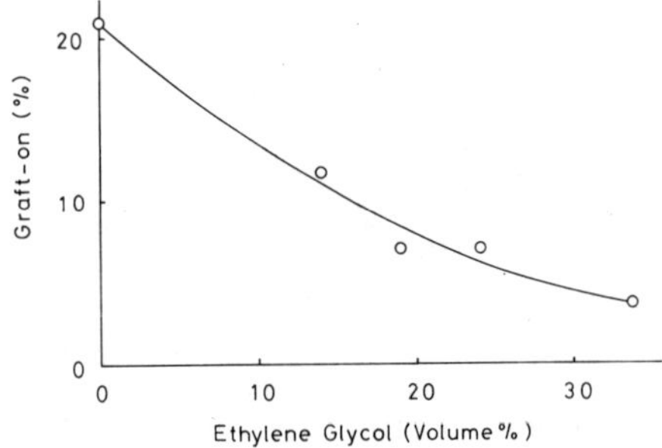

Figure 3. Apparent extent of grafting carried out in styrene–methanol–water–ethylene glycol reaction system at 30°C for 1 hr. Contents of styrene and methanol were 2% and 64%, respectively, and composition of water and ethylene glycol was changed.

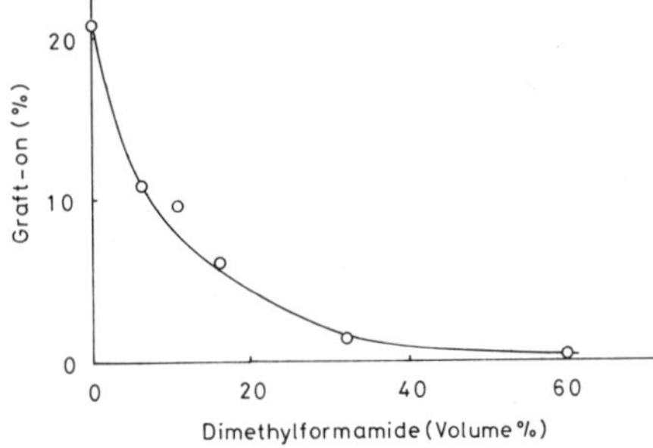

Figure 4. Apparent extent of grafting carried out in styrene–water–methanol–dimethylformamide reaction system at 30°C for 1 hr. Contents of styrene and water were 2% and 34%, respectively, and composition of methanol and dimethylformamide was changed.

Table II Reaction systems

reaction system	styrene	water	methanol	ethylene glycol	dimethyl-formamide
A	2 %	20 %	78 %		
B	2 %	34 %	64 %		
C	2 %	15 %	64 %	19 %	
D	2 %	34 %	58 %		6 %

content in styrene-water-methanol reaction systems (reaction system A and B), the apparent extent of grafting increased as water content increased. When ethylene glycol was added to the reaction system composed of 2 % styrene, 64 % methanol, and 34 % water, in place of water, grafting reaction proceeded slowly, but showed the same apparent extent of grafting as reaction system B without ethylene glycol after long reaction time. From these results, it is considered that the swelling of trunk polymer is very important factor to promote grafting reaction, but the rate of reaction is dependent on other factors. In the reaction system D, containing 6 % dimethylformamide, the rate of grafting reduced as compared with reaction system B composed of 2 % styrene, 34 % water,and 64 % methanol.

Amount of Reacted Cellulose. The relation between amount of reacted cellulose and reaction time is shown in Figure 6. The amount of reacted cellulose was the greatest and it was almost the same value even if reaction time proceeded in the reaction system B containing 34 % water and 64 % methanol. In other reaction systems, the amount of reacted cellulose was almost the same at the initial stage of reaction . But in the reaction system A composed of 2 % styrene, 20 % water, and 78 % methanol, the amount of reacted cellulose was unchanged with reaction time, and showed the lowest value in four reaction systems. In the reaction system C and D containing ethylene glycol, or dimethylformamide, the amount of reacted cellulose increased gradually as reaction time proceeded. These results may be caused by the difference of viscosity of these reaction solutions. Comparing these results with apparent extent of grafting, the increase of apparent extent of grafting is dependent on the increase of amount of reacted cellulose. From these facts, it is infered that reaction sites are decided at the initial stage of reaction in the reaction system containing strong swelling reagents with rapid diffusion to cellulose, but are made gradually with increase of reaction time in the reaction system containing

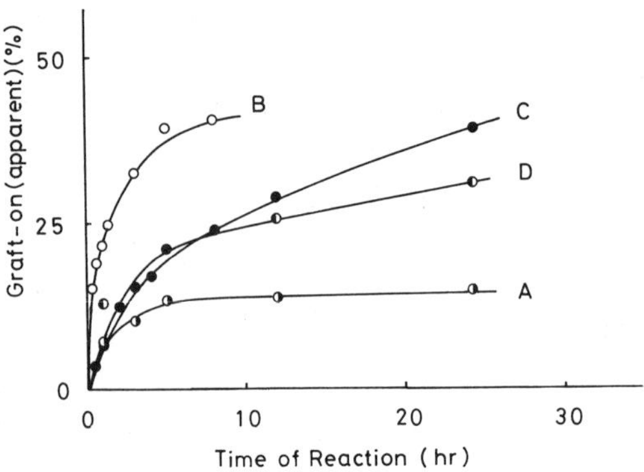

Figure 5. Apparent extent of grafting plotted against reaction time

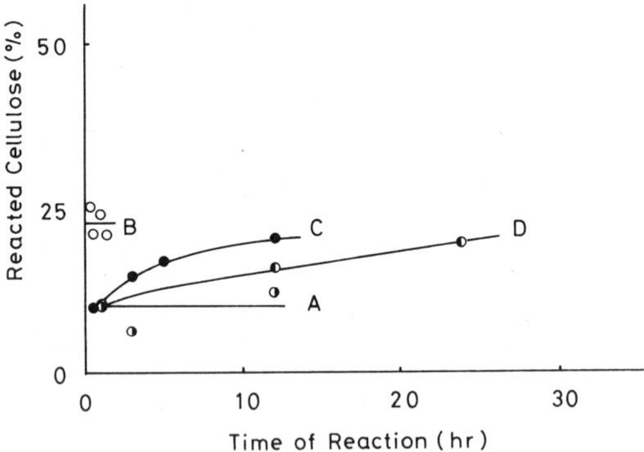

Figure 6. Relation between amount of reacted cellulose and reaction time

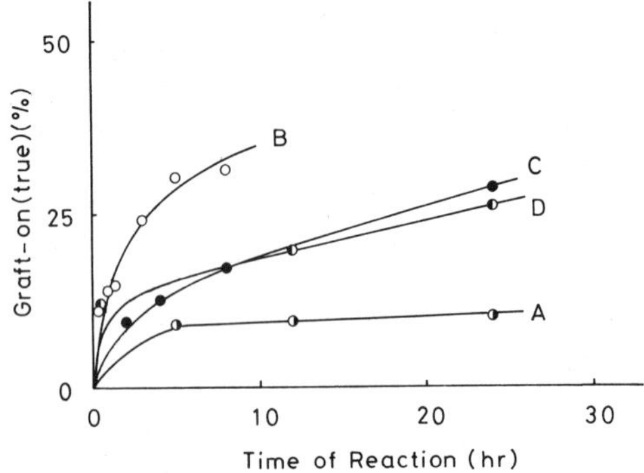

Figure 7. True extent of grafting plotted against reaction time

mild swelling reagents with mild diffusion to cellulose.

True Extent of Grafting. True extent of grafting is plotted against reaction time and shown in Figure 7. The true extent of grafting showed the same tendency as apparent extent of grafting, and it was the greatest in the reaction system B containing 34 % water and 64 % methanol.

Graft Efficiency. Graft efficiency, the rate of true extent of grafting to apparent extent of grafting, is shown in Figure 8. Graft efficiency was the greatest in the reaction system D containing 58 % methanol, 6 % dimethylformamide, and 34 % water, and was the lowest in the reaction system A containing 78 % methanol and 20 % water. In other reaction systems B and C containing the same amount of methanol, graft efficiency was almost the same. As it is evident that graft efficiency is affected by the amount of methanol from these results, it may be considered that methanol is a kind of chain transfer reagents.

Molecular Weight. Viscosity average molecular weight against reaction time is shown in Figure 9. Molecular weight in the reaction system B containing 34 % water and 64 % methanol increased remarkably as reaction time proceeded, but in the case of reaction system A containing 20 % water and 78 % methanol, molecular weight did not change with reaction time. Molecular weight in the reaction systems C and D containing ethylene glycol, or dimethylformamide increased gradually as reaction time proceeded. The relation between viscosity average molecular weight and true extent of grafting is shown in Figure 10. In the reaction system B containing 34 % water and 64 % methanol, molecular weight increased markedly with increase of true extent of grafting, especially when true extent of grafting exceeded about 25 %. In the reaction system containing dimethylformamide, molecular weight increased in a monotone as true extent of grafting increased. But in the reaction system C containing ethylene glycol, molecular weight increased gradually as true extent of grafting increased. Such results may be caused by the facts that the Trommsdorff effect of methanol for growing graft polymer radicals is greater than that of dimethylformamide, and that of water is greater than ethylene glycol.

Reaction Sites. Number of reaction sites is plotted against reaction time and shown in Figure 11. There was a maximum of reaction sites in the reaction system B containing 34 % water and 64 % methanol. In the reaction systems C and D containing ethylene

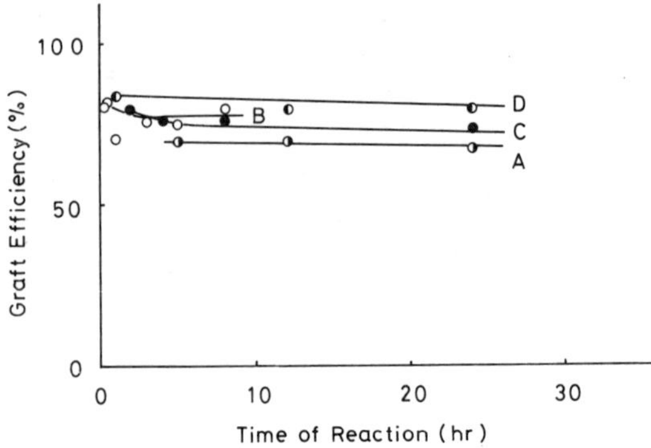

Figure 8. Graft efficiency vs. reaction time

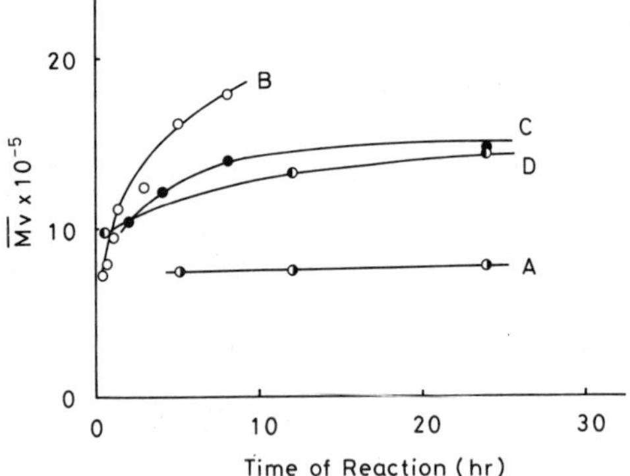

Figure 9. Viscosity average molecular weight plotted against reaction time

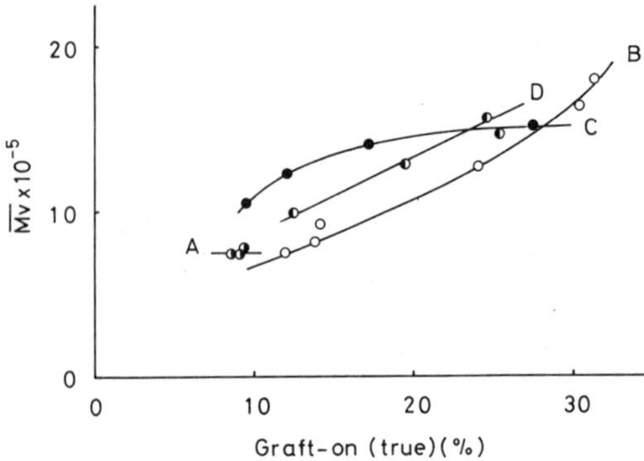

Figure 10. Relation between viscosity average molecular weight and true extent of grafting

glycol, or dimethylformamide, the number of reaction sites increased as reaction time proceeded, but in the reaction system A with 78 % methanol and 20 % water, the number of reaction sites was almost unchanged. In Figure 12, the relation between reaction sites and true extent of grafting is shown. In the reaction system B containing 34 % water and 64 % methanol, reaction sites increased at the initial period of reaction and decreased after true extent of grafting exceeded about 25 %. In other reaction systems C and D, number of reaction sites increased as true extent of grafting. But in the reaction system A, the number of reaction sites did not changed. The details will be discussed later. The generation of maximum number of reaction sites in the reaction system B is probably due to change of termination mechanism of growing graft polystyrene radicals as grafting reaction proceeded. Comparing these results with the results of amount of reacted cellulose (Figure 6), it is considered that the formation of reaction sites is due to the rate of diffusion of styrene, which is controled by swelling of trunk polymer by reaction solution, onto cellulose radicals generated by irradiation.

Decay of Cellulose Radicals. In Figure 13, the decay of cellulose radicals in various reaction solutions determined by ESR is shown. The decay of cellulose radicals was the slowest in the reaction system C containing ethylene glycol, but in other reaction systems the decay was almost the same. The decay of cellulose radicals is thought to express the rate of diffusion of reaction solution onto cellulose radicals for initiation of grafting reaction by styrene, and for scavenging by solvents. Therefore, the amount of cellulose radicals decayed does not necessarily mean the number of reaction sites. In comparison with the amount of reacted cellulose and the number of reaction sites (Figures 6 and 11), it is infered that in the reaction system C containing ethylene glycol, grafting reaction proceeded slowly because monomer diffuses onto cellulose radicals slowly. In the reaction system B containing 34 % water and 64 % methanol, grafting reaction proceeded rapidly for fast diffusion of monomer onto cellulose radicals. From these results mentioned above, it is considered that swelling of trunk polymer is one of the important factors to promote grafting reaction. But the rate to swell cellulose is also important factor. The rapid decay of cellulose radicals in the reaction system A containing 20 % water and 78 % methanol is guessed to be due to the termination of cellulose radicals by addition of solvents

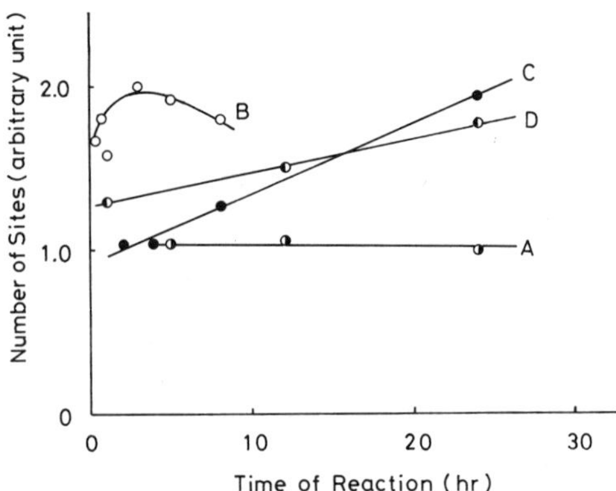

Figure 11. Number of reaction sites vs. reaction time

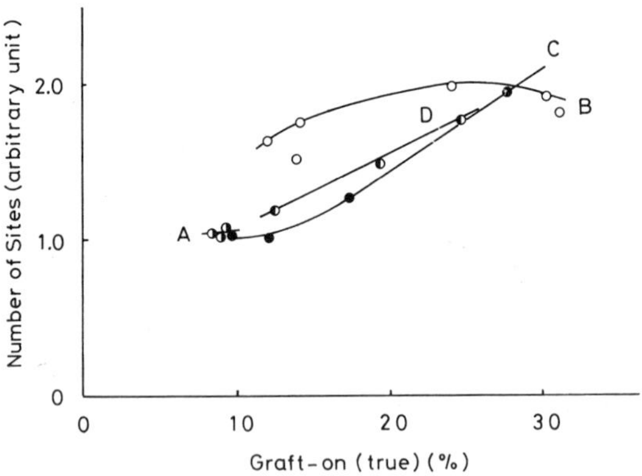

Figure 12. Relation between number of reaction sites and true extent of grafting

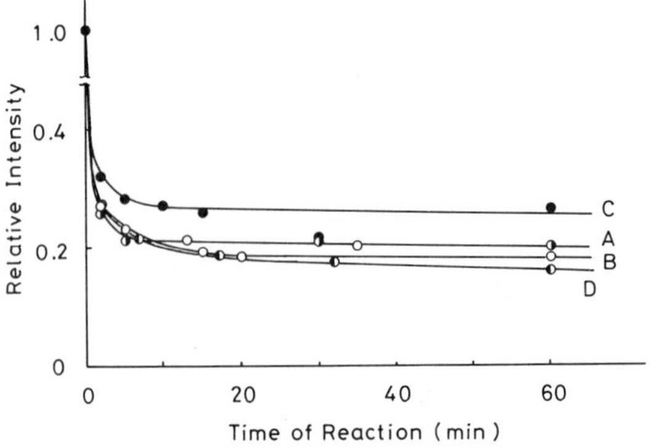

Figure 13. Decay of cellulose radicals in various reaction systems determined by ESR plotted against reaction time

and not due to the initiation of polymerization by styrene. In the reaction system D containing dimethylformamide, reaction sites and the amount of reacted cellulose were less than those in the reaction system B containing 34 % water and 64 % methanol. This result is guessed to be due to the difference of influence to graft polymer radicals between methanol and dimethylformamide.

Conclusion

In view of the results mentioned above, the most important factor to control grafting reaction onto the irradiated cotton cellulose is considered to be the rate of diffusion of monomer onto cellulose radicals which is dependent on reaction systems and have effect on the number of reaction sites and the amount of reacted cellulose. The number of reaction sites and the amount of reacted cellulose are also affected by the degree of swelling of trunk polymer. The Trommsdorff effect of solvent for growing graft polymer radicals is important factor to promote grafting reaction by increase of molecular weight of graft polymer.

Literature Cited

(1) Sakurada,I., Okada,T., and Kugo,E., Doitai To Hoshasen,(1960),3, 35
(2) Sakurada,I., Okada,T., and Kugo,E., Doitai To Hoshasen,(1959),2, 297
(3) Sakurada, I., Okada, T.,and Kugo,E.,Doitai To Hoshasen,(1959),2, 306
(4) Sakurada,I., Okada,T., and Kugo,E., Doitai To Hoshasen, (1959),2, 316
(5) Sakurada,I.,Okada,T., Uchida,M., and Kugo,E., Doitai To Hoshasen,(1959),2, 581
(6) Dilli,S., and Garnett,J.L., J. Polym. Sci. A-1,(1966),2323
(7) Dilli,S., and Garnett,J.L., J. Appl. Polym. Sci., (1967),11, 839
(8) Dilli,S., and Garnett, J.L.,J. Appl. Polym. Sci., (1967),11, 859
(9) Dilli,S., and Garnett,J.L., Aust. J. Chem., (1968), 21, 397
(10) Dilli,S., and Garnett,J.L., Aust. J. Chem.,(1968), 21, 1827
(11) Dilli,S. and Garnett,J.L., Aust. J. Chem.,(1970), 23, 1163
(12) Dilli,S., and Garnett,J.L., Aust. J. Chem.,(1970), 23, 1767
(13) Dilli,S.,and Garnett,J.L., Aust. J. Chem.,(1971), 24, 981
(14) Alexander,W.J.,and Mitchell,R.L., Anal. Chem., (1949),21, 4497

(15) Sobue,H., and Migita,N.,"Cellulose Handbook" 301, Asakura Shoten, Tokyo, 1958
(16) Sakurada,I., Okada,T., and Kaji,K., paper presented at 161st ACS National Meeting, Los Angels, March to April, 1971

22

Interaction of Radiation with Cellulose in the Solid State

GLYN O. PHILLIPS

North Wales Institute of Higher Education, Clwyd, Wales

P. J. BAUGH and J. F. McKELLAR

Department of Chemistry and Applied Chemistry, University of Salford, England

C. VON SONNTAG

Max-Planck Institute, Mulheim, Germany

When pure cellulose is irradiated with ultraviolet light of wavelength below 3000 Å, degradation readily occurs as a result of the direct absorption of the radiation. Intense irradiation with near-ultraviolet radiation results in a loss of tensile strength, a reduced degree of polymerisation and an increase in number of carboxyl and carbonyl groups[1]. Far ultraviolet (2537Å) irradiation of hydocellulose gives the gaseous products H_2, CO and CO_2[2].

Energetically ca 100 Kcal/mole is required to break a carbonhydrogen bond. The energy requirement for the degradation of cellulose by direct photolysis is met by radiation of wavelengths 3400Å or shorter. A prerequisite, however, is that the radiation must be absorbed by the moelcule. It must be inferred from the induced photochemical reaction that such absorption occurs, although the mechanism of energy absorption is uncertain.

High energy ionizing radiations similarly degrade cellulose most effectively as demonstrated by the extensive studies of J. C. Arthur Jr., and co-workers[3]. Cellulose closely resembles other solid state carbohydrates showing a remarkable susceptibility to degradation by γ-radiation. Polymeric carbohydrate systems degrade with -G (number of molecules destroyed/100 ev) ca.15. Subsequently α-lactose monohydrate was found to be even more radiation -labile and G-values of 40-60 were found[4]. Crystalline D-fructose behaves similarly with -G values 40-60[5]. The generality of this unusual behaviour of solid-state carbohydrate systems on γ-irradiation is being now constantly demonstrated in several laboratories[6]. To date the most remarkable is crystalline 2-deoxy-D-ribose where -G values > 650 have been encountered[7]. It is important to add that such transformations show these high yields directly in the solid state, and are not in any way dependent on work-up conditions.

Only isolated examples are known of such extreme radiation susceptibility and certainly no group of compounds resembles solid carbohydrates in this respect. Even more unusual and valuable is that degradation leads to a limited number of products. For α-lactose monohydrate, for example, three types of transformation have been recognised[8].

ROUTE 1

(A)

Water Elimination

(B)

Radical produced on γ-irradiation

Radical (B) abstracts H atom from C_1 to give radical (A) via a chain process and the product: 2-deoxy-lactobiono-lactone (G = 20)

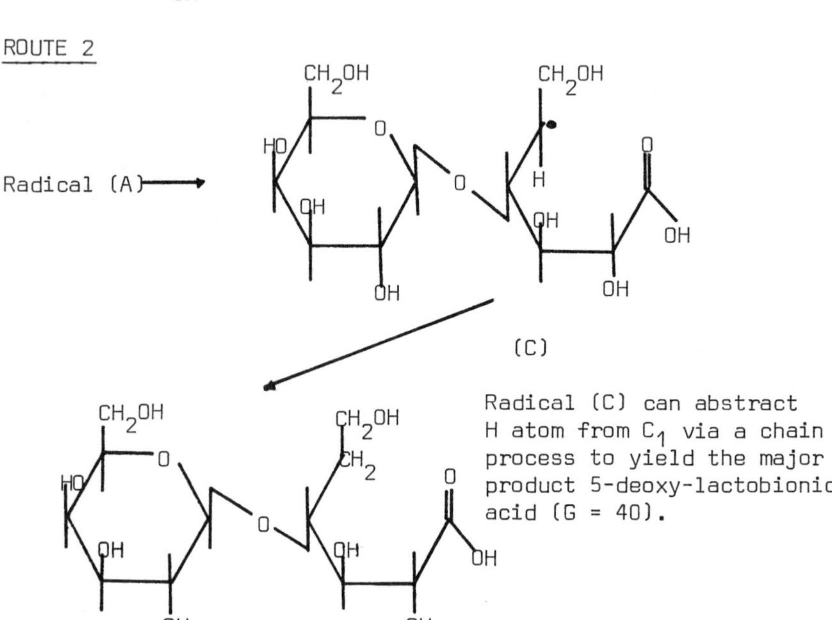

ROUTE 2
Radical (A)
(C)
Radical (C) can abstract H atom from C_1 via a chain process to yield the major product 5-deoxy-lactobionic acid (G = 40).

ROUTE 3

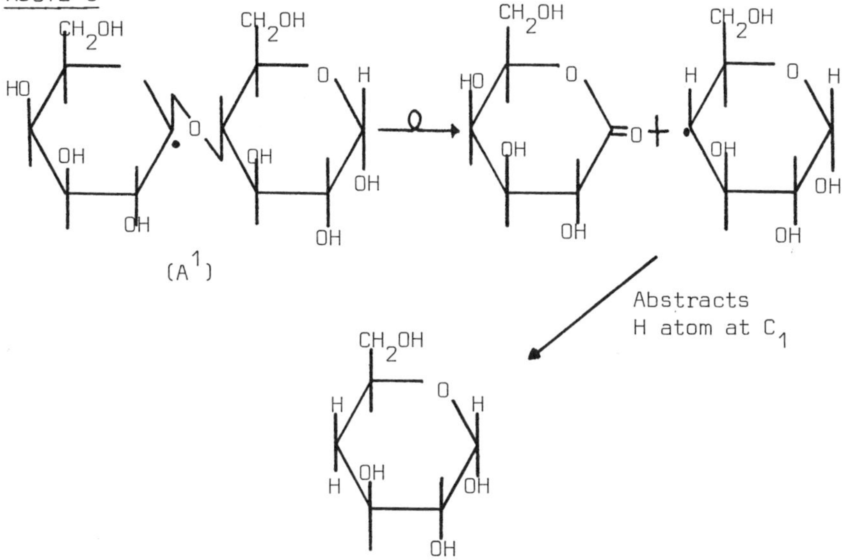

These transformations are encountered in the other systems also. For example, 2-deoxy-D-ribose via ROUTE 2 yields 2,5-dideoxy-D-erythropentonic acid (E)

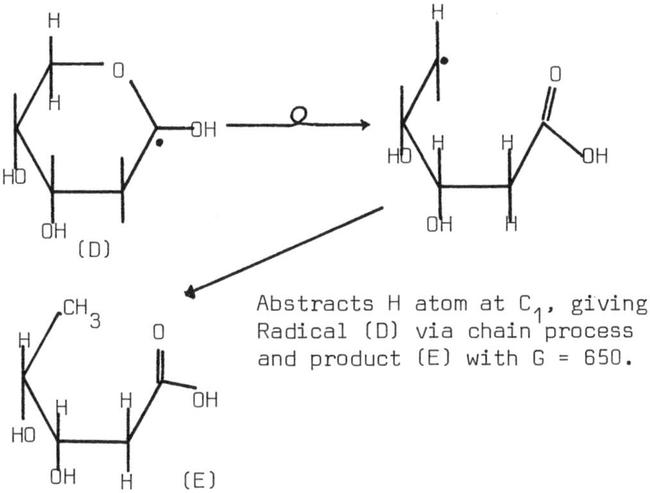

Similarly, D-fructose yields 6-deoxy-D-threo-2, 5-hexodiulose with G = 60 at a dose of $0.8 \times 10^{20}eVg^{-1}$. As would be anticipated, on the basis of a radical chain reaction occurring, product yields are dose rate dependent. Using this procedure, gram-yields of the products can readily be prepared.

After 25 years of intensive study, only now using the combination of gas chromatographic and mass spectral techniques can the products of irradiation of mono and disaccharides be identified. Moreover, a regular chemical pattern is emerging which allows products of γ-irradiation from particular carbohydrates to be predicted. Here we shall describe an extension of this chemical study to oligosaccharides, which enable the products from cellulose irradiation to be predicted.

Modification of the cellulose can considerably change the behaviour to cellulose towards radiation. The presence of vat dyes on cotton greatly increases the rate of photodegradation and represents an important technological problem[9]. Normally the objective is to ensure minimum degradation because of the unsirable effects of phototendering of the cellulose fibre. However, this effect has also been utilised to assist with waste paper disposal[10]. On the other hand, certain aromatic groups when introduced into the cellulose molecule greatly stabilise the polymer to radiation damage[11]. This procedure provides a procedure for reducing the susceptibility of cellulose to radiation damage and greatly improves its behaviour when ionizing radiation is used to sterilize cellulosics.

This paper considers the interaction of visible/ultraviolet and ionizing radiations with cellulose, and particularly the contrasting mechanisms by which protection and sensitization can be induced in relation to the radiation action.

Products after γ-Irradiation of Oligosaccharides

The high molecular weight of the oligo - and polysaccharides makes direct gas chromatographic (GC) analysis of radiation products difficult and identification by GC - MS completely impossible. As a model for cellulose and other biological polysaccharides we have utilised the cylic oligosaccharides - the cycloamyloses. These have the advantage that they do not possess a free reducing group, which generally dominates the chemical changes following irradiation of low molecular weight carbohydrates. Thus, it is possible to study main-chain chemical reactions, which are mainly responsible for the changes which occur when cellulose is treated with γ -radiation.

Samples of the oligosaccharide in hydrated and freeze-dried forms were irradiated to doses of $\sim 10^{21}eVg^{-1}$, sufficient to produce ca 4% conversion to products, based on the absolute -G values of 6.6 and 14.2 determined respectively, for these states by Phillips and Young[12]. The hydrolysate from the irradiated oligosaccharide was fractionated by column chromatography, reduced with $NaBH_4$ or $NaBD_4$, trimethylsilylated (before and after reduction) and identified by GC or GC - MS as previously described[13]. The chemical modifications induced in the oligosaccharide molecule on γ-irradiation are shown in Table 1. Only modification 6 appears to result in a monosaccharide (5-deoxy-glucose), which has not previously been identified as a product from carbohydrate irradiations.

With the precise knowledge of the products formed, a valid attempt can be made to locate the initial sites of attack and to propose reaction mechanisms which account for the functional group changes.

Two conclusions can be drawn from the monosaccharides identified. First there is a predominance of monosaccharides modified at the C1, C4 and C5 positions and secondly hydrolysate products resemble those identified for irradiated cellobiose[8]. In both instances there is little evidence of modiciations of the anhydroglucose unit (a.g.u.) at the C2, C3 or C6 positions.

It is not possible to be specific about the manner in which the radicals are initially formed in a solid carbohydrate matrix. The fate of the geminate ion-pair is unknown and it is assumed that in the ordered H-bonded solid, recombination occurs giving a predominance of excited molecules. One fate of such molecules would be to split off hydrogen atoms from the most reactive sites to form radicals[14].

By analogy with previous work, it can be assumed that the locations of radiation-induced keto groups in the a.g.u. coincide with the sites of initial attack, i.e. the primary radical sites. These groups arise in irradiated solid carbohydrates by reaction steps involving (a) hydrogen removal from an α-OH group during disproportionation (b) radical rearrangement (c) water eliminations and (d) radical hydrolysis. Deoxy groups are formed adjacent to keto functions via step (e) followed by H-atom addition during disproportionation. To explain the formation of keto and deoxy functions, here, similar reaction steps must also be considered.

Schemes I to III illustrate the radiation-induced radical reactions which lead to the reaction products identified. The last step in each pathway is hydrolysis to the monosaccharide identified in the hydrolysate mixture.

The radical reaction processes leading to functional group changes resulting from initial attack at C1 are outlined in SCHEME I. A gluconic Acid δ-lactone function is formed either after a radical hydolysis reaction followed by hydrogen atom removal or preferably radical rearrangement, C1→C4. In the latter instant the transfer of the radical site to C4 yields a 4-deoxy function after hydrogen atom addition. The modified straight chain oligosaccharide would thus contain these two functions on terminal a.g.u. and be drived from malta-hexa- or heptaose.

Other minor reactions involving loss of CO and H_2O from C1 radicals are proposed to account for the formation of five carbon

TABLE I Chemical Modifications Deduced from Monosaccharides identified in hydrolysate mixtures isolated from γ-irradiated α-and β-cycloamylose hydrates.

No.	Chemical Modification to a.g.u.	Site attacked[a] in a.g.u.	Compound[b] determined	Precursor in hydrolysate mixture
1	δ-lactone	C1	gluconic acid δ-lactone	same
2	4-deoxy	C1 → C4 (C5 → C1)	4-deoxy-glucose	4-deoxy-glucose
3	five-carbon	C1	arabitol	arabinose
4	five carbon	C1	2-deoxy-ribitol	2-deoxy-ribose
5	4-keto-	C4 C1	galactitol	4-keto-glucose
6	5-deoxy-	C4 C1	5-deoxy-hexitol	5-deoxy-glucose
7	3-deoxy-4-keto	C4	3-deoxy-galactitol-	3-deoxy-4-keto-glucose
8	5-keto-	C5 C1	iditol	5-keto-glucose
9	6-deoxy-5-keto	C5	6-deoxy-hexitols	6-deoxy-5-keto-glucose
10	5-deoxy-dialdo-	C_6	5-deoxy-hexitol	5-deoxy-xylo-hexodialdose
11	5-carbon	unknown	ribitiol	ribose
12	5-carbon	unknown	xylitol	xylose
13	5-carbon	unknown	3-deoxy-pentitol	3-deoxy-pentulose
14	4-carbon	unknown	erythritol	erythrose
15	4-carbon	unknown	threitol	threose

a. C4 → C1 refers to initial attack at C4 followed by rearrangement of radical site to C1.

b. Determined after reduction with NaBD4 (GC-MS) or $NaBH_4$ (GC).

Scheme I. Chemical modifications following radical formation at C1 of the cycloamylose anhydroglucose unit

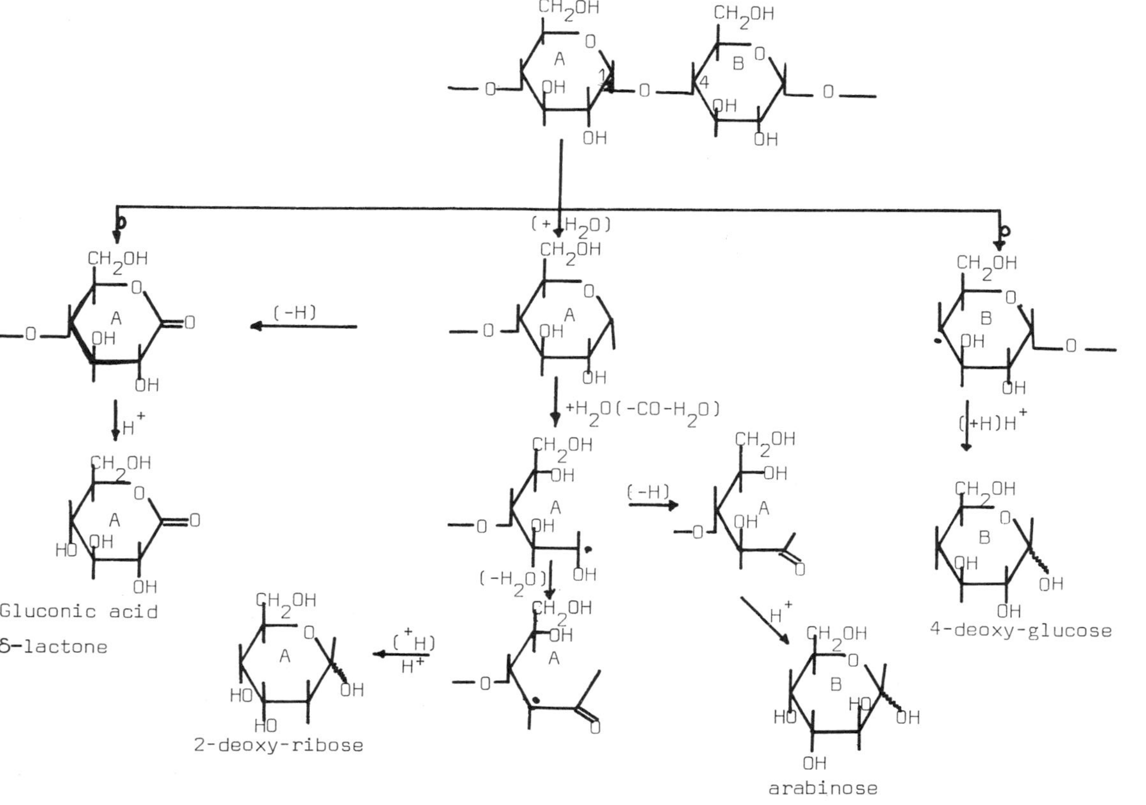

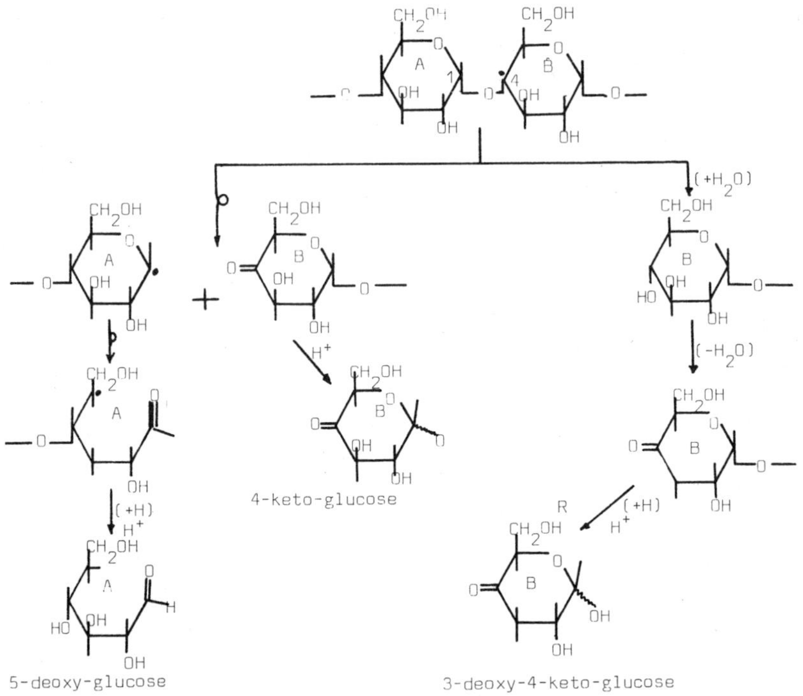

5-deoxy-glucose

4-keto-glucose

3-deoxy-4-keto-glucose

Scheme II. Chemical modifications following radical formation at C4 in the cycloamylose anhydroglucose unit

5-keto-glucose

4-deoxy-glucose

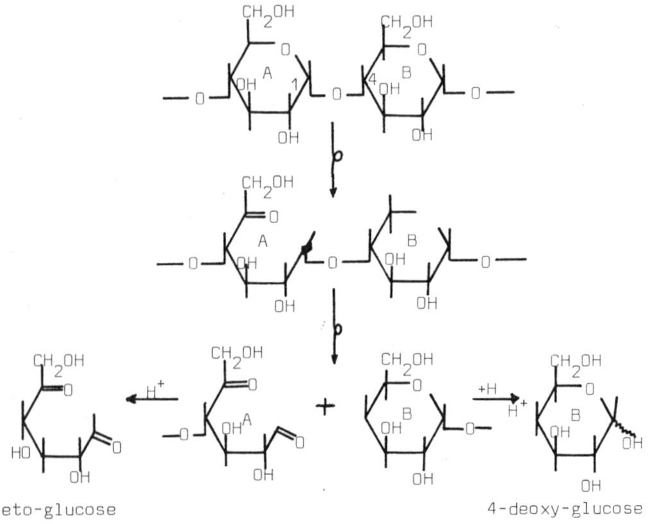

Scheme III. Chemical modifications following radical formation at C5 in the cycloamylose anhydroglucose unit

sugars, arabinose and 2-deoxy-ribose (see also SCHEME I).

A radical rearrangements, C4→C1, considered to be the most probable step following initial attack at C4 gives rise to a 4-keto function. A 5-deoxy function is also formed after further rearrangement of the radical site, C1→C5, followed by subsequent hydrogen addition as illustrated in SCHEME II. 3-Deoxy-4-keto function is formed via a radical hydrolysis step followed by water elimination and hydrogen addition (see also SCHEME II).

SCHEME III illustrates the radical reactions leading to chemical modifications identified following initial attack at C5. A radical rearrangement, C5→C1, accounts for the formation of a 5-keto-function. A 4-deoxy function is produced after further rearrangement of the radical site, C1 to C4, followed by hydrogen atom addition. Thus formation of a 4-deoxy functional group is possible via initial attack at C1 or C5. The straight chain doubly modified oligosaccharides of the maltahexa or heptaose type can also possess 4-deoxy and 5-keto-functions on terminal a.g.u. A 6-deoxy-5-keto function has also been identified and the radical reaction process leading to this modificiation is considered to commence at C5 (see SCHEME III) since the keto group is located at this position. A radical reaction process analogous to that leading to formation of this function is aqueous solution is difficult to envisage here.

The formation of a 5-deoxy-xylohexadialdose function is also difficult to explain. It is probably formed via initial radical attack at C6 since its formation from a C6 D-glycosyl radical has been proposed and proceeds via a rearrangement, C6 to C5, followed by water elimination and hydrogen atom addition. The oligosaccharide modified in this way, however, should remain in its cyclic form unless ring opening renders the modified a.g.u. unstable.

No satisfactory explanation can be offered at this time for certain products (transformations 10-15). However, one vital point emerges from the diversity of the products. Once formed within the solid state carbohydrate matrix, the radicals can undergo facile reactions, with chain processes common and leading to unusually high decomposition yields.

The generality of these chemical transformations, induced by ionizing radiations in solid state carbohydrates, from mono to disaccharides, and cellobiose in particular, along with the present evidence that the processes must also occur in γ-irradiated oligosaccharides allows extrapolation of the results to cellulose.

Protection of Cellulose towards Ionizing Radiations

Such chemical changes are positively undesirable when cellulosic materials are being sterilized by γ-irradiation, for example, packaging material and cotton sutures. Once produced, the radicals by the processes described, lead to irreversible chemical changes in the material. Thus any protective process, to be effective, must be introduced prior to radical formulation or the radical chain process must be prevented. It is on this basis that we developed the energy scavenging method for the radiation protection of cellulose when irradiated in the solid state. Facile singlet- singlet resonance energy transfer can occur to an aromatic group, either associated with or substituted into the carbohydrate, provided the first singlet level (E_1) of the associated molecule ~4 eV. Using fibrous cotton cellulose, estimates of the distance over which energy transfer can occur may be obtained. Here the degree of substitution (d.s.) of the aromatic groups can be varied and protection studied as a function of d.s. The measured parameter was the breaking strength of the fibrous benzoylated cellulose polymer. Protection continues as low as d.s. 0.2 of benzoyl groups. Assuming that the benzoyl groups are randomly substituted on the cellulose molecule, the maximum distance between benzoyl groups may be calculated as a function of the degree of substitution and as a binomial distribution. In the experimental case, the assumption of random substitution of benzoyl groups will not be exactly true, since the physical structure of the cellulose will limit the accessibility of highly ordered regions to reactions with benzoyl chloride. Therefore, in the experimental case, maximum spacing of benzoyl groups will tend to be greater than those indicated. Since protection is afforded at d.s. 0.2 when the calculated maximum spacings of benzoyl groups on the cellulose molecule is 7.2 - 8.2nm, an energy transfer over this distance in cotton cellulose is clearly possible. The method has proved a practicable method of protecting cotton cellulose towards γ-radiation[11]. Scavenging of H atoms by the aromatic groups could also be an important factor in reducing decomposition by preventing the propagation of radical chain processes.

Reactions of Cellulose promoted by ultraviolet and visible Radiations

The purest cotton can be degraded by 265 nm ultraviolet radiation[15, 16]. For D-glucose in aqueous solution $n \rightarrow \sigma^*$ absorption at the lactol oxygen leads to the step-wise degradation of the molecule via pentose, tetrose, diose to CO_2, which is, in effect, a reversal of photosynthesis. In cotton cellulose yarn, there can be observed in addition, using diffuse reflectance measurements, a broad absorption at ~270 nm (4.6 eV), which is a characteristic of the solid matrix[17].

It is this absorption which is the dominant process leading to direct photolysis of cellulose.

However, when cellulose is dyed, for example, with anthraquinonoid dyes, visible light initiates a greatly accelerated degradation of the cellulose and is frequently accompanied by fading of the dye. The mechanism of this process has been elucidated[18-20].

$$\begin{array}{ccccccc}
\text{Cell-H} & + & A & \longrightarrow & A^* & + & \text{Cell-H} \\
 & O_2 & \uparrow & H_2O & & & \downarrow \\
 & & AH_2 & \xleftarrow{\text{Cell-H}} & AH\cdot & + & \text{Cell}\cdot + H\cdot \\
 & & & & & & \downarrow \text{degradation}
\end{array}$$

where A represents the ground-state dye

A* the excited dye

AH· dye semiquinone

AH_2 dye anthhydroquinone

when the dye is aggregated, the initial excitation may be additionally dissipated by

$$A^* + A \longrightarrow (A\cdot^+ \text{-------------} A\cdot^-)$$

$$(A\cdot^+ \text{------} A\cdot^-) + O_2 \longrightarrow A\cdot^+ + A + O_2^-$$

$$A\cdot^+ + OH^- \longrightarrow A + OH$$

$$.OH + A \longrightarrow \text{hydroxylated dye}$$

$$.OH + \text{cell} \longrightarrow \text{Cell}\cdot + H_2O$$

$$\text{Cell}\cdot \longrightarrow \text{degradation}$$

where $(A\cdot^+ \text{------} A\cdot^-)$ represent the ion pair formed by the semi-reduced and semi-oxidised form of the dye. Thus light, oxygen, water and dye exert a profound influence on the course of the phototendering.

Protection of Cotton Cellulose towards light induced processes

Two approaches have been utilized to eliminate the degradation of cellulose sensitized by the dye. First, to establish the factors in relation to the dye responsible for the tending process and secondly remove the excitation energy from the dye before it is able to initiate degradation.

For a considerable period it was not explicable why two dyes of almost identical structure differed vastly in their ability to sensitize the degradation of cellulosics. It is now possible to explain this difference by reference to the behaviour of 1- and 2-piperidino anthraquinones[21]. Although the 1-piperidino derivative is more readily photoreduced in solution, it is significantly less active than the 2-piperidino in sensitizing the photodegradation of the yarn. It has been possible on the basis of the hydrogen - abstraction theory of phototendering to explain this apparent anomaly and to establish a necessary spectroscopic characteristic of the 'active' dye[22]. It was found that the long wavelength band of both piperidino derivatives was shown to be charge-transfer in nature. On protonation of the 1-piperidino compound, the charge-transfer band disappeared leaving a shoulder attributable to the $n\pi^*$ band. In agreement with this assignment, the protonated derivative abstracted hydrogen from the substrate very much faster than the unprotonated form. Thus, a nominally 'poor' sensitizer had been converted by protonation to a 'good' sensitizer due to the removal of the charge-transfer band and enabling the then lowest excited state ($n\pi^*$) to abstract hydrogen. With these materials also, another mechanism of photodegradation could be envisaged. The photochemistry of the two derivatives proved quite different in alkaline solution. On photolysis of the 1-piperidino derivative, the dianion (A^{2-}) is formed from the semiquinone radical from the reactions:

$$AH^* + OH^- \longrightarrow A^{\cdot -} + H_2O$$

and

$$2A^{*-} \longrightarrow A^{2-} + A$$

With the 2-piperidino derivative however, the final product was the semiquinone radical anion ($A^{\cdot -}$). This is formed by an electron transfer process from hydroxide and alkoxide ions to the photoexcited charge-transfer state of the quinone:

$$A^* + OH^- \longrightarrow A^{\cdot -} + OH$$

Such a process of direct electron abstraction by the 2-piperidino derivative explains its greater sensitizing action, on a textile, particularly in a highly polar environment. The resultant OH radicals readily promote degradation through abstraction processes.

These observations on model systems have been confirmed by a study of selected commercial vat dyes in solution and on cellulosic substrates[23-25]. Initially the triplet - triplet absorption spectra of two commercial vat dyes (C1 67300 and C1 60515) were characterised by laser flash photolysis and the rate constants of oxygen quenching determined[25]. Immediately an apparent anomaly of phototendering was removed. Why was it possible for excited (probably triplet) states able to initiate reactions in oxygen? Indeed, consumption of oxygen is one of the main characteristics of these reactions. We have found that the lifetime of the triplet states of these dyes is very short ($t_{\frac{1}{2}} \sim 1\mu s$). Thus the triplet plus-oxygen reactions on a fabric will not be capable of competition with triplet-plus-substrate reactions. Both dyes show similar phototendering of cellulose, but I sensitizes the photo-oxidation of tetralin via a process of H-atom abstraction 30 times faster than II.

Three different types of experiment involving these two dyes on cellulosic substrates were carried out:

(a) I and II were dyed on cotton fabric and portions were treated with a triplet quenching nickel ketoxime chelate[26]. On irradiation a distinct protective affect towards photodegradation of the cotton was observed only with I.

(b) Fabrics dyed with I and II were irradiated in a stream of oxygen with and without furan present. If a singlet oxygen mechanism were involved protection should occur. No significant protection occurred despite the fact that both gave a high quantum yield of singlet oxygen in solution.

(c) Cellulosic films freed from plasticiser and dyed with I and II were flashed in air using a nanosecond laser flash of wavelength 347nm. Very rapid formation of transient species was observed in both instances. By comparison of their spectra with those observed from the flash photolysis of the dyes in alcoholic solutions it is seen that the transient from I resemble the semiquinone radical AH. While that from II the semiquinone radical anion.

We conclude, therefore, that although singlet oxygen is undoubtedly formed on irradiation of the two vat dyes, this is not involved in the major primary process for initiating phototendering. For I the degradation occurs by the reaction:

$$A^* + \text{Cell-H} \longrightarrow AH\cdot + \text{Cell.}$$

For dye II the initial reaction is:

$$A^* + OH^- \longrightarrow A\cdot^- + OH$$

followed by $\text{Cell-H} + OH \longrightarrow \text{Cell}\cdot + H_2O$

By considering the two different mechanisms suitable protective procedures can be devised. For the former process, applications of the triplet-quenching nickel ketoxime chelate to the dyed fabric ensures considerable protection. In the second mechanism, where direct abstraction is slow, it is important to remove the polar electron donating environment.

These dyes can be usefully used on cellulose, but should not be used on a more polar substrate, e.g. nylon.

Dye Fading associated with Phototendering

In addition to our studies on the sensitizing effect of anthraquinone dyes on cellulosic substrates in connection with the problem of phototendering, we have recently been examining the effect of various substituent groups in the dye structure that can affect its photostability. This work has yielded some remarkable results which show that even relatively minor changes in dye structure can markedly affect light-fastness (LF) on a fabric. This is a finding of considerable technological interest since it guides the dyestuff chemist to the synthesis of new dyes of commercially desirable light stability. The reasons for the effect of these structural changes on LF has thus led us to our current studies of the character and properties of the photoexcited states of the dyes when they are in solution and also when dyed on commercial fabrics.

Our interest in this aspect of dye photochemistry developed from a discovery, purely by chance, that simple benzanthrone disperse dyes on polyester showed remarkable difference in L.F. with only minor changes in the nature of the substituent R (Figure 1) in the 6-position. For example, when R is an anilino group then no fluorescence, phosphorescence or photochemical activity (as indicated by flash photolysis) is observed - and the L.F. of the dye is very high. Laser flash studies of the type illustrated in Figure 2 have enabled us to detect triplet formation with many benzanthrone derivatives, but not with the 6-anilino derivatives. These flash experiments, taken in conjunction with the fluorescence and phosphorescence experiments, have led us to the conclusion that the 6-anilino substituent induces a very fast, radiationless deactiveation from the first excited singlet state. Simply stated - if the first excited singlet state is so short lived that it cannot emit fluorescence, undergo intersystem crossing or indeed react with any species in its environment (unexcited dye molecule, oxygen, substate etc) then it is being very efficiently returned to its ground state,

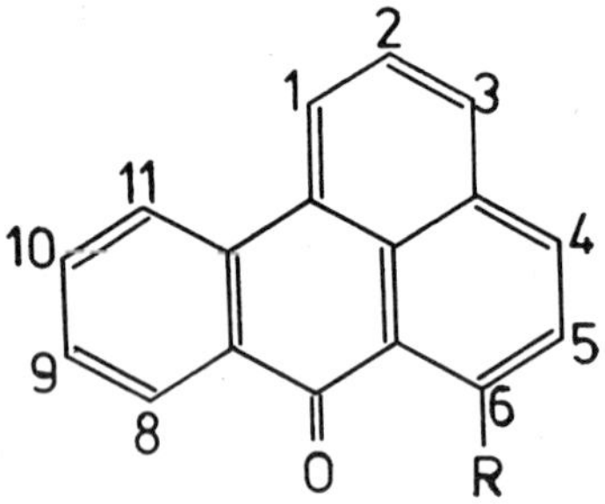

Table

R	Transient[a] species observed	Fluorescence[b] and/or phophorescence	Light[c] fastness ISO scale
H	Yes	Yes	<1
OH	Yes	Yes	1-2
NH_2	Yes	Yes	2
$NHCH_2CH_2OH$	Yes	Yes	2
NHC_6H_5	No	No	6-7
$NHC_6H_4OCH_3$	No	No	6-7

[a] Photoflash energies of 400*J* using a conventional flash photolysis apparatus;[8] [b] measurements made with an Aminco Bowman Spectrophotofluorimeter. Fluorescence measurements at room temp. using a range of solvents. Phosphorescence measurements at 77K using mixed alcoholic glasses; [c] measurements made on polyester fabric as specified by British Standard 1006[9] using a 'Xenotest' 150 fading lamp

Figure 1

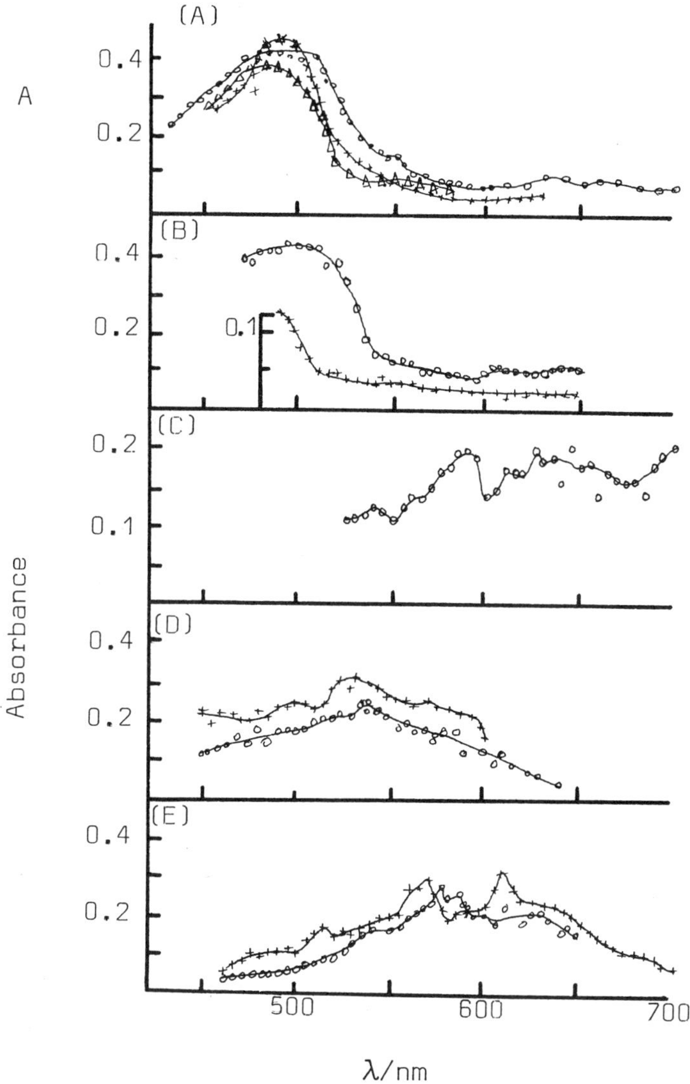

Figure 2. Absorption spectra of the transients observed at the end of the pulse on laser flash photolysis of: benzanthrone (A); 3-methoxybenzanthrone (B); 4-anilinobenzanthrone (C); 6-hydroxybenzanthrone (D); and 6-aminobenzanthrone (E)—in benzene O; 1,1,2-trichlorotrifluoroethane △; methanol X; or 2-propanol +.

Light Fastness of Anthraquinone Disperse Dyes on Poly(ethylene terphthalate)

Anthraquinone	m.p. °C	Standard depth 1/3	1/1	2/1	Anthraquinone	m.p. °C	Standard depth 1/3	1/1	2/1
1-hydroxy-	193		>8		2-*N*-piperidino-	165		<1	
1-methoxy-	169		<1RD		1-amino-2-methyl-	205		5-6	
1-amino-	254		4-5	5	1,2,4;triamino-	>330		3-4	
1-methylamino-	167		4		1,4-dihyroxy-	193-4		>8	
1-acetylamino-	220-1		6-7	7	1-amino-4-chloro-	176		1R	
1-chloroacetylamino-	217		7	7	1-amino-4-bromo-	178		1-2R	
1-benzoylamino-	256-7	7	7		1-amino-4-hydroxy-	207-8		6	
1-anilino-	148		7-8		1-amino-4-methoxy-	166-7		3-4	3
1-*p*-toluidino-	266-7	7	7		1,4-diamino-	267		5	5
1-*N*-methylanilino-	130-1	4	5		1-amino-4-methylamino-	192-3		3RD	3-4RD
1-*N*-piperidino-	121		3-4		1,4-bismethylamino-	224		3RD	
2-hydroxy-	309-10		<1RD		1,4-bis-*p*-toluidino-	321-2	8		
2-methoxy-	196-7	<1Y			1-amino-5-chloro-	219		4-5	
2-amino-	307	1	1		1,5-diamino-	317-8		3RD	
2-methylamino-	235		<1RD		1,5-di-*N*-piperidino-	205		4	
2-acetylamino-	265-6	<1			1,8-di-*N*-piperidino-	185		3	
2-dimethylamino-	183-4		<1RD	<1RD					

Figure 3

chemically unchanged i.e. the result will be that the dye will be very fast to light.

Developing this concept to the anthraquinone dyes has required work along two distinct lines. In the first we have measured that L.F. of a wide range of highly purified, relatively simple dyes in the hope that any structural features influencing the L.F. may become apparent. Figure 3 shows the interesting feature that those dyes substituted by hydroxy or amino substituents in the 1-position have significantly higher L.F. than their counterparts substituted in the 2-position (cf 1 and 2-hydroxy; 1-and 2-amino; 1 and 2-piperdino derivatives). Our second line of investigation, therefore, has been to study the photoexcited states of those pairs of dyes of contrasting L.F. to see if this can. . ad to rational explanations for the difference. As yet, no clear explanation has emerged but some progress has been ma e particularly with the hydroxy derivates.

The factors con rolling the L.F. of the amino derivatives certainly appear to b e complex, particularly when it is noted that the difference i L.F. between the 1- and 2- derivatives is not nearly so great a that between the corresponding hydroxy compounds. However, c ne interesting feature has emerged. It has been known for some ti ne that the 1- substituted derivatives there is an empirical relation between the basicity of the amino group and the L.F. of he dye. Figure 4 shows that this is indeed correct from ou data. However, the figure also shows that two derivatives dc not conform to this relationship and that both are <u>anilino</u> derivatives; these dyes both possess higher L.F. than their basicity would indicate. Remarkably, this is the same 'anilino group influence' as we found with the benzanthrone dyes. Evidently this group when ortho to a carbonyl group,

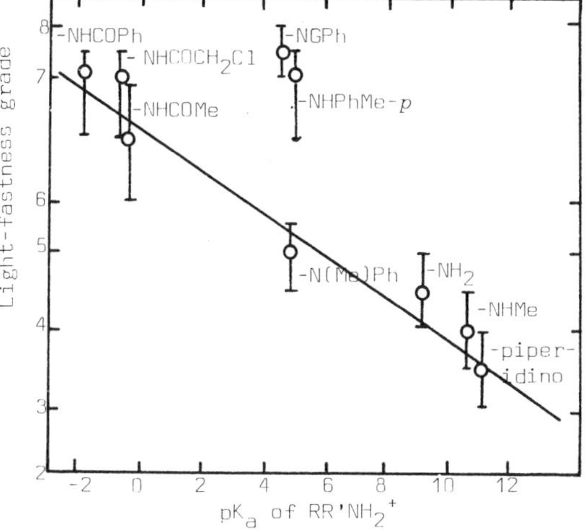

Figure 4. Relationship between the light-fastness grade of 1-RR'N-anthraquinones on polyester and the pK$_a$ of RR'NH$_2^+$. The light-fastness grades are spaced according to the fastness ratio between successive standards (31), and the vertical lines on the experimental points define the limits of experimental uncertainty in the observed light-fastness grades.

Fluorescence and phosphorescence maxima and quantum yields for hydroxyanthraquinone in relation to their light fastness.

Anthraquinone	Fluorescence		Phosphorescence			Light fastness
	λ/nm	Φ_f	λ/nm	Φ_p	τ_p/secs	1/1 depth
2- hydroxy	472,500	$\sim 10^{-5}$	515	0.06	0.05	<1RD
2,6-di "	463,505	"	532(s),560	0.23	0.16	1
1- "	570,597(s)	0.004	NO.	-	-	>8
1,2-di "	570(s), 590	"	"	-	-	7-8
1,4-di "	536,565	0.05	"	-	-	>8
1,5-di "	575,605(s)	0.006	"	-	-	>8
1,8-di "	530,570	0.015	"	-	-	7
1,4,5-tri "	543,572	0.073	"	-	-	>8
1,2,5,8-tetra	550,561,591	0.004	"	-	-	8

τ_p = phosphorescence lifetime; NO = not observed; RD = redder; (s) = shoulder.

Figure 5

enhances L.F. and we have found that this effect also occurs with more structurally complex commercial dyes.

Thus we can conclude that the deactivating effect of this structural feature may be quite widespread in dye photochemistry, giving us access to the synthesis of more highly lightfast dyes.

Again, using our second line of investigation, the factors controlling the L.F. of the hydroxy derivatives have proved to be more easily related to the photoexcited states of the dye. Figure 5 summarizes the fluorescence and phosphorescence quantum yields of non 1- and 2- substituted hydroxy derivatives and gives the corresponding L.F. of the dyes on polyester fabric. By far the most interesting feature is the clear cut difference in the data (yantum yield and L.F.) between those dyes with a hydroxyl group in the 1- position and those without. From the data in Figure 5 and other sources we have constructed a potential energy diagram for the 1- and 2- hydroxy derivatives (Figure 6) which we believe assists our interpretation of these differences.

For the 1- hydroxy derivative it is seen that the $^3n\ \pi^*$ state lies above the lowest $^1\pi\ \pi^*$ state so intersystem crossing will not occur. This conclusion is supported by the fact that no phosphorescence was detected within the wavelength range of our instrument (sensitive up to 1000 nm) and that a large potential energy gap must therefore exist between the $^1\ \pi\ \pi^*$ state and any low lying $^3\ \pi\ \pi^*$ state, if indeed one exists. Thus no potentially photoactive triplet state is formed, and even the flourescence quantum yield values are low enough to indicate rapid and efficient radiationless deactivation from the first excited singlet state. Unlike the 'anilino deactivating effect' whose mechanism we do not understand our knowledge of ortho hydroxy substituted carbonyl compounds indicated that the mechanism is one of rapid proton exchange in the lowest excited singlet state.

For the 2-hydroxy derivative proton exchange deactivation is much less probable so the LF of the dye should be low. However, the relative protential energy levels of this dye indicate a

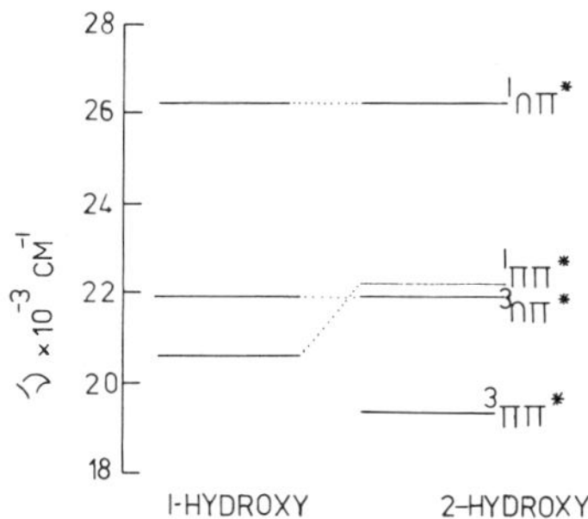

Figure 6

further reason for low LF which would thus explain the very large difference between the two derivatives. In Figure 6 it is seen that the $^3n\ \pi^*$ state lies just below that of the $^1\ \pi\ \pi^*$ state so that inersystem crossing will be strongly favoured by two factors (a) the small size of the potential energy difference between the two states and (b) the fact that the crossing involves state of differing configuration. These two factors clearly account for the very low fluorescence quantum yield. Following this very efficient inter-system crossing to the $^3n\ \pi^*$ state, further rapid deactivation will occur by internal conversion within the triplet manifold to the $^3\pi\ \pi^*$ state. At this point however, further radiationless deactivation to the ground state is likely to be inefficient on both multiplicity and configuration grounds, and this would explain the substantial quantym yields and relatively long lifetimes of the phosphorescence from the 2-hydroxy derivatives. Thus these compounds exist for long periods in a potentially photoactive triplet state which strongly enhances their probability of chemical reaction with their environment. If this reaction changes the chemical constitution of the dye in any way such that a change of colour is observed by the eye (the human eye is extremely sensitive to small changes in shade) dye fading will be the result.

ABSTRACT

Cellulose is particularly susceptible to the action of ultraviolet and ionizing radiations in the solid state. The high degree of order, inherent in the solid state system, is a prerequisite for the efficient energy transfer processes and radical chain reactions which are triggered by the radiation action. Mechanisms of radiation initiated processes have been elucidated and will be reviewed.

Dyes promote sensitized photodegradation of cellulose, particularly anthraquinonoid systems. The factors responsible for the behaviour of the sensitizer have been related to the spectroscopic characteristics of the dyes.

Based on the mechanisms elucidated, efficient procedures have been developed to protect dyed and undyed cellulosic systems to light and ionizing radiations.

Literature Cited

1. REINE, J. & ARTHUR, J.C. Jr., Text Res.J.,(1970) 40, 90.
2. BOS, J. & BUCHANAN, C. J. Polymer Sci., A1, (1973) 11 833.
3. BLOUIN, F.A. & ARTHUR, J.C. Jr., Textile Res.J.,(1963) 33, 727.
4. AHMED, N.V. BAUGH, P.J. & PHILLIPS, G. O. J.Chem.Soc. Perkin 11, (1972), 1305.
5. LOGROTH, G. Int.J.Rad,Phys.Chem. (1972) 4, 277.
6. GEJVALL, T. & LOFROTH, G. Acta Chem.Scand. (1973) 27, 1108.

7. SONNTAG, C.VON, NEUWALD, K. & DIZDAROGLU, M. Radiation Res., (1974), 58, 1.
8. SONNTAG, C.VON, & DIZDOROGLU, M. Z.Naturforsch, (1973) 286 367.
9. McKELLAR, J.F. Radiation Res.Rev., (1971), 3, 141.
10. ESKINS, E. BUCHER, S. & SLONEKER, T. Photochem.Photobiol. (1973), 18, 195.
11. ARTHUR, J.C. Jr., BLOUIN, F.A. & PHILLIPS, G.O. U.S. Patent 3, 519, 382 (1970).
12. PHILLIPS, G.O. & YOUNG, M. J.Chem.Soc., A, (1966), 383.
13. DIZDAROGLU, M. HENNEBERG, D. SCHOMBERG, G. & SONNTAG, C.VON, Z.Natursforsch, (1975), 306, 416.
14. HARTMAN, V. SONNTAG, C.VON, & SCHULTE-FROHLSLINDE, D. Z.Natursforsch, (1970), 256, 1394 + others.
15. PHILLIPS, G.O. & ARTHUR, J.C. Jr. Textile Res.J. (1964), 34 572-579.
16. BAUGH, P.J. & PHILLIPS, G.O. in Cellulose and Cellulose Derivatives, Wiley (1971), 5, Part 5, 1015-1047.
17. THOMAS, G. Ph.D. Thesis, University of Salford, (1972).
18. BAUGH, P.J. PHILLIPS, G.O. & WORTHINGTON, N.W. J.Soc.Dyers Colourists,(1969), 85, 241-245.
19. BAUGH, P.J., PHILLIPS, G.O. & WORTHINGTON, N.W. J.Soc.Dyers Colourists, (1970), 86 19-24.
20. BAUGH, P.J., PHILLIPS, G.O. & WORTHINGTON, N.W. Proceedings 1st Symposium International de la Recherche Textile Cotonniere (Institut Textile de France), Paris, (1970) P.767-779.
21. DAVIES, A.K., McKELLAR, J.F. & PHILLIPS, G.O. Proc.Roy.Soc. London, (1971), A323, 69-87.
22. BENTLEY, P., McKELLAR, J.F. & PHILLIPS, G.O. Review of Progress in Coloration and Related Topics. J.Soc.Dyers & Colourists, (1974) 5, 33-48.
23. DAVIES, A.K., FORD, R., GEE, G.A., McKELLAR, J.F. & PHILLIPS, G.O., J.Chem.Soc., Chem.Comms., (1972) 873-874.
24. DAVIES, A.K., GEE, G.A., McKELLAR & PHILLIPS, G.O., Chemistry and Industry, (1973), 431-432.
25. GEE, G.A., PHILLIPS, G.O. & RICHARD, J.T., J.Soc.Dyers and Colourists, (1973), 89, 285.
26. BRIGGS, P.J. & McKELLAR, J.F., Chem.& Ind. (1967), 622.

23

Grafting of Monomers to Cellulose Using UV and Gamma Radiation as Initiators

JOHN L. GARNETT

The University of New South Wales, Kensington, N.S.W. 2033, Australia

Grafting of monomers to cellulose initiated by gamma radiation has previously been extensively studied by a wide variety of groups throughout the world (1-8). By comparison, equivalent data from the analogous photosensitized process is less comprehensive (9-13). In the present manuscript further more detailed work for UV copolymerization to cellulose will be discussed. These results will be compared with corresponding data from grafting with ionizing radiation. In the latter system, the use of additives, particularly certain polycyclic aromatic hydrocarbons, has been shown to be valuable in the enhancement of grafting (4). More recently, in preliminary communications (14, 15), the presence of inorganic acids has been reported to increase grafting yields dramatically under certain experimental conditions in the gamma ray system. The present paper includes a comprehensive study of this acid catalyzed grafting enhancement with ionizing radiation. Finally the value of both UV and ionizing radiation techniques as a means for producing useful copolymers will be critically discussed.

Experimental Grafting Procedures

The grafting procedure used in the ionizing radiation work has been outlined previously (4); however for copolymerization in the presence of acid, a number of modifications to the earlier methods were necessary (14,15). In particular, because of the problems associated with the effect of trace impurities on the radiation chemistry of methanol (16), careful purification of this solvent was carried out before grafting as previously recommended (16). For runs which were performed under deoxygenated conditions (17), the monomer solutions were purged with nitrogen for 30 minutes, then stoppered under nitrogen.

For the UV work, the predominant number of runs utilized a Philips 90W high pressure mercury vapour lamp fitted with a quartz envelope. Samples to be grafted were strips (5 x 4 cm) of Whatman 41 filter paper, which were immersed in a solution (25 ml)

consisting of purified monomer, solvent and photosensitizer, all contained in test-tubes which were stoppered or sealed under vacuum after several freeze-thaw cycles. Lightly stoppered pyrex tubes were predominantly used for simplicity in these experiments since, although evacuated quartz tubes give higher grafting efficiencies (18), reasonable rates of copolymerization can still be obtained under the former conditions. Tubes were held in a rotating rack and were irradiated at a distance of 12 cm, unless otherwise stated, from the source. Actinometry was performed with the uranyl nitrate-oxalic acid system, although some calibrations were also carried out with potassium ferrioxalate. After irradiation was complete, the paper strips were extracted and treated in the same manner as described in the preceding ionizing radiation technique.

Grafting with Ionizing Radiation

In the simultaneous technique for grafting to cellulose using initiation by ionizing radiation, in particular gamma rays, the role of solvent is important (1,3,5,6,7,19) and has been reviewed (4). For copolymerization with most monomers the low molecular weight alcohols are generally the most useful solvents. Using styrene as representative monomer, typical copolymerization behaviour in the alcohols (4) shows that grafting virtually cuts out at *n*-butanol. The presence of an accelerated copolymerization or Trommsdorff effect is also found. This is only observed at certain monomer concentrations and is also radiation dose and dose-rate dependent.

Effect of Acid Additives on Grafting. Preliminary studies (14,15) have shown that the presence of certain mineral acids can lead to an appreciable increase in copolymerization especially when styrene is being radiation grafted to cellulose. More detailed studies outlined in this manuscript show that for the above grafting reaction, sulfuric acid is the most efficient mineral acid when cellulose is the trunk polymer (Table I). The enhancement predominates at monomer concentrations up to 40% when methanol is used as representative low molecular weight alcohol for styrene. Hydrochloric acid is also active although phase separation problems limit the versatility of this additive. Nitric acid is marginally satisfactory but only with 10% monomer solutions and at acidities no higher than 0.2M. At higher acidities nitric acid appears to attack and oxidise the cellulose before sufficient protective grafting can occur.

If the radiation dose and dose rate are kept constant at 200 krad and 27 krad/hour, respectively, the effect of molarity of sulfuric acid on styrene grafting in methanol can be evaluated (Table II). The relationship between acidity and graft is complicated. However, marginal enhancement in copolymerization is observed at least up to 60% monomer concentration containing

1.1×10^{-3}M H_2SO_4. The effect of sulfuric acid on the grafting reaction at different radiation doses and dose-rates is shown in Table III.

TABLE I. Effect of Mineral Acid on Radiation-Induced Grafting of Styrene in Methanol to Cellulose *a*

% Styrene (by volume)	Graft (%) in					
	No Acid	H_2SO_4 1M	HCl 2M	HNO_3 2M	HNO_3 0.2M	H_3PO_4 .67M
10	6.2	27.8	26.5	1.0	7.4	8.5
20	30.3	82.7	-*b*	7.5	29.1	25.8
30	41.3	33.7	18.3*b*	9.1	37.4	30.6
40	-	19.7	19.1*b*	11.2	30.6	28.7
50	38.6	-	-	-	-	-
60	-	25.5*b*	15.1*b*	20.6	34.8	30.4
80	56.1	20.6*b*	8.6*b*	8.0	28.5	31.5

a Dose rate = 2.64×10^4 rads/hr. Total dose = 0.20×10^6 rads.
b Phase separation observed.

TABLE II. Effect of Molarity of Sulfuric Acid on Radiation Grafting of Styrene in Methanol to Cellulose at Constant Dose and Dose-Rate *a*

Styrene (%) in Methanol	Graft (%) in Concentration of Sulfuric Acid (M)						
	0.0	1.1×10^{-3}	1.1×10^{-2}	5.4×10^{-2}	1.1×10^{-1}	5.4×10^{-1}	1.1
10	1.2	5.4	7.5	8.8	11.0	19.4	21.2
20	8.9	22.6	28.7	38.4	48.2	58.5	74.7
30	13.9	41.0	44.7	47.5	49.6	39.3	27.0
40	16.3	27.5	17.8	34.0	30.2	20.2	16.1
60	29.8	31.2	29.1	26.8	23.0	16.5	15.1
80	34.1	24.8	26.5	19.3	21.3	12.4	10.1

a Irradiation in evacuated vessels at a dose rate of 2.73×10^4 rads/hr to a total dose of 2×10^6 rads.

At representative low doses and dose-rates (25 krads, 18.9 krads/hour), grafting is enhanced at low acidities; however, as the dose and dose-rate are increased, copolymerization is accelerated at higher acidities. A detailed study of this acid effect (Table IV) at very low dose rates (4.8 krad/hour) and total doses (2.3 krads), but at high constant acidity (1.0M), shows that the magnitude of the grafting increase with H_2SO_4 is favoured at the lower monomer concentrations (20%). As the dose and dose-rates are increased at this higher acidity (Tables V and VI), the general pattern of the acid enhancement in grafting is complicated by the onset of the gel or Trommsdorff effect (4) which is both dose and dose-rate dependent. This Trommsdorff peak in grafting is superimposed on the simple acid enhancement and will be discussed separately in a later section of this paper.

TABLE III. Effect of Molarity of Sulfuric Acid on Radiation Grafting of Styrene in Methanol to Cellulose at Different Doses and Dose-Rates.[a]

Dose Rate (krad/hr)	18.9	18.9	47.1
Dose (krad)	25	50	50
Acidity (M)	Graft (%)		
0	0.4	9.3	4.6
10^{-2}	-	8.1	1.6
10^{-1}	7.2	11.1	-
5×10^{-1}	1.2	-	-
1.0	2.4	16.7	7.2

[a] Concentration of solutions 20% (v/v). Results in duplicate, irradiations in de-oxygenated solution under nitrogen.

TABLE IV. Grafting of Styrene in Methanol to Cellulose in the Presence of Sulfuric Acid at very Low Dose-Rates.[a]

Conditions		Graft (%) at							
Dose Rate (krad/hr)		4.8		4.8		9.1		9.1	
Total Dose (krad)		2.3		53		28		43	
Acid (M)		0	1.0	0	1.0	0	1.0	0	1.0
Concentration of Monomer (% v/v)	20	7.8	16.2	28.2	50.0	8.0	14.2	13.3	24.3
	22	7.0	11.9	27.6	42.2	8.0	15.8	12.8	27.9
	24	14.5	15.4	40.4	42.9	11.0	16.9	14.9	25.6
	26	10.3	15.8	40.6	38.3	10.8	15.9	14.5	11.4
	28	10.9	12.0	48.6	31.4	13.5	10.5	17.6	21.9
	30	12.1	14.8	48.2	27.5	12.3	16.1	18.2	20.6

[a] Results in quadruplicate, irradiation in de-oxygenated solution under nitrogen.

TABLE V. Grafting of Styrene in Methanol to Cellulose in the Presence of Sulfuric Acid at Low Dose Rates. *a*

Conditions		Graft (%) at									
Dose Rate (krad/hr)		18.7		18.7		39.6		39.6		64.4	
Total Dose (krad)		28.1		65.6		29.7		69.3		32.2	
Acid (M)		0	0.5	0	0.5	0	0.5	0	0.5	0	0.5
Conc. of	20	3.1	5.2	15.0	22.5	3.0	2.1	7.1	12.1	2.1	2.0
Monomer	25	5.5	5.6	20.2	25.7	1.7	5.0	9.3	15.0	1.7	1.3
(% v/v)	30	-	-	-	-	3.4	6.9	11.3	14.8	2.3	3.2

a Results in quadruplicate, irradiations in de-oxygenated solution under nitrogen.

TABLE VI. Grafting of Styrene in Methanol to Cellulose in the Presence of Sulfuric Acid at Medium Dose Rates. *a*

Conditions		Graft (%) at							
Dose Rate (krad/hr)		94.5		96		97		105	
Total Dose (krad)		70		100		30		100	
Acid (M)		0	0.1	0	1.0	0	0.1	0	0.1
Concentration of Monomer (% v/v)	10	0.3	1.5	0.7	5.1	0	0.2	1.1	2.1
	15	1.3	2.3	2.6	14.3	1.3	0.8	3.1	4.4
	20	2.7	4.3	5.9	17.5	3.4	4.7	4.7	7.3
	25	4.3	8.4	8.8	18.3	5.5	7.4	6.9	10.1
	30	5.5	9.5	10.4	18.6	7.0	9.4	8.6	12.2
	35	7.0	10.0	11.7	17.2	9.2	10.7	10.0	13.5
	40	8.7	11.7	12.7	16.5	9.9	9.2	11.0	13.3

a Results in quadruplicate, irradiations in de-oxygenated solution under nitrogen.

Grafting in this system at low dose-rate and low total doses is important for two reasons. Firstly, for many cellulose copolymer applications, grafts of 10-20% are usually sufficient, thus low radiation doses are generally needed for this purpose, especially if an additive will accentuate the copolymer yield. Secondly, at low dose-rates, oxygen effects become particularly significant. Previous studies (17) have shown that very bad scatter in radiation grafting occurs at low dose-rates (< 3000 rads/hour). For this reason, a study of the effect of oxygen on the acid enhancement in grafting was essential (Tables

VII and VIII). It is also the reason why all low dose-rate work reported in this paper in earlier tables has been under de-oxygenated conditions (14,15,20).

TABLE VII. Effect of Air on Grafting of Styrene in Methanol to Cellulose in the Presence of Acid at High Dose Rates

Conditions	Graft (%) at							
Dose Rate (krad/hr)	19.5[a]		405[a]				501	
Total Dose (krad)	200		200				70	
	Vacuum[b]		Air		Vacuum[b]		Vacuum[b]	
Acid (M)	0	1	0	1	0	1	0	1
	43.0	76.8	36.9	41.1	25.8	44.5	6.9	8.0

[a] 30% solution of monomer, remainder 50%. Results in quadruplicate. For runs in air, tubes were stoppered under atmospheric conditions.

[b] Evacuated vessels (10^{-3}Torr) after three freeze thaw cycles.

TABLE VIII. Effect of Air on Grafting of Styrene in Methanol to Cellulose in the Presence of Sulfuric Acid at Medium Dose Rates.[a]

% Styrene	% Graft			
	No Acid		0.5M H_2SO_4	
	Vacuum	Air	Vacuum	Air
10	10.1	4.3	25.5	11.3
20	20.4	10.1	56.6	18.4
30	32.0	17.9	45.4	22.4
40	35.8	23.4	35.0	20.4

[a] Dose rate 4×10^4 rads/hr to 0.2×10^6 rads.

The data in Table VII show that at the high dose rate (405 krad/hour) there is acid enhancement in the vacuum irradiated samples particularly at 200 krads total dose, whereas the acid effect is marginal for the irradiation in air under the high dose rate conditions used. At medium dose rates the effect of acid in air irradiations is shown in Table VIII where the enhancement in graft at a dose rate of 40,000 rads/hour at 0.2 megarads is significant (and reproducible); however, grafting in the corresponding vacuum irradiated samples is higher.

Effect of Acid on the Trommsdorff Peak in Grafting. One particular property which is of both fundamental and preparative significance in a radiation copolymerization system is the occurrence of a gel or Trommsdorff effect (4) In this respect the addition of mineral acid can affect a radiation grafting system in two ways. Thus acid can (a) induce a Trommsdorff peak in a system (Tables I and II) or (b) enhance the magnitude of the effect if already present (Table IX). The value of acid enhancement is obvious if samples of cellulose copolymers are desired, since the total radiation dose required to give a particular graft is much lower in the presence of acid and thus possible radiation-induced decomposition of the trunk polymer is minimised (4,21).

TABLE IX. Enhancement in Trommsdorff Effect with Mineral Acid in Radiation-induced Grafting of Styrene in Methanol in De-oxygenated Solutions.[a]

% Styrene (by Volume)	Graft (%) in Concentration (M) of Sulfuric Acid				
	0.0	$0.9x10^{-4}$	$0.9x10^{-3}$	$0.9x10^{-2}$	$0.7x10^{-1}$
15	35.1	38.4	31.5	50.8	78.7
20	61.9	69.3	65.6	87.2	109.8
25	99.3	112.5	110.8	123.3	162.7
30	104.5	113.9	98.8	109.6	v.high
35	100.9	102.8	80.0	90.7	124.0
45	80.5	52.6	42.0	48.4	82.5

[a] Dose rate = 6.77 x 10 rads/hr. Time of irradiation, 17 hr. Results in quadruplicate, irradiation in de-oxygenated solution under nitrogen.

The data in Table IX are for irradiations in nitrogen. Similar Trommsdorff effects are observed for irradiations in air (Table X). Comparison of these data with the air irradiations in Table VIII illustrate two significant features of gamma ray grafting to cellulose. Firstly, the irradiations in air without acid, in both Tables, demonstrate the very good reproducibility of the procedure for the 10-40% monomer concentration range. The inclusion of 0.5M acid (Table VIII) gives a different type of grafting behaviour to the inclusion of 0.1M acid (Table X), again emphasising the role of acidity in these reactions.

Under the present radiation grafting conditions, cellulose itself contains approximately 6% residual moisture (23°C, 65% RH). Addition of further water, without acid being present, to a solution where the Trommsdorff peak is operative leads to a progressive decrease in graft (15). Addition of acid to these solutions increases the graft for a specific water content up to 3% v/v water. By contrast with acid enhancement in grafting, inclusion of alkali (15) leads to a progressive decrease in copolymerization.

TABLE X. Effect of Mineral Acid on Trommsdorff Effect in Radiation-induced Grafting of Styrene in Methanol to Cellulose in Air.[a]

Styrene (% v/v)		10	20	30	40	50	60	80
Graft (%)	No Acid	3.2	8	16	22	25	28	34
	0.1 M H_2SO_4	4.2	24	33	31	27	25	16

[a] Dose-rate of 4 x 10^4 rads/hr to 0.2 x 10^6 in air.

Acid Effects in Mixed Solvents. The role of low molecular weight alcohols is particularly important in radiation grafting (4). Thus, for the styrene system, copolymerization virtually cuts out at *n*-butanol and no graft is achieved with the longer chain alcohols (4). If, however, a longer chain alcohol, such as *n*-octanol, is added to methanol (1:1), there is a marked enhancement in grafting to cellulose, particularly for monomer concentrations of from 20-40% (Table XI). Addition of mineral acid to such a mixed solvent system increases the magnitude of the graft even further. This observation is important both mechanistically and for preparative work. Thus, in the synthesis

TABLE XI. Gamma Ray Induced Grafting of Styrene Using Methanol and Methanol/Octanol in a Fixed Ratio (1:1) as Solvents with and without Acid. [a]

Styrene (% v/v) in Solvent	Graft (%) in		
	Methanol	Methanol/Octanol (1:1)	
		No Acid	0.1M H_2SO_4
10	3.2	3.4	5.0
20	7.7	11	20
30	17	23	30
40	22	30	27
50	25	33	23
60	28	33	19
80	34	34	10

[a] Solutions of styrene in solvent irradiated at 4 x 10^4 rads/hr to 0.2 x 10^6 rads total dose in air.

of grafted celluloses, the lower the radiation dose required to give a particular degree of copolymerization the better, since cellulose itself is known from ESR studies (10,22,23) to be degraded by even relatively small doses of ionizing radiation.

Grafting of Monomers Other than Styrene - The Vinylpyridines. Extensive data from the radiation copolymerization of monomers other than styrene to cellulose has been reported (4), particularly the role of solvent in such reactions (4). In this respect the vinylpyridines are unique and preliminary results (24) of solvent phenomena with these monomers suggest the participation of monomer-solvent complexes in the grafting reaction. The vinylpyridines normally require relatively high doses of radiation for significant grafting, and thus any solvents which yield enhanced copolymerization with these monomers would be valuable for preparative purposes.

For convenience, in this vinylpyridine work, a number of arbitrary classifications which assist subsequent interpretation will be proposed (24). Thus a promoting solvent is one which dissolves the monomer and facilitates grafting when cellulose and solution are mutually irradiated. Where the promoter is not a solvent and both monomer and promoter are dissolved in a third compound which is not, itself, a promoting solvent, this latter compound is called a co-solvent. Butanol and dioxane are used for this purpose.

TABLE XII. Comparison of Cellulose Graft from Three Vinylpyridine Monomers with Eleven Selected Solvents.[a]

Solvent	Graft (%) for Monomer[b]		
	2-Vinylpyridine	2-methyl, 5-Vinylpyridine	4-Vinyl-pyridine
2,2,2-Trichloroethanol	0	0	1.1*
2,2,2-Trifluoroethanol	0.5	3.6	5.3*
2-Methoxyethanol	27.8	33.0	28.0*
Ethanol	26.0	60.1	24.8*
n-Propanol	0	0.8	2.9*
3-Picoline	35.1	1.4	3.5
Pyrrolidine	10.1	8.6	10.9
1,2-Diaminoethane	38.0	18.1	26.2
Tri-*n*-butylamine	16.4	0.2	2.7
Dimethylformamide	32.8	93.3	19.4*
Dimethylacetamide	43.7	54.9	16.1*

[a] All solutions were 30% (v/v) in monomer.

[b] Radiation dose was 5 x 10^6 rads at 0.1 x 10^6 rads/hr except for those marked with an asterisk, which were 2 x 10^6 rads at the same dose-rate.

Three vinylpyridine monomers are useful in radiation grafting. These are 2-vinylpyridine, 2-methyl, 5-vinylpyridine and 4-vinylpyridine. From the data in Table XII, the radiation grafting behaviour of all three monomers is similar in the representative series of eleven solvents examined. Thus, the best grafting solvent for one monomer will generally be the best for all three monomers, although the magnitude of the graft may vary. The one exception in the eleven solvents appears to be 3-picoline, which is very good only for 2-vinylpyridine.

Because of the uniform solvent grafting property of the three vinylpyridine monomers, one monomer can be used to study the representative behaviour of all three in grafting reactions. Surface area of cellulose is also known not to influence vinylpyridine grafting (24). With 2-vinylpyridine, the low molecular weight alcohols, glycols and methylcellosolve, are the best of the simple hydroxy compounds for radiation grafting (Table XIII). With the same monomer, in amine type solvents (Table XIII), the mono-, di- and tri-butylamines are all effective as is allylamine although the latter gives a smaller graft than *n*-butylamine. Diamines, like diols, are effective promoters, the effectiveness tending to decrease with chain length. Progressive methyl substitution on the amine also significantly reduces grafting progressively. Of the aromatic amines studied, only benzylamine gave any graft. Pyridine and 3-picoline are the best heterocyclic solvents for grafting with 2-vinylpyridine, the remaining picoline isomers being much less effective. Hydrogenation of pyridine to piperidine results in a marked reduction in graft. Dihydropyridine also leads to lower grafting, thus pyridine, in contrast to benzene, depends markedly on its aromatic nature for its graft promoting properties. Fusion of a benzene ring to pyridine as in quinoline and quinaldine completely inhibits grafting. Of the remaining miscellaneous nitrogen-containing compounds used as promoting solvents for grafting 2-vinylpyridine, dimethylacetamide, dimethylsulfoxide, dimethylformamide and formamide are the most effective.

When both NH_2 and OH functional groups are incorporated into the one solvent molecule in the correct positions, they act in a synergistic fashion, giving enhanced grafting with 2-vinylpyridine (Table XIV). Of the solvents examined with NH_2 and OH incorporated, 2-aminoethanol, N-methyl-2-aminoethanol and 3-aminopropanol were the most effective; however, the most reactive of all was 3-aminopropane-1,2-diol, although being a solid at room temperature this compound could only be used in solution with co-solvents such as *n*-butanol. If the OH and NH_2 were substituted on the same carbon atom in the molecule, the solvent was poor for grafting. Similarly, dimethylation of 3-aminopropanol completely eliminated the copolymerization properties of the parent solvent.

Effect of pH of Solvent on Grafting Vinylpyridines. All three vinylpyridine monomers are promoted by alcohols with pH

TABLE XIII. Promoting Solvents for Grafting 2-Vinylpyridine.[a,b,c]

Alcohols, Heterocyclics	Graft (%)	Amines, Derivatives	Graft (%)
Methylcellosolve	27.8	Ethylenediamine	38.0
Ethanol	27.0	Dibutylamine	31.0
Methanol	27.0	Tributylamine	16.4
Glycol	20.0	1,4-Diaminobutane	14.2
Propan-1,2-diol	20.0	*n*-Butylamine	12.2
Digol	13.5	N,N-Dimethyldiaminoethane	11.8
Ethylcellosolve	9.0	1,3-Diaminopropane	11.0
3-Picoline	35.1	Dimethylacetamide	43.7
Pyridine	35.0	Dimethylsulfoxide	35.5
Pyrrolidine	10.0	Dimethylformamide	32.8
		Formamide	25.5

[a] Concentration of monomer in solvent, 30% (v/v) Radiation dose 5 x 10^6 rads at 0.1 x 10^6 rads/hr.

[b] The following compounds were poor promoters (< 9% graft): Trimethyleneglycol, 2-chloroethanol, allyl alcohol, furfural, *t*-butanol, trigol, 2,2,2-trifluoroethanol, N-methyldiaminoethane, N,N-dimethyldiaminomethane, allylamine, triethylenetetramine, benzylamine, N,N,N'-trimethyldiaminoethane, N,N,N',N'-tetramethyldiaminoethane, 2-picoline, piperidine, pyrrol, 4-picoline, acetonitrile, nitrobenzene.

[c] The following compounds were non-promoters: Propan-1-ol, propan-2-ol, 2-methylpropan-1-ol, 1-methylpropan-1-ol, 2,2,2-trichloroethanol, 2-butoxyethanol, cyclohexanol phenol, phenylmethanol, 2-phenylethanol, 1,2-butanediol, 2,3-butanediol, 2,6-lutidine, 2,4,6-collidine, quinoline, 2-methylquinoline, 2-aminopyridine, chloropyridine, 4-hydroxypyridine, aniline, N,N-dimethylaniline, phenylhydrazine, nitromethane, cyclohexane, benzene, chloroform, carbon tetrachloride, trichlorethylene, chlorobenzene, thiophen, tri-n-butylphosphine, acetone, 2-pentanone, 3-pentanone, acetophenone, 2,3-butandione, acetylacetone, isophorone, propanal, furfural, benzaldehyde, tetrahydrofuran, acetic acid, methyl acetate, 1,4-dioxan, epichlorhydrin and hexamethylphosphoric triamide.

values from 12.4 to 16.0 (Table XV), the range of promotion generally being 4VP > MVP > 2VP. Water with a pH of 15.5 acts as a strong promoter when used with a co-solvent (24). The unsaturated derivative, allyl alcohol, with a pH of 15.5 yields poor grafting. For the methyl pyridines, the pH of promoting solvents is lowered to 5.2-5.9 (Table XVI). In addition to pH and interdependence with it are swelling and homopolymer formation (Table XVII). Pyridine, aminoethanol, methanol and 3-picoline are the most attractive grafting solvents from these data. Some solvents, *e.g.* 2-aminoethanol, pyrrol, pyrrolidine and allylamine caused almost complete disintegration of the paper when added to the paper without monomer. However, in the presence of monomer, swelling was reduced to an extent where the experiment could be satisfactorily carried out. Finally the data in Table XVII show that there is no direct relationship between homopolymerization and graft for the vinylpyridines.

TABLE XIV. Grafting Yields of Derivatives of 2-Aminoethanol Used as Solvents for 2-Vinylpyridine.[a]

Compound	Graft %	Compound	Graft %
2-Aminoethanol	55.3	2-Hydroxy-1-aminopropane	19.8
N-Methyl-2-aminoethanol	67.6	2-Aminopropan-1-ol	14.9
N-t-Butyl-2-aminoethanol	0[b]	2-Amino-2-methylpropan-1-ol	2.9
3-Aminopropanol	53.1	2-Aminobutan-1-ol	10.1
N,N-Dimethyl-3-aminopropanol	0	2-Methoxy-1-aminoethane	3.8
3-Aminopropan-1:2-diol	high (15.0)[b]		

[a] Concentration of monomer in solvent - 30% (v/v). Radiation dose of 5 x 10^6 rads at 0.1 x 10^6 rads/hr.

[b] 10% solution in *n*-butanol.

TABLE XV. Effect of pK Alcohols as Solvents on the Grafting of the Three Vinylpyridines.[a]

R[b]	pK[c]	Graft (%)		
		2VP	MVP	4VP
CCl_3	12.24	nil	nil	1.1
CF_3	12.37	0.5	3.6	5.3
CH_2Cl	14.31	3.6	-	-
CH_3OCH_2	14.8	27.8	33.0	28.0
$HOCH_2$	15.1	20	-	-
H	15.5	26.0	60.1	24.8
$CH_2=CH$	15.5	2.1	-	-
CH_3	15.9	27.0	20.6	8.1
$CH_3.CH_2$		nil	0.8	2.9

[a] All solutions contained 30% monomer (v/v). Radiation dose was 5 x 10^6 rads for 2VP and MVP but 2 x 10^6 rads for 4VP at a dose-rate of 0.1 x 10^6 rads/hr.

[b] R is the substituent in the molecule RCH_2OH.

[c] See ref. (24).

TABLE XVI. Effect of pK of Methylpyridines as Solvents in Grafting of 2VP [a,b]

Compound	pK	Graft (%)
Pyridine	5.2	35.0
2-Picoline	5.9	7.8
3-Picoline	5.7	35.1
4-Picoline	6.05	1.4

[a] Concentration of monomer in solvent, 30% (v/v). Radiation dose of 5 x 10^6 rads at 0.1 x 10^6 rads/hr.

[b] The following lutidines with pK > 6.4 gave no graft:- 2,4-lutidine, 3,4-lutidine, 3,5-lutidine, 2,6-lutidine and 2,4,6-collidine.

TABLE XVII. Comparison of Selected Solvents for Grafting 2VP in Terms of Swelling of Paper, Homopolymer Formed and Graft Obtained on Cellulose.[a]

Solvent	Swelling[b]	Homopolymer[b]	Graft (%)
Methanol	+	+	27
Butanol	-	-	0
Ethyleneglycol	+	+++	20
Allylalcohol	+	+++	2.1
t-Butanol	+	+++	1.6
2-Phenylethanol	-	++	2.0
Benzylalcohol	-	++	0
Diacetal	-	+++	0
Pyrrol	++	++	6.4
Pyrrolidine	+++	+++	10.0
Pyridine	-	+	35.0
Ethylene diamine	++	-	38.0
Allylamine	+++	-	6.5
Butylamine	++	-	12.2
Aniline	-	+	0
Aminoethanol	+++	-	55.3
Formamide	+	+++	25.5
Cyclohexane	-	+++	0
Heptane	-	+++	0
3-Picoline	-	-	35.1
2-Picoline	-	+	7.8
Carbon Tetrachloride	-	+++	0
Dioxan	-	++	0

[a] Monomer concentration in solution, 30% (v/v) Total radiation dose 5 x 10^6 rads at 0.1 x 10^6 rads/hr.

[b] Response assessed at nil (-) to very strong (+++).

Mechanism of Grafting with Ionizing Radiation

The precise nature of the role of solvent in radiation grafting using the simultaneous technique has already been reviewed (4) and is determined by a number of interdependent factors. Thus it should be a good solvent for both monomer and homopolymer but solvency alone is inadequate to describe solvent activity in grafting. Many good solvents (*e.g.* benzene) are ineffective. Physical properties such as wetting, swelling and dielectric constant should also be considered in any complete solvent discussion. In terms of wetting, a solvent must wet the surface of the polymer chain to which grafting is occurring. Hence, although styrene can be grafted to cellulose in non-wetting solvents such as hexane (4), high radiation doses are required in this hydrocarbon. Wetting solvents require much lower doses for comparable grafting. With respect to swelling, ability of solvent and monomer to penetrate the individual fibres of the cellulose is important. Stamm and Tarkow (25) showed that primary penetration of solvent into the inner layers of the cellulose fibre can be related to the molecular volume and dielectric constant of the solvent. Thus water, with a low molecular volume and a dielectric constant over 15, penetrates deeply and quickly. Methanol and formamide are good penetrating solvents. Once having penetrated and swollen the fibrous structure of the cellulose, molecules of greater molecular volume, *i.e.* monomers, can penetrate the swollen structure and graft.

Acid Enhancement Effect on Monomer and Solvent. Physical parameters, alone, do not satisfactorily explain the large enhancement in graft observed when acid is used as additive, especially at low concentrations (Tables II, IX). Mineral acids, particularly sulfuric, are significantly better than organic acids such as acetic. The possibility that the favourable effect of acid is due exclusively to improved accessibility of styrene to the cellulose structure is also not tenable since hydroxide ion also possesses the same property and alkali retards grafting (15). These acid and alkali results suggest that the acid enhancement is chemical in nature. Swelling and accessibility of acid would probably make a small contribution to the increased graft, since acid is known to promote uncoiling of cellulose chains during hydrolysis.

The predominant acid enhancement effect observed in the present work appears therefore to be due essentially to radiation chemistry phenomena. In terms of this model the acid additive interferes with the precursors of the radicals produced from radiolysis in the grafting system. Acid additives should thus advantageously affect both the propagation rate of styrene polymerization and also the number and nature of the sites in the cellulose available for grafting.

If the radiation grafting of styrene in methanol to cellulose is used as representative system, the radiolysis of all three components in the presence of acid needs to be considered. In the radiolysis of solvent methanol, the yield of the major product, hydrogen ($G(H_2)$), is considerably increased under certain conditions by mineral acid (16,26,27). Preliminary studies attributed the enhancement in $G(H_2)$ to an increase in G(H). However, since the radiolysis of methanol leads to the formation of ions (28), free radicals and excited states, all three pathways are considered to contribute to $G(H_2)$. In the basic radiation chemistry of methanol, CH_3OH^+ and $CH_3OH_2^+$ are the protonated species formed. Neutralization of $CH_3OH_2^+$ by electron capture (29) yields hydrogen atoms and excited methanol molecules which decompose to radicals and molecular products. Solvated electrons are also produced in methanol. Sulfuric acid also affects the yields of radiolysis products from methanol (11,26,27,30), these results being interpreted in terms of the affinity of hydrogen ions for the solvated electron. The $\Delta G(H_2)$ found with sulfuric acid addition during methanol radiolysis is consistent with electron scavenging by hydrogen ions as in Equation 1, thus enhancing G(H) and hence $G(H_2)$ yields. The concentration of $CH_3OH_2^+$is higher in

$$CH_3OH_2^+ + e \rightarrow CH_3OH + H \qquad (1)$$

the presence of acid; hence yields of excited states, radicals and molecular products are also increased by acid.

Thus, with the addition of acid to the methanol-styrene radiation grafting solution, there is an increase in the concentration of excited states, radicals and ions, particularly hydrogen atoms. Styrene, being a strong radical scavenger, readily reacts with the above species, particularly the hydrogen atoms, leading to both higher polymerization rates and grafting. Increased H atom yields also explain the observed acid effects associated with the Trommsdorff effect in grafting.

Acid Enhancement Effect on Cellulose. The presence of acid can further increase grafting efficiencies in the present system by influencing the concentration of radicals and grafting sites formed in the trunk polymer, cellulose. It has previously been suggested (4,14) that radiolytically produced hydrogen atoms are important in the mechanism of the present grafting system since cellulose macroradicals can be formed by hydrogen abstraction. Inclusion of mineral acid in the grafting solution leads to the formation of species such as (I). Capture of secondary electrons by species (I) leads to more H atoms and excited molecules (II), which then decompose to yield macroradicals (III) for further grafting and also additional H atoms. This is one mechanism for the formation of alkoxy radical sites (III) for copolymerization, whereas alkyl radicals (IV) are usually obtained preferentially

$$\gt C\lt^{H}_{OH_2^+} \;;\quad \left[\gt C\lt^{H}_{OH}\right]^* \;;\quad \gt C\lt^{H}_{O^\bullet} \;;\quad \gt C^\bullet\!\!-OH$$

(I) (II) (III) (IV)

by energetic hydrogen abstraction reactions. Electron spin resonance studies (10,22,23) have been used to separate such processes. Hence, it can be seen how inclusion of acid in a grafting solution can increase the number of macroradicals in the cellulose and thus enhance the grafting.

Effect of Air on the Acid Enhancement. When acid is excluded from the grafting reaction, the data show that copolymerization in air, at 405 krad/hour and 200 krads dose, is slightly better than grafting in vacuum, thus confirming the earlier work of Dilli and Garnett (19) for dose rates above 70 krad/hour. Grafting to cellulose in air at dose-rates lower than this value is restricted by poor reproducibility especially at dose-rates $<$ 20 krad/hour (17). In previous preliminary studies (14,15,20), a marginal increase in graft was observed for air irradiations in the presence of acid at the relatively high dose rate of 405 krad/hour (20). In this earlier work (15,20), some runs in the low dose rate region of 19.5 krad/hour suggested that acid might retard grafting only in the presence of air. However, because of severe oxygen scavenging and poor reproducibility, insufficient runs were carried out to permit accurate statistical analysis as Garnett and Martin (17) had previously done when acid was absent. The earlier acid results at 19.5 krad/hour in the presence of air were thus inconclusive. In the corresponding runs in vacuum, large increases in copolymerization are found with the inclusion of acid. Mechanistically, these data suggest that oxygen scavenges the precursor to the grafting and that, at low dose rates, this scavenging is more efficient than the copolymerization. If the dose rate is increased to 40 krad/hour, then oxygen scavenging effects are minimized and an increased graft with inclusion of acid is observed even with air irradiations. The magnitude of the acid enhancement in air, however, remains less than the corresponding increase for vacuum irradiations when acid is included (Table VIII).

Specific Grafting Mechanism in Acid. For the radiation grafting of styrene in methanol to cellulose in the absence of acid, Dilli and Garnett (4,31) developed a charge-transfer theory for the copolymerization which was applicable also to the grafting of a wide range of monomers to other trunk polymers. The present acid effects observed in grafting are consistent with the charge-transfer theory. The basic principle of the theory is that radiation-induced trapped radicals are available for bonding in the trunk polymer. Charge-transfer adsorption of monomer or growing polymer to the trunk polymer, as in Equation 2,

$$2\dot{P} + C_6H_5\text{-}CH{=}CH_2 \longrightarrow \underset{\downarrow \dot{P}}{C_6H_5}\text{-}\underset{\downarrow \dot{P}}{CH{=}}CH_2 \qquad (2)$$

facilitates subsequent grafting. With styrene and irradiated cellulose as model, the complex in Equation 2 is formed, showing the delocalized π-bonding between styrene and the free valencies of the irradiated cellulose. From this intermediate charge-transfer complex, a number of specific grafting mechanisms can be developed involving σ-bonded species. These mechanisms have already been discussed in detail elsewhere (4,15) and are applicable in the present system. Further, from recent fundamental studies of the pulse radiolysis of styrene and styrene solutions (32), it appears that under certain radiolysis conditions, hydrogen atoms can add with equal propability to either side-chain or ring of styrene to give species such as (V) and (VI). Thus styrene in grafting reactions can be copolymerized via intermediates, involving π-olefin complexing through the side-chain with

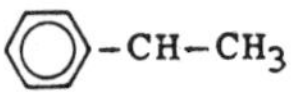

C_6H_5–CH–CH_3 (V) H–C_6H_5–CH=CH_2 (VI)

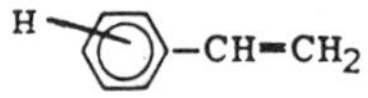

multiple cross-linking through the radical sites on the aromatic ring.

The present charge-transfer theory is also satisfactory for interpreting acid effects observed in the current work. Thus an increase in G(H) yield in the presence of acid leads to increased grafting sites in cellulose by hydrogen atom abstraction processes. Increased hydrogen atom yields with acid are also relevant to monomer reactivity, since, under certain radiolysis conditions, hydrogen atoms can add with equal probability to either side-chain or ring of styrene molecules (32). Finally, the acid dependency of the Trommsdorff effect is also consistent with the Dilli and Garnett (4,31) mechanism. Thus the increase in grafting yields at the Trommsdorff peak, as the acid concentration is increased to ≃1M, parallels Sherman's data (26) for the effect of acidity on $G(H_2)$ in the radiolysis of methanol. Since the position of the Trommsdorff peak depends predominantly on dose and dose-rate effects, the significance of the present acid dependency of the dose and dose-rate effect is obvious. The possibility that mechanisms other than the present radical processes, *e.g.* energy transfer and ionic processes, are significant in the acid enhanced grafting work should also be considered; however, present data (30) indicate that effects due to these competing processes are small.

Grafting of Polar Monomers - the Vinylpyridines. The acid enhancement effects observed in the preceding sections in the styrene system again draw attention to the possibility of ionic contributions to the general grafting mechanism, especially since ions are known to be formed in a gamma radiation process. The unique properties observed in radiation copolymerization with the vinylpyridines necessitate modifications to the grafting mechanism already proposed (4,15). Unfortunately, it is difficult to discover whether acid enhancement effects, analogous to those found with styrene, also occur in vinylpyridine grafting since the heterocyclic nitrogen of the vinylpyridine is protonated by acid and the resulting salt is usually insolubilized in most common copolymerization solvents.

Grafting in the vinylpyridines reflects the significance of polar effects in these reactions since the most desirable solvents for these reactions contain oxygen or nitrogen, the best actually containing both types of atoms as in the alkanolamines (Table XIV). A further common factor generally present in solvents which promote vinylpyridine grafting is the donor/acceptor property relative to protons. The affect of the donor/acceptor property is modified by other structural properties of the solvent. Thus, size of solvent molecule is important; a progressive increase in molecular weight of a promoting species, *e.g.* alcohols, results in a decrease and ultimate elimination of promoting properties.

Specific Grafting Mechanism for Vinylpyridines. Radical processes as previously discussed for the radiation grafting of styrene are also relevant to the grafting of the vinylpyridines. The retarding effect of conventional free radical scavengers such as ketones and aldehydes support this conclusion. However, other experimental evidence presented here suggests that polar intermediates are also important in the copolymerization, thus ionic species would appear to be significant and may even provide the predominant grafting pathway. For copolymerization reactions it is difficult to unequivocally demonstrate the role of ions. However, polar solvent effects found such as pH dependency on graft and also the donor/acceptor property exhibited by many solvents for vinylpyridine grafting, support such a theory (24).

In the vinylpyridine system, it thus appears that an association between solvent and monomer is advantageous to grafting (24). Suitable solvents also possess the necessary physical properties already described, such as satisfactory wetting and swelling of the trunk polymer. Under these conditions, it is envisaged that the role of the solvent-monomer complex in the grafting is to transport monomer to the cellulose (either surface or bulk), thus lowering the activation energy for the attack of monomer on cellulose due to the compatibility of the functional groups in both solvent and trunk polymer. With this model, the additional important role of the solvent is to accept a proton from the cellulose, thus facilitating ionization of the hydroxyl group of the trunk polymer, which then becomes available for bonding to the monomer via the vinyl group as in Equations 3 through 7. Such a scheme not only fulfils the role of solvent as a proton acceptor, but also provides an anion ($cello^-$) with which the vinyl group of the monomer can react. This is analogous to the proposal by Goutiere and Gole (33) for grafting to carbanions. In the proposed mechanism, the intermediate suggested in Equation 5

$$\text{celloH} + \text{S} \rightarrow \text{cello}^- + \text{SH}^+ \quad (3)$$

$$\text{cello}^- + 2\text{VP} \rightarrow \text{cello-CH}_2\text{-}\bar{\text{C}}\text{H-Py} \quad (4)$$

$$\text{SH}^+ + \text{cello-CH}_2\text{-}\bar{\text{C}}\text{HPy} \rightarrow \text{cello-CH}_2\text{-}\underset{\text{SH}^+}{\underset{|}{\bar{\text{C}}\text{HPy}}} \rightarrow \text{S} + \text{cello-CH}_2\text{-CH}_2\text{Py} \quad (5)$$

$$\text{or SH}^+ + e^-_{\text{solv}} \rightarrow \text{S} + \text{H} \quad (6)$$

$$\text{H} + \text{H} \rightarrow \text{H}_2 \quad (7)$$

may be important since the ability of the SH^+ species, *i.e.* the protonated solvent molecule, to associate with the pyridyl moiety of the anion may be essential for high grafting efficiency with most solvents. Analogous to this intermediate (Equation 5) is the species formed during initial association between solvent and monomer which may well be molecular association similar to that referred to by Plesch (34) as a co-catalyst in cationic polymerization. Thus, in the radiation grafting of the vinylpyridines, ionic processes as well as free radical intermediates may be important mechanistically.

Grafting with UV

The simultaneous grafting technique using UV initiation is directly analogous to the gamma ray process. Similar types of solvents and monomers can be used in both systems; however, sensitizers are required for the UV process to achieve maximum grafting efficiency within a reasonable time of irradiation. By contrast with the ionizing radiation system, little detailed work has been published for the photosensitized grafting process (9-11). Oster and Yang (35) have reviewed the types of photosensitizers used in simple photopolymerization. No systematic study of the effect of solvent structure on the UV grafting reaction has previously been reported. These data are needed for a direct comparison of the UV process with the gamma ray system.

Alcohols as Solvents for Styrene Grafting. A model system incorporating a satisfactory sensitizer is necessary before a critical evaluation of the alcohols as solvents in this copolymerization can be achieved. Uranyl nitrate, manganese pentacarbonyl, Michler's ketone, the disodium salt of anthraquinone-2,6-disulfonic acid, biacetyl and benzoin ethyl ether have all previously been used in simple photopolymerization reactions. However, the last three only have been utilized as photosensitizers in UV grafting (12,13,18). Of the group, uranyl nitrate, biacetyl, then benzoin ethyl ether are the most efficient for UV copolymerization of styrene in methanol to cellulose (12,18). Grafting in quartz is more effective than in pyrex; however the level of copolymerization in pyrex is still satisfactory for general use (12,13,18).

If styrene in methanol is grafted for short irradiation times (3 hours), the yield of copolymer gradually builds up to a maximum at 90% monomer concentration (12,13,18). With the other alcohols (Table XVIII), there is a progressive decrease in grafting yield with increasing alcohol chain length, copolymerization virtually cutting out with *n*-butanol. A dramatic increase in

TABLE XVIII. Photosensitized Grafting of Styrene to Cellulose Using the Simple Alcohols as Solvents, also Methanol/Isobutanol and Methanol/Octanol in Fixed Ratio (1:1).[a]

Monomer Conc. (% v/v)	20	40	60	80	90
Alcohol	% Graft				
Methanol	13	28	34	53	64
Ethanol	9	17	22	50	70
n-Propanol	5	12	18	30	20
iso-Propanol	5	7	5	14	8
n-Butanol	5	5	5	5	5
iso-Butanol	5	5	5	5	5
t-Butanol	5	5	5	5	5
n-Octanol	5	5	5	5	5
Methanol/*iso*-Butanol (1:1)	9	26	45	67	
Methanol/*n*-Octanol (1:1)	35	49	60	78	

[a] Solutions contained 1% w/v of uranyl nitrate and irradiated for 24 hr at 24 cm from 90W high pressure UV lamp.

TABLE XIX. Photosensitized Grafting of Styrene to Cellulose Using Methanol/Octanol in Different Ratios as Mixed Solvents.[a]

% Octanol in Solvent	0	10	25	33	40	50	55	60	66	75	80	85	90	100
Styrene (%)	% Graft													
80	53	58	61	-	68	76	80	82	90	78	32	16	5	5
60	34	38	48	-	-	54	60	75	89	51	-	-	-	4
40	28	-	30	33	29	48	41	50	50	32	-	-	-	3

[a] Experimental conditions as in Table XVIII.

copolymerization efficiency is achieved if methanol, an active solvent, is mixed (1:1) with a poor grafting solvent such as *iso*-butanol or *n*-octanol. If the grafting is optimised in the mixed solvent (Table XIX), a Trommsdorff effect is observed at 66% of *n*-octanol in methanol, the position of this peak being virtually independent of the monomer concentration for the range of styrene in methanol examined (40-80%).

TABLE XX. UV Grafting of Acrylates, Methacrylates, Acrylonitrile, 4-Vinylpyridine and Vinyl Acetate in Methanol to Cellulose.[a]

Monomer[b]	Grafting (%) in		
	1 hour	2 hours	3 hours
AA	-	-	15.3
AMA	1.7	10.2	-
MMA	-	-	296.0
TEGDM	8.7	34.0	-
EGDM	4.4	-	-
EGDA	4.9	*c*	-
DEGDM	6.2	14.5	-
DEGDA	3.6	-	-
1,6-HDD	4.0	-	-
1,3-BGDM	7.6	-	-
CMA	2.2	-	-
PETA	13.1	16.3	-
PEGDM	14.3	-	-
TMPTA	-	22.3	-
ACN	-	-	15.8
VA	-	-	2.0
4-VP	-	-	10.1

a Solution of acrylate (30% v/v) in methanol containing uranyl nitrate (1% w/v) and cellulose irradiated at 24 cm from a 90W high pressure UV lamp.

b AA = acrylic acid; AMA = allyl methacrylate; MMA = methyl methacrylate; TEGDM = triethylene glycol dimethacrylate; EGDM = ethylene glycol dimethacrylate; EGDA = ethylene glycol diacrylate; DEGDM = diethylene glycol dimethacrylate; DEGAA = diethylene glycol diacrylate; 1,6-HDD = 1,6-hexane diol diacrylate; 1,3-BGDM = 1,3-butylene glycol dimethacrylate; CMA = cyclohexyl methacrylate; PETA = pentaerythritol triacrylate; PEGDM = polyethylene glycol dimethacrylate; TMPTA = trimethylol propane triacrylate; ACN = acrylonitrile; VA = vinyl acetate; 4-VP = 4-vinylpyridine.

c Homopolymer formation is severe in all runs reported in this Table; however most of the copolymers can be recovered, after grafting, for experimental purposes, although this was not possible with this EGDA sample.

Effect of Monomer Structure. The ability to UV graft monomers other than styrene is both of fundamental and practical significance. Methyl methacrylate in methanol grafts more rapidly than styrene (Table XX) using uranyl nitrate as sensitizer. However, homopolymer formation can be a problem with the former monomer. Homopolymerization is also severe with copolymerization of the very reactive multifunctional acrylates (Table XX) but, after short exposure times which yield grafts of 5-15%, the copolymer can usually be recovered and purified from homopolymer. Acrylonitrile and acrylic acid also undergo severe homopolymerization concurrent with copolymerization.

If the level of copolymerization is not too high (20%), the grafted sample can be recovered and separated from homopolymer. With vinyl acetate, both grafting and homopolymer yields were low, whereas with 4-vinylpyridine, reasonable copolymerization was achieved with virtually no homopolymer formation.

Mechanism of UV Grafting

Both inorganic (*e.g.* uranyl nitrate) and organic (benzoin ethyl ether, biacetyl) compounds can sensitize UV grafting to cellulose, hence copolymerization is capable of being performed in both aqueous and non-aqueous media. Thus the role of water in these reactions can be studied, water being a useful co-solvent in UV grafting (24). Solvent effects are important in photosensitized copolymerization with cellulose using the simultaneous technique and, in particular, physical properties such as wetting and swelling are essential for efficient grafting. The relationship between solvent structure and homopolymerization is also critical, since if the homopolymer is not soluble in the grafting solvent, the homopolymer precipitates and the resulting turbidity either terminates further copolymerization or leads to erratic grafting. Using the alcohols as representative solvents, it is obvious that a small molecule such as methanol, which can both wet and swell the trunk polymer, leads to significant grafting. As chain length and degree of branching of alcohol increase, there is a corresponding decrease in grafting efficiency, copolymerization virtually cutting out with *n*-butanol, which is a relatively poor swelling solvent for cellulose.

A most unusual feature of the alcohol data is that copolymerization is enhanced when a poor grafting solvent (*iso*-butanol or octanol) is added to a solvent where copolymerization readily occurs (methanol). The data are consistent with the fact that the longer chain alcohol would be a better solvent than methanol for styrene homopolymer. Thus, in the mixed methanol-octanol solvent, homopolymer would be maintained in solution, the turbidity effect which inhibits grafting would be eliminated and ultimately efficient copolymerization achieved. Presumably the optimum in enhancement occurs at a particular octanol concentration due to a compromise in the role of methanol, *i.e.* sufficient methanol must remain in the solvent to permit efficient swelling and wetting, and thus allow grafting.

However, the data in Table XVIII for the effect of chain length of alcohol on the grafting enhancement, show that *iso*-butanol is less effective than *n*-octanol. Thus turbidity effects and solubilization of homopolymer, alone, do not fully explain the magnitude or the relative alcohol reactivity of the enhancement effect. The trend in the alcohol data may be explained more satisfactorily if complex formation between monomer and alcohol is proposed as a grafting intermediate. Spectroscopic evidence is available (36) to show that complexes between alkanes and aromatics are readily formed, the alkane approaching to within 4 Å of the aromatic in mixtures of the two, as in solutions of hexane and benzene. Linear alkanes are very effective in the formation of such complexes, the alkane lying horizontally across the aromatic ring. In the grafting reactions, it is proposed that the alkyl portion of the alcohol and the aromatic ring of the monomer are complexed. The formation of such complexes facilitates grafting and contributes to the enhancement effect. Thus, the resulting complex can diffuse into the cellulose which has already been pre-swollen by methanol. The polar hydroxyl group on the longer chain alcohol, *e.g.* octanol, would assist diffusion of this alcohol into the swollen cellulose. The complexing properties of the alkyl part of the longer chain alcohol would also assist diffusion of monomer into cellulose. In terms of this model, octanol alone as solvent, would not swell cellulose sufficiently to yield appreciable grafting, consisting with observations. It is thus plausible to suggest that monomer/solvent complexes are mechanistically important in photosensitized copolymerization reactions. The present UV system would then constitute a further example of the participation of charge-transfer complexes as intermediates in general polymerization reactions (4,31,34,37).

Specific Mechanism for Photosensitized Grafting to Cellulose. Processes that occur when both inorganic and organic photosensitizers are used to graft monomers to cellulose are broadly analogous; however, mechanistically there are specific aspects associated with each system which suggest that it is more convenient to discuss inorganic and organic systems separately. Styrene will be used as representative monomer for the mechanisms discussed.

(i) Inorganic Systems - Uranyl Salts. The photochemistry of the uranyl ion and its role as a photosensitizer have been reviewed (38). In photosensitized copolymerization, two predominant reactions are important: (a) intermolecular hydrogen atom abstraction and (b) energy transfer. In a typical grafting system consisting of styrene/methanol/cellulose, intermolecular hydrogen abstraction can promote copolymerization in several ways. Radicals can be formed in solvent methanol (Equations 8 and 9).

$$UO_2^{2+} + h\nu \rightarrow (UO_2^{2+})^* \tag{8}$$

$$(UO_2^{2+})^* + CH_3OH \rightarrow CH_3O\cdot + H^+ + UO_2^{2+} \tag{9}$$

In liquid methanol, the methoxy radical ($CH_3O\cdot$) is the principal species formed, whereas with aqueous methanol $\dot{C}H_2OH$ predominates (39). With other alcohols, R $\dot{C}HOH$ is the predominant species formed (39). These solvent radicals can then abstract hydrogen atoms from the trunk polymer to yield grafting sites.

In a similar manner sensitizer can diffuse into the alcohol preswollen cellulose and either directly abstract hydrogen atoms or rupture bonds (40) to form additional grafting sites. In the presence of air, these reactions may be further modified by peroxy radical formation. Energy transfer processes may also be involved in grafting (41).

(ii) Organic Systems. The mechanisms for UV grafting initiated by organic photosensitizers are generally similar to those already proposed for the inorganic systems, both radical and energy transfer processes being possible. Radicals can be formed by homolytic cleavage in the sensitizer. Hydrogen abstraction by these radicals from trunk polymer then yields grafting sites. Direct hydrogen abstraction from trunk polymer is also possible (Equation 10).

$$AB \xrightarrow{h\nu} AB^{*} \xrightarrow{\text{celloH}} ABH + \text{cello}^{\bullet} \qquad (10)$$

The essential mechanistic difference in the mode of operation of the various organic photosensitizers is predominantly the relative emphasis of reactions depicted by Equation 10 and also the nature of the radicals formed in homolytic cleavage. With the two most successful organic photosensitizers used in the present work, benzoin ethyl ether and biacetyl, the types of radicals formed are shown in Equations 11 and 12.

$$C_6H_5 - \overset{O}{\overset{\|}{C}} - \underset{H}{\overset{OEt}{\overset{|}{C}}} - C_6H_5 \xrightarrow{hv} C_6H_5\overset{O}{\overset{\|}{C}}{}^{\bullet} + C_6H_5\underset{H}{\dot{C}} - OEt \qquad (11)$$

$$CH_3 - \overset{O}{\overset{\|}{C}} - \overset{O}{\overset{\|}{C}} - CH_3 \xrightarrow{hv} 2CH_3 - \overset{O}{\overset{\|}{C}}{}^{\bullet} \qquad (12)$$

The stability of the resulting radical and steric factors then predominantly determine the relative efficiencies of the two processes.

Comparison of UV and Gamma Ray Grafting Systems

There are a number of properties common to both gamma ray and photosensitized UV copolymerization systems when cellulose is the trunk polymer in the simultaneous irradiation procedure. Each process is predominantly free radical in nature with a possible contribution from energy transfer. With the gamma ray system only, ionic processes could also play a small but significant mechanistic role. Solvent structure is important since only those solvents possessing the necessary physical properties required to swell and wet the trunk polymer are the most suitable for efficient grafting by both methods of initiation.

In both gamma ray and UV systems, analogous enhancement effects in grafting are observed in the presence of certain additives. With the former system, mineral acid and a co-solvent effect can significantly increase the grafting yield. A similar co-solvent effect is observed in UV copolymerization, thus grafting is higher in a swelling-nonswelling solvent mixture (methanol-octanol) than in methanol alone.

Again, in both radiation initiated grafting systems, Trommsdorff effects are observed; however with gamma radiation this gel peak usually occurs at approximately 30% styrene in methanol, whereas in the UV process the peak is found at 80-90% monomer concentration. The reason for this difference in position of the peaks may be associated with the nature of the formation or activation of radicals in the trunk polymer. With ionizing radiation, complete penetration of the solution and cellulose occurs during grafting. Thus radicals are directly formed in the trunk polymer and are immediately available for termination and therefore grafting when monomer diffuses to the site. By contrast, with the UV system, radical sites in the trunk polymer are predominantly only formed after sensitizer and/or solvent radicals have diffused into the polymer and abstracted hydrogen atoms. Because of this diffusional limitation relevant only to the UV system, chain length of grafted polymer will be increased before termination. Since the higher molecular weight chain to be grafted will be more soluble in a solvent containing high percentages of monomer, the Trommsdorff peak for the UV grafting of styrene in methanol to cellulose is shifted to the 80-90% monomer concentration region.

The relationship between monomer structure and ease of copolymerization is also similar for both radiation systems. Thus homopolymer formation is always a competing reaction to graft. With styrene, homopolymer formation is minimal in both radiation grafting systems, especially when grafting in methanolic solution. However, with reactive monomers such as methyl methacrylate, acrylonitrile, acrylic acid and the polyacrylates, homopolymerization can be serious and methods are at present being developed to overcome the problem (12).

The final property common to both radiation systems is the actual mechanism of the grafting. For copolymerization, radicals must be formed in the trunk polymer. With gamma rays, this can occur from the effect of direct primary radiation on the polymer, as well as by secondary reactions involving H atoms. In UV initiation, the former process is less efficient and predominantly radical formation is by diffusion of sensitizer to the cellulose site followed by reaction. Thus in both radiation systems, monomer can diffuse to site and graft by the charge-transfer mechanism already discussed. Consistent with this theory, there is accumulating data, again from both radiation systems, to show that monomer/solvent complexes are important intermediates in the charge-transfer grafting mechanism (4,31).

The fact that many properties are common to both gamma and UV simultaneous irradiation processes for grafting is valuable not only mechanistically, but also from a preparative viewpoint. Thus any developments in one radiation system can usually be directly applied to the other with only minor modifications to the experimental procedure. There is also no necessity to use gamma irradiation facilities to prepare experimental quantities of cellulose copolymers, since, in some instances, the corresponding UV process is even simpler and can be more convenient with the use of conventional small laboratory photochemical equipment. Finally, the idea of using developments interchangeably in UV and gamma radiation grafting to cellulose is now being extended to other trunk polymers such as the polyolefins, polyvinylchloride and wool with similar success (42,43). Extension of the principles developed for gamma ray grafting to EB work is also being explored (43).

Acknowledgement. The author thanks the Australian Institute of Nuclear Science and Engineering and the Australian Atomic Energy Commission for the irradiations. Financial support from the Australian Wool Corporation and the Australian Research Grants Committee is also gratefully acknowledged. The author thanks the following co-workers in this cooperative project: Drs. Davids, Davis, Dilli, Fletcher, Phuoc, Reid, Rock and Schwarz.

Literature Cited

1. Krassig, H.A., Stannett, V.T., Advan.Polymer Sci. (1965), 4, 111.
2. Moore, P.W., Rev. Pure Appl. Chem. (1970), 20, 139.
3. Arthur, J.C. Jr., Advan. Chem. Ser. (1971), 99, 321.
4. Dilli, S., Garnett, J.L., Martin, E.C., Phuoc, D.H., J. Polymer Sci. C (1972), 37, 57.
5. Usmanov, Kh. U., Aikhodzhsev, B.I., Azisov, U., J. Polymer Sci. (1961), 53, 87.
6. Okamura, S., Iwasaki, T., Kobayashi, Y., Hayashi, K., I.A.E.A. Conf. Appl. Large Radiation Sources Ind. especially Chem. Proc., Warsaw, 1959.
7. Huang, R.Y.M., Rapson, W.H., J. Polym. Sci. Fourth Cellulose Conference (1963), 169.
8. Guthrie, J.T., Haq, Z., Polymer (1974), 15, 133
9. Geacintov, N., Stannett, V.T., Abrahamson, E.W., Hermans, J.J., J. Appl. Polymer Sci. (1960), 3, 54.
10. Arthur, J.C. Jr., Polym. Preprints (1975), 16, 419.
11. Kubota, H., Murata, Y., Ogiwara, Y., J. Polymer Sci. (1973), 11, 485.
12. Davis, N.P., Garnett, J.L., Urquhart, R.G., Polymer Letters, in press.

13. Davis, N.P., Garnett, J.L., Urquhart, R.G., J. Polymer Sci., in press.
14. Dilli, S., Garnett, J.L. Phuoc, D.H., Polymer Letters, (1973), 11, 711.
15. Garnett, J.L. Phuoc, D.H., J. Polymer Sci., in press.
16. Ekstrom, A., Garnett, J.L., J. Chem. Soc. A (1968), 2416.
17. Garnett, J.L., Martin, E.C., J. Macromol. Sci. Chem. (1970), A4, 1193.
18. Davis, N.P., Garnett, J.L., unpublished data.
19. Dilli, S., Garnett, J.L., J. Appl. Polymer Sci. (1967),11,839.
20. Garnett, J.L., Phuoc, D.H., Airey, P.L., Sangster, D.F., Aust. J. Chem., in press.
21. Bosworth, R.C.L., Ernst, I., Garnett, J.L., Int. J. Appl. Radiation Isotopes (1961), 11, 152.
22. Dilli, S., Ernst, I.T., Garnett, J.L., Aust. J. Chem. (1967), 20, 911.
23. Florin, R.E., Wall, L.A., J. Polymer Sci. (1963), A-1, 1163.
24. Garnett, J.L., Martin, E.C., Polymer Letters, in press.
25. Stamm, A.J., Tarkow, H., J.Phys.Colloid Chem. (1950), 54, 745.
26. Sherman, W.V., J. Phys. Chem., (1967), 71, 4245
27. Baxendale, J.H., Mellows, F.W., J. Am. Chem. Soc. (1961), 83, 4720.
28. Taub, I.A., Harter, D.A., Sauer, M.C., Dorfman, L.M., J. Chem. Phys. (1964), 41, 479.
29. Baxendale, J.H., Sedgwick, R.D., Trans. Faraday Soc. (1961) 57, 2157.
30. Fletcher, G., Garnett, J.L., unpublished work.
31. Dilli, S., Garnett, J.L., J. Polymer Sci. (1966), 4, 2323
32. Sangster, D.F., Davison, A., J. Polymer Sci. (1975), 49, 191.
33. Goutiere, G., Gole, J., Bull. Soc. Chim. France, (1965), 153.
34. Plesch, P.H., "Progress in High Polymers", J.G. Robb, F.V. Peeker, Eds., Vol. 12, p.139, Heywood, London, 1968.
35. Oster, G., Yang, N.L., Chem. Rev. (1968), 68, 125.
36. Lamotte, M., Jousoot-Julien, J., Mantione, M.J., Claverie, P., Chem. Phys. Letters (1974), 27, 515.
37. Gaylord, N.G., Deshpande, A.B., Dixit, S.S., Maiti, S., Patnaik, B.K., J. Polymer Sci. (1975), A-1, 13, 467.
38. Burrows, H.D., Kemp, T.J., Chem. Rev. (1974), 3, 139.
39. Ledwith, A., Russell, R.J., Sutcliffe, L.H., Proc. Roy. Soc. (1973), A332, 151.
40. Greatorex, D., Hill, R.J., Kemp, T.J., Stone, T.J., J.C.S. Faraday I (1972), 68, 2059.
41. Venkatarao, K., Santappa, M., J. Polymer Sci. (1970), A-1, 8, 1785, 3429.
42. Garnett, J.L., This Conference, in press.
43. Fletcher, G., Garnett, J.L., Schwarz, T., unpublished work.

24

Some Biological Functions of Matrix Components in Benthic Algae in Relation to Their Chemistry and the Composition of Seawater

INGER-LILL ANDRESEN, OLAV SKIPNES, and OLAV SMIDSRØD
Institute of Marine Biochemistry, The University of Trondheim, 7034 Trondheim-NTH
KJETILL OSTGAARD
Institute of Biophysics, The Norwegian Institute of Technology, 7034 Trondheim-NTH
PER CHR. HEMMER
Institute of Theoretical Physics, The Norwegian Institute of Technology, 7034 Trondheim-NTH

The Organizing Committe must first be thanked for inviting me to participate in this interesting meeting and to give a lecture here today. When the invitation came about a year ago, the proposed title of the lecture was made so broad as to cover several possible directions and developments in our research within the general area of obtaining some molecular understanding of the biological functions of extracellular components in Benthic algae. The lecture today will be confined to their function in cementing the cells together and giving the plants sufficient rigidity to tolerate the large stresses caused by the rapid movements of the sea. A further narrowing of the scope of this lecture is made in that only brown algae will be considered.

Most brown algae contain alginate as the major intercellular substance and only some 1-10% of the dry matter as cellulose (1,2), the common structure substance in most plants. It has, therefore, been assumed by several authors (3,4,5) that alginate contributes to the mechanical properties of brown algal tissue, but almost no experimental data has so far been collected to verify this assumption.

Some indications have, however, been obtained from a study of the cation composition of algal tissue (4,6). It is well known that sodium alginate is soluble in water and in aqueous sodium chloride, and that it forms rigid gels in the presence of divalent metal ions, in particular calcium and strontium ions (5). The extracellular alginate in algal tissue is in rapid ion-exchange equilibrium with sea-water (6,7), and analysis has shown (6) that it is partly neutralized by calcium and strontium ions in spite of the large excess of sodium and magnesium ions in sea-water as shown in

Table I. This is due to the fact (5,8) that alginate has a much higher affinity to calcium and strontium ions than to sodium and magnesium ions. The alginate should, therefore, exist in the plant as gels of some rigidity.

TABLE I. CONCENTRATIONS IN EQUIVALENTS PER 1000 g OF SOME MAJOR CATIONS IN SEA-WATER AND A SECTION OF L. HYPERBOREA STIPE RINSED IN WATER TO REMOVE FREE SALTS.

	Sea-Water*	Stipe+
Na^+	0.47	0.05
Mg^{++}	0.11	0.11
Ca^{++}	0.02	0.14
Sr^{++}	0.0002	0.006

*From ref. 24 +From ref. 10

If alginate is responsible for the rigidity of brown algal tissue some correlation should exist between the chemical composition of alginate and the strength of the tissue.

Alginate is a binary heteropolymer containing 1,4-linked β-D-mannuronic acid and α-L-guluronic acid residues arranged along the chain in homopolymeric "blocks" of the two acid residues together with blocks of an alternating sequence (4). The relative amounts of the three types of blocks vary between species and between different type of tissue within one species (9).

It is well established that the strength of calcium alginate gels increases markedly with an increasing fraction of L-guluronic acid in the alginate, the reason being that the homopolymeric blocks of L-guluronic acid are responsible for the formation of junctions in the gels (5). Very little is known, however, about a possible correlation between the strength of brown algal tissue and alginate composition. The only evidence for the existence of such a correlation stems from the fact, mentioned by Haug in his review article on cell-wall polysaccharides (4), that the only known occurrence of an alginate with a composition approaching polymannuronic acid is in the medulla of the receptacles of certain species in the order Fucales where it forms a viscous solution, while the most guluronic-rich alginate is found in the old stipes of *Laminaria hyperborea* which have a very rigid structure.

We have performed experiments with tissue from the two algae *Laminaria digitata* and *Laminaria hyperborea* to look for such a correlation. These algae were selected for two reasons: Firstly, they have a well-defined stipe which easily can be cut into sections making it possible to determine stress-strain diagrams by compressing the sections. The leaves, on the other hand, can be cut into strips, and stress-strain diagrams can be determined by stretching. Secondly, these two algae contain very little fucan and sulfated polysaccharides, leaving alginate as the dominating matrix polysaccharide (8).

In Table II some results from compressing stipe sections of L. hyperborea are shown. The sections were cut at different distances from the leaves. The rigidity was obtained from an initial linear part of the stress-strain diagram. This yields an easily obtained

TABLE II. MODULUS OF RIGIDITY OF SECTIONS OF L. HYBERBOREA STIPE CUT AT DIFFERENT DISTANCES FROM TOP.

SERIES A		SERIES B	
Distance from top, cm	G, $^{kp}/cm^2$	Distance from top, cm	G, $^{kp}/cm^2$
10	131	6	103
15	189	8	180
24	253	21	268
26	258	23	258
34	287	41	346
36	290	43	381
44	316	57	388
46	312	59	360
52	303	72	403
54	480	74	541

M/G determinations of alginate from sections of the stipe cut at 12 and 48 cm from the leaf yielded M/G=0.9 and M/G=0.6, respectively. The corresponding alginate concentrations were 4.9 and 4.5 g per 100 cm^3 of fresh tissue, respectively.

number which can be used for comparison of different sections, but of course it is not a complete description of the mechanical properties of the algal tissue. Table II indicates a very significant tendency for an increase in modulus with increasing distance from the leaf. Haug (8) has shown that the ratio between D-mannuronic and L-guluronic acid residues in alginate from this algae decreases downwards in the stipe. The M/G-determinations given in the table II support this finding. The results from extraction studies given in the same table, show that the alginate concentration does not increase downwards in the stipe. In addition we have compared the strength of stipe and leaf in L. hyperborea, and the strength of different type of tissue in L. hyperborea and L. digitata (10). Without exception we find the same correlation as in Table II: The modulus increases with increasing fraction of L-guluronic acid residues in the alginate of the plant tissue.

These experimental results, indicating that the correlation between strength and chemical composition of alginate is the same in algal tissue as in calcium alginate gels, suggest that alginate contributes crucially to the mechanical properties of the algal tissue in *L. hyperborea* and *L. digitata*.

The strength of alginate gels is, as already mentioned, strongly dependent on the type of counterions neutralizing the polyanions (5). Some further understanding of the role of alginate in contributing to the strength of the algal tissue should, therefore, be obtained when we compare how algal tissue and alginate gels respond to changes in the composition of the counterions.

Changing the "native" counterions for sodium ions should, if alginate is responsible for the rigidity, result in a marked reduction in modulus. By dialyzing algal tissue to exhaustion against 0,1M NaEDTA in 0,2M aqueous NaCl the alginate is completely transferred to the sodium form (10). At this high ionic strength the alginate is insoluble and is not extracted from the plants (10). Table III shows that a marked drop in modulus is the result of this cation-exchange. Therefore alginate must obviously be responsible for the rigidity of the plants.

TABLE III. MODULUS OF RIGIDITY OF A SECTION OF *L. HYPERBOREA* STIPE BEFORE AND AFTER TREATMENT WITH 0.1M NaEDTA IN AQUEOUS 0.2M NaCl.

BEFORE:	$G = 350\ kp/cm^2$
AFTER:	$G = 10.5\ kp/cm^2$

An interesting question is now: To what extent is it possible to restore the modulus by exhaustive dialysis against sea-water again? Table IV, Series A, contains the results of extensive dialysis of NaEDTA-treated sections against sea-water and other salt solutions. It is seen that only some 12% of the modulus is restored by dialysis against sea-water. The table also shows that even dialysis against lead nitrate, known (5) to give very high moduli in alginate gels, does not restore the modulus to more than 30% of the original value. It seems, therefore, that the alginate in plant tissue behaves differently from pure alginate, and a closer look at the effect of different cations was found necessary.

TABLE IV. MODULUS OF RIGIDITY, G, OF SECTIONS OF STIPE FROM L. HYPERBOREA IN PERCENT OF ORIGINAL VALUE.

DIALYSIS FOR SIX DAYS AGAINST:
A: FIRST Na-EDTA (0.1M) + NaCl (0.2M), THEN DIFFERENT IONIC SOLUTIONS
B: DIFFERENT IONIC SOLUTIONS (I=0.55)

	A	B
SOLUTION	G %	G %
SEA-WATER	12	100
Na-EDTA	3	3
$MgCl_2$	8	20
$CaCl_2$	13	76
$SrCl_2$	11	73
$BaCl_2$	8	116
$Pb(NO_3)_2$	30	127

In Table IV, Series B, the results of direct, exhaustive dialysis against different salt solutions are given. When the results of series A and B are compared, it is seen that direct dialysis of the sections of the stipe against the different salt solutions leads in general to higher moduli than the NaEDTA-treated sections. As expected, magnesium ions reduce the modulus markedly compared to sea-water (series B). It is not clear why both calcium and strontium ions also reduce the modulus somewhat. It may be that other cations in the sea, for example barium and lead ions, contribute to the stiffness of living tissues. The data in Table IV exclude the possibility that sodium and magnesium ions alone can give sufficient strength.

In Table V similar results are given for pure alginate prepared from old L. hyperborea stipes with M/G= 0.4. Alginate solutions were first dialyzed against sea-water and then against different salt solutions, and the results should, therefore, be compared with the results of series B in Table IV. It is seen that the alginate gel responds much more to the different salt solutions than does the algal tissue. The trends in the moduli are, however, the same, pointing again to the importance of alginate as contributing to the rigidity of the algal tissue. The same series of experiments has also been carried out with algal tissue and alginate from L. digitate (11) with very similar results: It is possible to reduce the modulus of the algal tissue by Na^+ - and Mg^{2+}-ions as in alginate gels dialyzed first against sea-water, but, in contrast to the alginate gels, it is not possible to raise the modulus very much above the "native" value.

TABLE V. MODULUS OF RIGIDITY, G, OF GELS MADE OF ALGINATE FROM *L. HYPERBOREA* STIPE (M/G=0.4) AFTER DIALYSIS OF SODIUM ALGINATE AGAINST SEA-WATER FOLLOWED BY DIALYSIS AGAINST DIFFERENT IONIC SOLUTIONS (I=0.55). ALGINATE CONCENTRATION BEFORE DIALYSIS 2 % (w/v).*

SOLUTION	G, kp/cm^2	% OF SEA-WATER
SEA-WATER	0.32	100
NaCl	0.01	3
$MgCl_2$	0.05	16
$CaCl_2$	2.0	620
$SrCl_2$	2.1	660
$BaCl_2$	2.5	760
$Pb(NO_3)_2$	4.6	1440

*The concentrations in the gels are somewhat higher due to shrinkage during gel formation.

So far we have only compared trends in moduli obtained by perturbation of the counterion composition in algal tissue and alginate gels. A comparison of the absolute value of the modulus of alginate dialyzed against sea-water with that of the corresponding algal tissue is much more difficult, because of the problem of accounting for the presence of the cells in the tissue. A very rough comparison is attempted in Table VI. We have assumed that the cells have the same modulus as the intercellular substance. This substance occupies about half of the volume of the tissue (12), making the alginate concentration in the intercellular space about twice that of the measured concentration based on whole, fresh tissue. The moduli of alginate gels are proportional to the square of the alginate concentration (13), and the moduli of the alginate gels were corrected according to this relationship, making possible a comparison of moduli at the same concentration. Table VI indicates, quite surprisingly, that the alginate gel has a modulus only about one percent of the algal tissue value.

In spite of the very rough assumptions behind the comparison it is now necessary to ask to what extent other intercellular components contribute to the rigidity. The answer is not known yet. The algae in question contain, as already mentioned, some cellulose. Although Preston (3) and others have studied the organization of cellulose in algal tissue, no data exist enabling a quantitative estimate of the contribution of cellulose to the rigidity. If, however, alginate should contribute with only 1% of the modulus and cellulose with 99% it is hard to see why the strength of the algal tissue should at all be dependent on the M/G-ratio and

TABLE VI. COMPARISON OF MODULUS OF A SECTION OF L. HYPERBOREA STIPE WITH A GEL OF ALGINATE FROM THE SAME SECTION DIALYZED AGAINST SEA-WATER.
Alginate content in algal section: 4.5 g per 100 cm^3 of fresh tissue.

$$c_{alginate\ gel} = 2\%, \qquad c_{stipe} \simeq 9\%$$

$$G_{alginate} = 0.15\ kp/cm^2, \qquad G_{stipe} = 350\ kp/cm^2$$

$$\frac{(G_{c=9})\ stipe}{(G_{c=9})\ alginate} = \frac{350}{0.15\ (\frac{9}{2})^2} \simeq 100$$

on different counterion composition. Some very peculiar synergistic effects would then have to be postulated to explain why, for example, magnesium ions lower the modulus by about 80% and lead ions raise the modulus by about 30%. (For L. digitata stipe the corresponding figures are (11) 95 and 110%, respectively.) It is possible, therefore, that the alginate is organized in another way in the plant tissue than in the gels prepared in the laboratory, making it a better structure-forming substance in the former case.

A detailed discussion of possible differences in structure will not be attempted here. The remaining part of the lecture deals mainly with some ion-exchange and electron-microscope work which will indicate that such a difference indeed exists, and presents some ideas as to how they can be explained on a molecular level. These ideas may act as working hypothesis for further studies.

Previously we have studied the ion-exchange between magnesium and calcium ions in alginate (5,14). The selectivity coefficients obtained from such experiments depend not only upon the chemical composition of alginate but also upon the structural arrangements of the molecular chains. We may therefore use the calcium-magnesium ion-exchange as a "probe" to test whether the structure is different in the gel and in the plant. Alginate prepared from L. hyperborea stipe was dialyzed extensively, first against sea-water and then against different calcium-magnesium chloride solutions in the standard way (14) for obtaining selectivity coefficients. The results are given in Fig. 1 together with similar results on sections cut near each other in stipes of L. hyperborea. The algal tissue is seen to have markedly the highest selectivity coefficients, k_{Mg}^{Ca}, at low calcium content of the tissue.

In an attempt to explain these differences quantitatively we must take a closer look at the ion-exchange properties of alginate. Fig. 2 shows some results obtained (5) for the calcium-magnesium exchange reaction

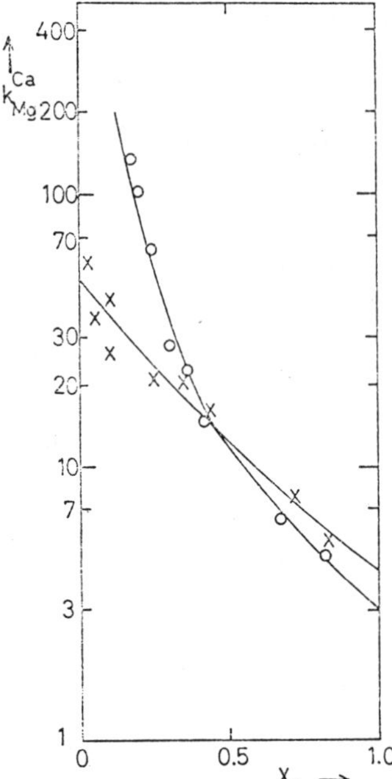

Figure 1. The selectivity coefficient k_{Mg}^{Ca} for alginate (M/G = 0.4) and sections of stipe from Laminaria hyperborea *vs. the fraction X_{Ca} of the polyelectrolyte sites involved in binding to calcium ions.* ×*: alginate, first dialyzed against sea-water;* ○*: sections of stipe.*

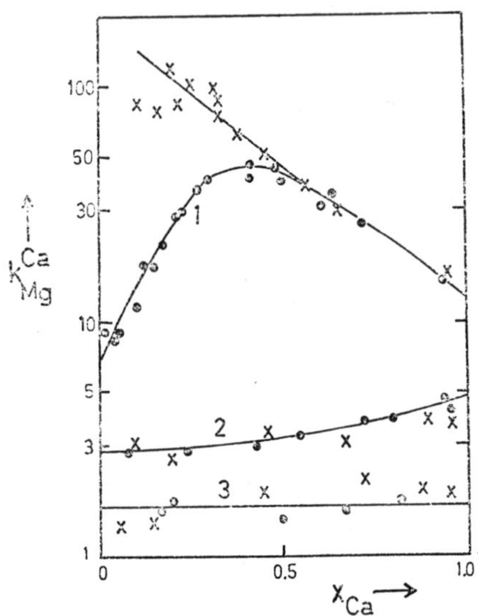

Figure 2. The selectivity coefficient k_{Mg}^{Ca} for alginate fragments vs. X_{Ca} (from Refs. 5 and 25). 1: Fragment with 90% guluronic acid (G.A.); 2: alternating fragment, 38% G.A.; 3: Fragment with 90% mannuronic acid (M.A.). ●*: dialysis of the fragments in the sodium form;* ×*: dialysis first against 0.2 M $CaCl_2$, then against the different mixtures of $CaCl_2$ and $MgCl_2$.*

on three fragments approaching in chemical composition the three types of "blocks" in alginate. The selectivity coefficients have been obtained in two ways: In one series the fragments were dialyzed directly against Ca/Mg-solutions. In the other series (crosses) the fragments were first dialyzed against calcium chloride and then against Ca/Mg-solutions. We see from the figure that the fragments rich in mannuronic acid residues ("MM-blocks") and the fragments rich in the alternating structure ("MG-blocks") are characterized by low selectivities, no hysteresis and no co-operativity, in contradistinction to the fragments rich in guluronic acid residues ("GG-blocks").

We have previously (14) tried to explain the co-operativity in the binding of calcium to "GG-blocks" by a theoretical model with nearest-neighbour co-operative effects, i.e. a model where binding of a calcium ion to one site in the chain favors the binding of a calcium ion in the neighbouring position so that sequences of calcium ions along the chains are formed. This model has now been modified for reasons to become clearer below, to incorporate the case where co-operativity can occur only in a fraction α of the binding sites in "GG-blocks". The rest of the sites bind calcium without co-operativity. The model (15) is defined in Fig. 3, and the theoretical calculations in Fig. 4 show that it can account for the maximum in the experimental curve relating the selectivity coefficients to the fraction X_{Ca} of the sites in the GG-fragments involved in binding to calcium ions (curve 1, Fig. 2).

We have done a least-square fitting with the model to experimental results for the Ca^{2+}-Mg^{2+} and Sr^{2+}-Mg^{2+} exchange reactions on GG-fragments. The best-fit curves are shown together with the experimental results in Fig. 5 and Fig. 6, and the corresponding model parameters are given in Table VII. We see that the Sr^{2+}-Mg^{2+} exchange reaction is characterized by a higher selectivity and a lower α-value than the Ca^{2+}-Mg^{2+} exchange reaction. We shall not comment on these results yet, but try to explain the hysteresis in Fig. 2, and the selectivities obtained for GG-fragments first dialyzed against Ca^{2+} or Sr^{2+} and then against mixtures of Ca^{2+}/Mg^{2+} and Sr^{2+}/Mg^{2+}.

We know that the high selectivities are results of ion-binding between chains (5,14), the inter-chain contact zones being the junctions in the alginate gels. If we now assume that dissociation of junctions is a slow process compared to the ion-exchange process, we can formulate a new model as shown in Fig. 7. The model assumes that all ions bound in junction zones

A. DEFINITION OF SELECTIVITY COEFFICIENTS

$$k_B^A = \frac{X_A}{X_B} \Big/ \frac{C_A}{C_B} = \frac{X_A}{X_B} \frac{1}{f}$$

B. ION-EXCHANGE WITHOUT CO-OPERATIVITY

$$X_A = \frac{k_o f}{1 + k_o f}$$

C. ION-EXCHANGE WITH NEAREST-NEIGHBOUR CO-OPERATIVE EFFECTS

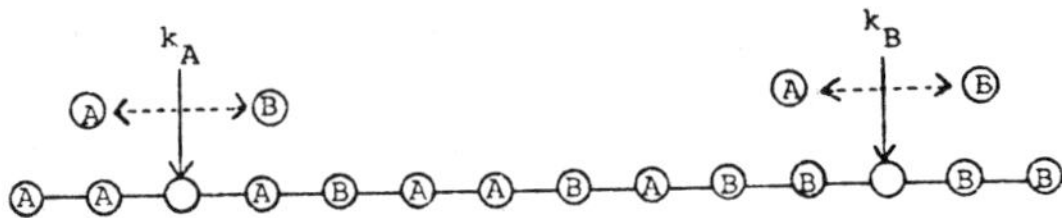

	Selectivity
Fraction α of GG-segment (g)	k_A, $\sqrt{k_A k_B}$, k_B between AA, AB, BB neighbours
Fraction 1-α of GG-segment	k_o
MM-segment (m)	k_m

$$(X_A)_{tot} = (X_A)_g g + (X_A)_m m$$

$$(X_A)_m = \frac{k_m f}{1 + k_m f}$$

$$(X_A)_g = \tfrac{1}{2}\alpha \left[1 + \frac{f\sqrt{k_A k_B} - 1}{\sqrt{(f\sqrt{k_A k_B} - 1)^2 + 4f\,k_B}}\right] + (1-\alpha)\frac{k_o f}{1 + k_o f}$$

Figure 3. Definition of selectivity coefficients, ion-exchange without cooperativity, and ion-exchange with nearest-neighbor cooperative effects. A. $k_B{}^A$ is the selectivity coefficient characterizing the exchange of B-ions for A-ions at a ratio, X_A/X_B, between the molfractions of A- and B-ions bound to the polyelectrolyte, when the ratio between the concentrations in the dialysate is $f = C_A/C_B$. B. Dependence of the molfraction, X_A, upon f and k_o, when k_o is the selectivity coefficient for ion-exchange without cooperativity. C. Expression for the molfraction, $(X_A)_{tot}$, for a fragment containing the fractions g and m of GG-segments and MM-segments, respectively. A fraction α of the GG-segments are characterized by nearest-neighbor cooperative effects. k_A and k_B are the selectivity coefficients for the exchange of B-ions for A-ions when both the neighbors are A and B, respectively.

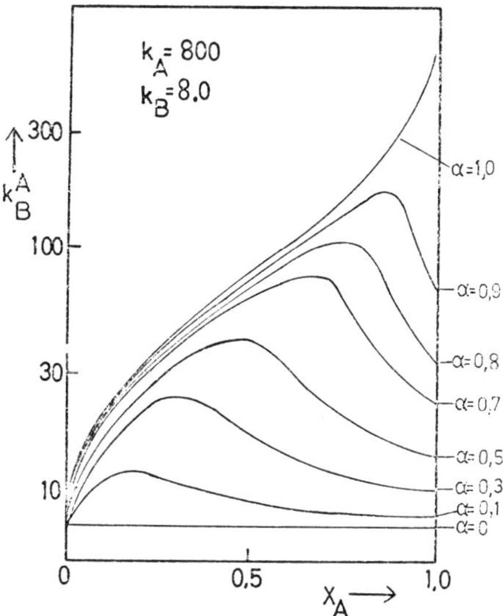

Figure 4. Theoretical calculation of $k_B{}^A$ *against* X_A *with the model defined in Figure 3 with* $k_B = k_o$

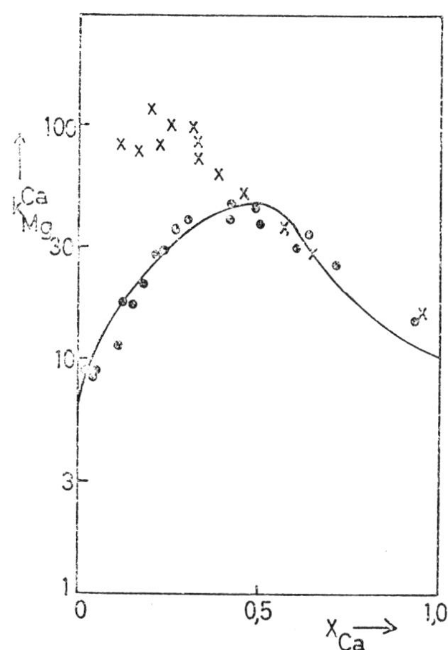

Figure 5. Experimental points for the Ca^{2+}*–*Mg^{2+} *exchange reaction and the curve giving the best fit with the model defined in Figure 3. Fragment containing 90% G.A. ×: dialysis of fragments in the sodium form; ●: dialysis of fragments in the calcium form.*

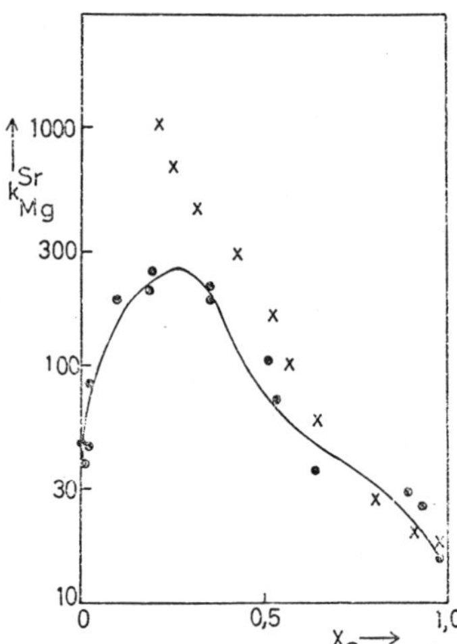

Figure 6. Experimental points for the Sr^{2+}–Mg^{2+} exchange reaction, and the curve giving the best fit with the model defined in Figure 3. Fragments containing 90% G.A. ×: dialysis of fragments in the sodium form; ●: dialysis of fragments in the strontium form.

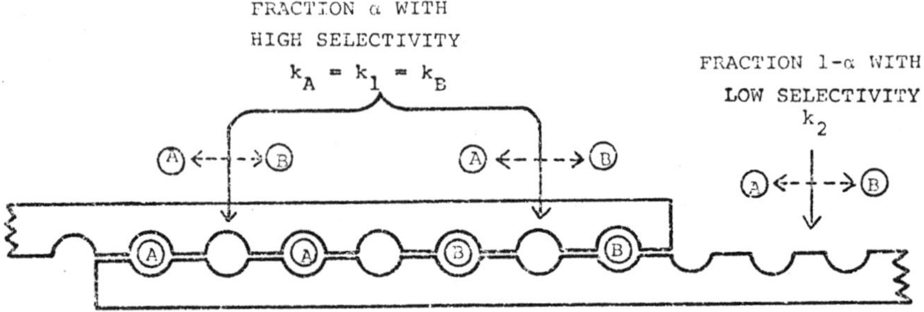

$$(X_A)_{tot} = \left(\frac{k_1 f}{1+k_1 f}\,\alpha + \frac{k_2 f}{1+k_2 f}\,(1-\alpha)\right) g + \frac{k_m f}{1+k_m f}\, m$$

Figure 7. Ion-exchange model assuming a fraction α of junction zones with high selectivity and a fraction (1 — α) of single chain segments with low selectivities. Compare Figure 3 for definition of parameters.

TABLE VII. ESTIMATION OF PARAMETERS IN THE MODEL WITH NEAREST NEIGHBOUR CO-OPERATIVE EFFECTS.[x] GG-FRAGMENTS CONTAINING 10% MANNURONIC ACID RESIDUES.

	α	k_A	k_B	k_o	k_m
Ca-Mg	0.6	830	8.0	7	1.7
Sr-Mg	0.37	8600	118	30	1.9

k_o OBTAINED FROM SOLUBLE GG-FRAGMENTS (14)
k_m OBTAINED FROM STUDIES OF MM-FRAGMENTS

[x]See definition of the model in Fig. 3.

have high selectivities regardless of the nature of the neighbouring ions, and that a fraction α of the sites in "GG-blocks" involved in junction formation is independent on the fraction of Ca^{2+} or Sr^{2+} neutralizing the sites. Least square-fitting of experimental results with this model (Figs. 8 and 9) yields the parameters shown in Table VIII. The conclusion emerging from this table is that the Sr^{2+}-Mg^{2+}-exchange reaction is again characterized by high selectivity coefficients but low α-value compared to the Ca^{2+}-Mg^{2+}-exchange reaction.

TABLE VIII. ESTIMATION OF PARAMETERS IN MODEL DEFINED IN FIG. 7.

	"GG-BLOCKS"				"MM-BLOCKS"	
	INTER-CHAIN		INTRA-CHAIN			
	α	k_1	$1-\alpha$	k_2	m	k_m
Ca-Mg	0.64	180	0.36	7	0.1	1.7
Sr-Mg	0.46	4200	0.54	30	0.1	1.9

It should be stressed that it is not the autors' opinion that these models can explain all details of the ion-exchange properties of alginate, but some indications that the models contain the essential features for such a description can be seen from the next two figures.

Firstly, the Sr^{2+}-Ca^{2+}-exchange should be characterized by the ratio between the selectivity coefficients k_{Mg}^{Sr} and k_{Mg}^{Ca}. Table VIII indicates that these ratios are about 20, 4 and 1, respectively, for the inter-chain binding, the intra-chain binding in "GG-blocks", and the intra-chain binding in "MM-blocks". These figures are close to those indicated by a direct study of the Sr^{2+}-Ca^{2+}-exchange reaction as seen in Fig. 10.

Secondly, it seems significant from Table VIII that the fraction of the sites in "GG-blocks" involved

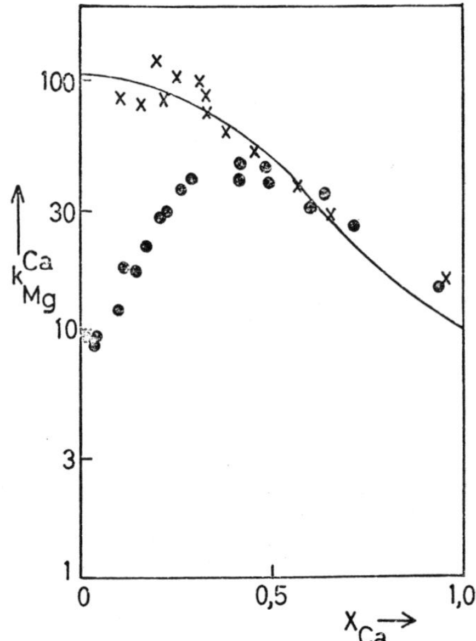

Figure 8. The same experimental points as in Figure 5, and the curve giving the best fit with the model defined in Figure 7

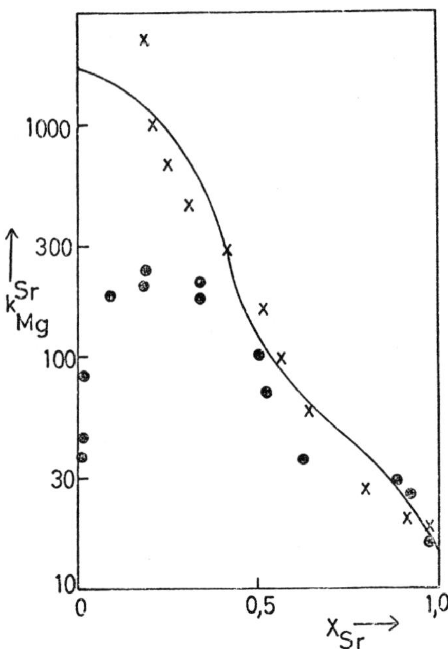

Figure 9. The same experimental points as in Figure 6, and the curve giving the best fit with the model defined in Figure 7

in inter-chain binding is lower in the Sr^{2+}-Mg^{2+} than in the Ca^{2+}-Mg^{2+} exchange reaction. This result is in agreement with what one would expect from the following consideration of a dimerization process of chain molecules: It is reasonalbe to assume that crosslinking starts between two monomer residues, one from each chain situated at random along the chains. If dissociation cannot occur before a contact zone of the maximum possible length is formed, a degree of overlap (corresponding to α in Table VII) of 0.5 will be obtained in a population of monodisperse chain molecules. For a Kuhn distribution of chain lengths this number will be 0.33 (16). To get a higher degree of overlap it is necessary for a series of dissociation and association steps to occur after the initial crosslinking. We know that dissociation of junctions in calcium and strontium alginate gels is a slow process. If we assume that dissociation of dimers is prevented for kinetic reasons when the contact zones contain more than a critical number of monomer residues, the degree of overlap will become higher, the higher this number is (16). This number would be expected to be higher for calcium than for strontium ions linking the chains together because of a stronger binding of the ions in between chains in the latter case. The difference between the α-values obtained in Table VIII is thus in agreement with this reasoning.

We can now come back to the ion-exchange experiments in Fig. 1, and try to see if we, with the selectivity coefficient and the α-value obtained from studying the fragments, can reproduce the experimental curve. Fig. 11 shows that we get a reasonably good fit for the alginate. By assuming that $\alpha = 1.0$ in the stipe, we can explain some of the differences at low X_{Ca}-values. The discrepancy at intermediate and high X_{Ca}-values may be due to the presence in the stipe of some sulphate groups without selectivity (27).

Calcium alginate gels are characterized by syneresis due to a slow crosslinking occuring after the gels have been formed (5,17). If $\alpha = 1$ in the algal tissue, such a crosslinking reaction should not occur, equilibrium already having been reached. In Fig. 12 the rate of syneresis is compared at 50°C for a calcium alginate gel and a stipe from *L. digitata* (11). The marked difference seen in the figure clearly indicates that the stipe represents more closely an equilibrium situation than the calcium alginate gel does.

If $\alpha = 1$ in the alginate of the algal tissue, and lower in alginate gels prepared at the laboratory by dialysis and "random crosslinking", then one would

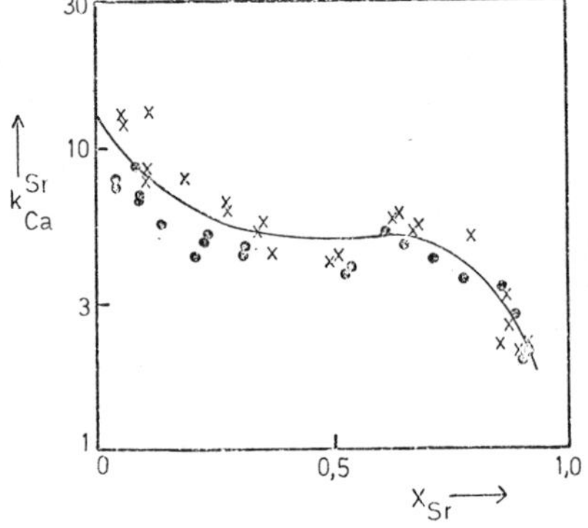

Figure 10. The selectivity coefficient, k_{Ca}^{Sr}*, vs.* X_{Sr} *for an alginate fragment containing 10% G.A.* ×*: dialysis of fragments in the strontium form;* ●*: dialysis of fragments in the calcium form.*

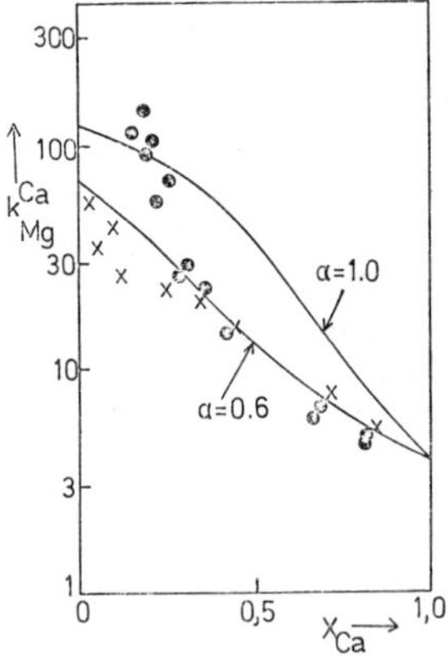

Figure 11. Experimental points and theoretical curves for k_{Mg}^{Ca} *vs.* X_{Ca} *for* L. hyperborea *stipe and alginate from* L. hyperborea. ×*: alginate;* ●*: stipe. Theoretical calculation with block composition*[26]*: GG-blocks 62%, MM-blocks 15%, MG-blocks 22%; alginate:* $\alpha = 0.6$*, stipe:* $\alpha = 1.0$*; selectivity coefficients from Table VIII, and* $k_{Mg}^{Ca} = 3.0$ *for MG-blocks from Figure 2.*

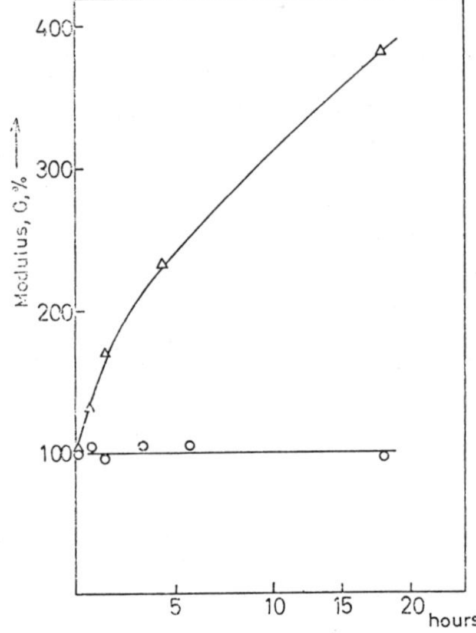

Figure 12. Sections of stipe from Laminaria digitata *and alginate from the same algae treated for different times at 50°C before determination of modulus* (11)*. Modulus in % of original value.* ○*: section of stipe from* L. digitata*,* △*: Ca-alginate* (M/G = *1.6*, c = *2%*).

expect to find a chain structure with more correlations between the chains in the former case. We have previously studied the chain structure in calcium alginate gels with alginate prepared from L. hyperborea stipe by electron microscopy (18). The main impression from several photographs is indeed that of a randomly crosslinked system with "pores" differing widely in size (see Fig. 13). Skipnes has also studied the intercellular substance in sections from Ascophyllum nodosum (12). Both by osmium staining and without staining, the presence of pores in the matrix substance is revealed. Fig. 14 shows an electron micrograph of algal tissue from L. hyperborea stipe. The presence of "pores" is clearly seen, but here they seem to be more narrowly distributed in size than in the alginate gel.

By measuring the diameters of 112 well-defined pores in the alginate gel and the algal tissue the distribution curves given in Fig. 15 are obtained. The "tail" in the distribution curve for alginate is absent in the algal tissue. More work is needed in this field, but an inescapable conclusion seems to be that the alginate gel in the algal tissue can not be a result of a random crosslinking process.

Much work has been carried out at this Institute to show that the "GG-blocks" in alginate are synthesized on the polymer level from mannuronic acid residues by a C_5-epimerase (19,20,21,22,23). Calcium ions, or other gel-forming divalent metal ions, are needed for the reaction to occur. Many possibilities exist for explaining the synthesis of a chain structure with some degree of order between the chains. The discussion above does not favor the mechanism pictured in Fig. 16; that the alginate is epimerized to its final chemical structure before crosslinking by sea-water occurs. The alternative, pictured in Fig. 17, seems more probable: It is suggested that crosslinking of the "GG-blocks" occurs simultaneously with their pairwise synthesis. A degree of overlap of unity is then obtained. The enzyme (or enzymes) must in this case be looked upon as a gelforming enzyme being able to catalyze a slow physical reaction in addition to catalyzing the more common slow chemical reaction. Work with this enzyme is currently being carried out at this Institute.

Literature Cited.

1. Black, W.A.P., J. Mar. Assoc. U.K. (1953) 29, 379.
2. Black, W.A.P., J. Mar. Assoc. U.K. (1954) 30, 49.
3. Preston, R.D., "The Physical Biology of Plant Cell Walls", Chapman and Hall, London (1974).

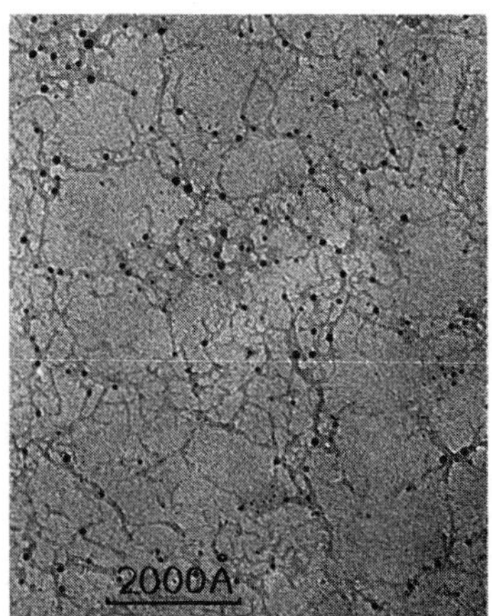

Figure 13. Electron microscope picture of calcium alginate gel (18). Gels are dialyzed against $Pb(NO_3)_2$ to yield good contrast. Some precipitation of $PbCl_2$ (from remaining $CaCl_2$ in the gel) is seen as black spheres. The threads are much thicker than in experiments without $PbCl_2$ (18, 25).

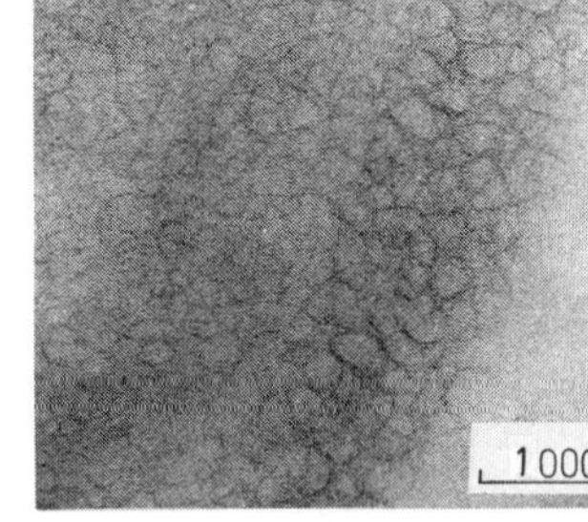

Figure 14. Electron microscope picture of a section of L. hyperborea, *stipe, outer cortex. The stipes are dialyzed against $Pb(NO_3)$ to give a good contrast. The white area to the right is part of a cell.*

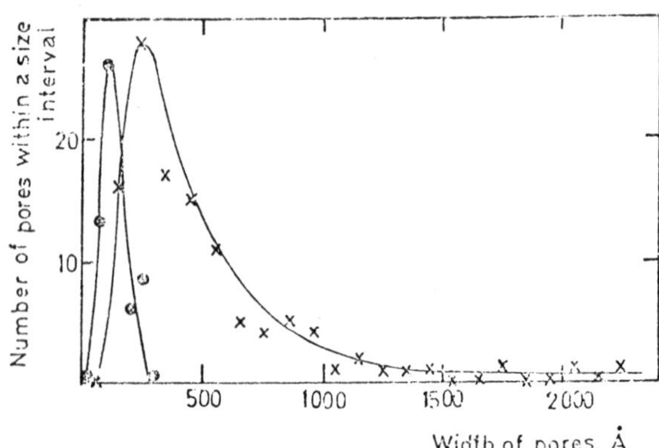

Figure 15. Distribution of pore size (diameter) in an alginate gel (18) and in a section of stipe from L. hyperborea. *●: stipe; ×: alginate.*

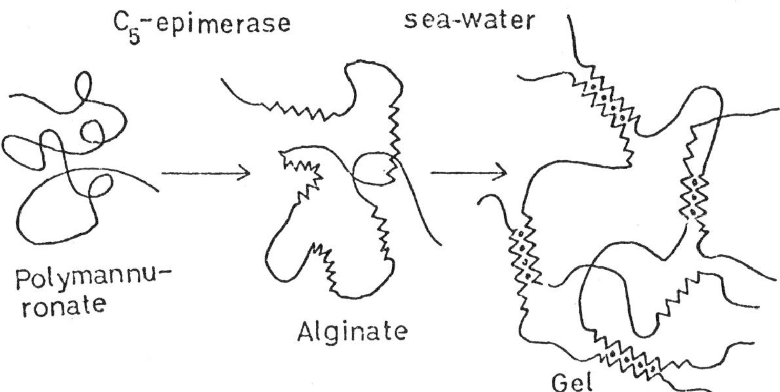

Figure 16. Model for random crosslinking of alginate epimerized to its final chemical structure before crosslinking. The degree of overlap, $\alpha < 1$. ∿∿∿ : GG-blocks; ∼ : MM and Mg-blocks.

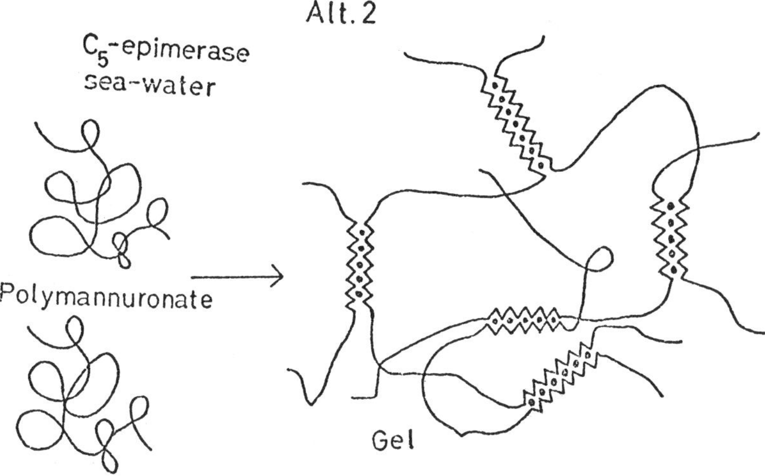

Figure 17. Model for crosslinking occurring simultaneously with the C_5-epimerization. The degree of overlap is unity.

4. Haug, A., in MTP International Review of Science, Volume 11, "Plant Biochemistry", p. 51, Butterworths London, University Park Press, Baltimore, (1974).
5. Smidsrød, O., "Some Physical Properties of Alginates in Solution and in the Gel State", Report No. 34, Norwegian Institute of Seaweed Research, Trondheim (1973).
6. Haug, A. and Smidsrød, O., Nature (London) (1967) 215, 1167.
7. Skipnes, O., Roald, T., and Haug, A., Physiol. Plant. (1975) 34, 314.
8. Haug, A., "Composition and Properties of Alginate", Report No. 30, Norwegian Institute of Seaweed Research, Trondheim (1964).
9. Haug, A., Larsen, B., and Smidsrød, O., Carbohyd. Res. (1974) 32, 217.
10. Andresen, I.L. and Smidsrød, O., to be submitted to Marine Chemistry.
11. Stokkenes, D., Andresen, I.L. and Smidsrød, O., unpublished results.
12. Skipnes, O., Thesis, Institute of Marine Biochemistry, University of Trondheim, Trondheim (1973).
13. Smidsrød, O. and Haug, A., Acta Chem. Scand. (1972), 26, 79.
14. Smidsrød, O. and Haug, A., Acta Chem. Scand. (1972), 26, 2063.
15. Østgaard, K., Thesis, Institute of Theoretical Physics and Institute of Marine Biochemistry, The University of Trondheim, Trondheim (1974).
16. Hemmer, P.C., personal communication.
17. Andresen, I.L. and Smidsrød, O., Carbohyd. Res. in press.
18. Skipnes, O. and Smidsrød, O., in Report No. 34, Norwegian Institute of Seaweed Research, Trondheim (1973).
19. Haug, A. and Larsen, B., Biochim. Biophys. Acta (1969) 192, 557.
20. Hellebust, J.A. and Haug, A., Proc. 6th Intern. Seaweed Symp. 1968, Madrid (1969) p. 463.
21. Madgwick, J., Haug, A. and Larsen, B., Acta Chem. Scand. (1973) 27, 3592.
22. Haug, A. and Larsen, B., Proc. Phytochem. Soc. (1974) 207.
23. Madgwick, J., Haug, A. and Larsen, B., to be submitted to Bot. Mar.
24. Harvey, H.W. "The Chemistry and Fertility of Sea-Water", p. 4, University Press, Cambridge, 1963.
25. Smidsrød, O., Faraday Disc. Chem. Soc. No. 57 (1974) 263.

26. Grasdalen, H., personal communication.
27. Myklestad, S., personal communication.

25

The Place of Cellulose under Energy Scarcity[1]

IRVING S. GOLDSTEIN

Department of Wood and Paper Science, North Carolina State University, Raleigh, NC 27607

Cellulose is the most abundant organic material on earth, and more important it is a renewable resource with some 50 billion tons being formed each year by land plants alone representing the fixation of 200 x 10^{18} calories of solar energy (1). The commercial forests of the U. S. at a conservative estimate are capable of producing 500 million tons of cellulose annually of which about 100 million tons are now consumed as just over twice that quantity of lumber, wood products and pulpwood (2). The approximately 50 million tons of wood pulp produced in the U. S. annually is for the most part cellulose, and amounts to about three times the total 1974 U. S. production of plastics, synthetic fibers and rubber (3).

So plentiful a material should be relatively inexpensive, and in fact wood costs from one to two cents a pound. The price of cellulose, however, depends on its purity and the extent to which it has been separated from the lignin and hemicelluloses with which it is naturally associated.

We use cellulose in the form of cotton, paper, regenerated cellulose and cellulose derivatives, and wood. All of these applications are based on the macromolecular structure of cellulose, which is a highly oriented, crystalline, linear polymer of D-anhydroglucopyranose units linked by β-1,4 glycosidic bonds with a degree of polymerization which may be as high as 10,000 in the native state.

Recent increases in the price of petroleum and natural gas have focused attention on the potential uses of cellulose for chemicals and energy. The direct combustion, gasification, or conversion to liquid fuels of cellulose in the form of agricultural or municipal residues or wood is in principle similar to the conversion of solid fossil material such as coal. Neither these processes nor the conversion of cellulose to chemical feed-

[1] Presented at Symposium on Macromolecules and Future Societal Needs, Macromolecular Secretariat.

stocks (4) are concerned with cellulose as a macromolecule, and although of potentially great significance for both energy and resources they will not be considered here. The place of cellulose as a macromolecule under energy scarcity will be considered separately for cotton, paper, cellulose derivatives and wood in the following sections. The CORRIM report of the National Research Council (5) contains a wealth of background material on this subject.

Cotton

The cotton fiber is essentially pure cellulose and has served man's textile needs for thousands of years. Although the relative importance of cotton in our total textile consumption has declined in favor of synthetics, more cotton is still consumed than any other fiber and the U. S. production has remained remarkably constant with about the same production of 3.3 million tons in 1920 and 1972 (2.9 million tons in 1974). However, the 1972 production was on only 14 million acres of land compared to almost three times as much in 1920. The increased yield per acre has resulted from irrigation, genetic improvements, mechanization, fertilization and pest control; most of these agronomic techniques are highly energy intensive.

Per capita consumption of cotton in the U. S. has declined from 26.5 pounds in 1920 to 18.5 in 1974. Projections by the U.S. Department of Agriculture indicate that this will drop to 12 pounds by 2000. Although the energy consumption for producing cotton and converting it to finished cloth is only about half that for producing cloth from synthetic fibers from petrochemicals (6), this factor alone would not be sufficient to reverse the decline in per capita consumption.

Our present cotton production is concentrated on land highly suited for growing cotton. Significant expansion to former levels of cotton acreage would require cultivation of additional land less desirable for cotton and/or displacement of other agricultural crops of perhaps higher value. Furthermore, the displacement of cotton has been in large part brought about by the more desirable durable press properties of polyester blends.

However, because cotton is a renewable resource it will continue to play an important role in the future. The magnitude of this role can be increased by successful research to accomplish greater efficiency in production, handling and processing, and to attain new performance qualities to compete with man-made fibers in such areas as durable press and flame retardance.

Paper

Paper and paperboard production in the U. S. is now at a level of about 60 million tons annually with over 80% from virgin pulp and less than 20% recycled material. American Paper

Institute projections indicate total production of about 90 million tons in 1985 and almost 140 million tons in 2000 with no major increase in the proportion of recycled material. This annual growth rate of 3.5% is based on population growth and projections of past production increases and may be excessive if consumer attitudes towards the use and reuse of paper change in response to increasing energy costs and decreasing wood supplies. Nonetheless, paper and paperboard will still represent a steadily growing use of cellulose.

Paper pulps increase in price from about $0.10/lb. for groundwood to $0.17/lb. for bleached chemical pulps, but at all levels paper is regarded as disposable. Even during the period of decreasing petrochemical prices paper maintained its growth with only occasional market penetrations by synthetic polymers. Anticipated large scale replacement of cellulose by synthetic pulps for printing papers did not materialize and are not likely to do so in the U. S. now that petrochemical prices have risen so markedly. In economies where both petroleum and cellulose must be imported local conditions may provide an improved competitive position for synthetic pulps.

Most pulp mills are essentially self-sufficient in energy in that they derive their process energy from that portion of the wood which is removed in the pulping process. The high energy demands of the papermaking stage could also be met from wood at the expense of potential raw material. A forest resource based industry can operate without any fossil fuels at all.

Since a major use of paper products is in packaging it is significant that both the energy content of the materials and the energy of manufacture are greater for plastic milk containers and plastic bags than their paper counterparts (5). Energy considerations will militate against further market shifts toward plastics and favor paper. At the same time combinations of synthetic polymers and cellulose in the form of blends, composites and coatings will become increasingly important in meeting packaging and communications needs at low cost/effectiveness ratios.

Regenerated Cellulose and Cellulose Derivatives

In contrast to the growth of wood pulp for paper applications the expensive high purity chemical cellulose or dissolving pulp is produced to the extent of less than 2 million tons per year in the U. S. and has been facing declining markets (7). Chemical cellulose is the starting material for rayon and acetate fiber, cellophane, cellulose ester plastics and cellulose ether gums. The high purity demands of these applications require additional processing steps for purifying the wood pulp and result in lower product yield. Chemical pulps as a consequence cost as much as $0.23 per lb.

The growth of petrochemically derived completely synthetic

fibers has been based not only on population growth and replacement of natural fibers, but also at the expense of rayon and acetate which instead of growing with the economy have started to decline. An important factor in this trend has been the higher energy consumption in the production of man-made cellulosic fabrics than non-cellulosic especially in fiber production (5). Despite an advantage in the energy content of the raw materials the total energy consumption of 34.7 kWh/lb. of fiber for cellulosics is greater than the 30.6 value for non-cellulosics primarily because production of cellulosic fibers requires 22.1 kWh/lb. as compared to only 11.4 for non-cellulosics.

Now that petrochemical prices are rising rapidly, cellulose and its derivatives hold the potential for assuming a more important role in our organic materials picture because of the lower energy content of the raw material. However, this will not take place spontaneously, but will require improvements in cellulose technology which has been neglected for decades.

The cost and performance of cellulosics relative to petrochemical polymers can be improved in four areas:

(1) production and yield of chemical cellulose
(2) conversion of cellulose into derivatives
(3) regeneration of cellulose and shaping of products
(4) properties of the cellulose-derived products.

Unless their cost and performance compare favorably with polymers based on petroleum or coal, cellulosics will not experience a resurgence. Since much of the present cellulose technology is highly energy intensive, the mere availability of cellulose as a renewable resource will not assure increased utilization of cellulosic polymers, unless their total cost becomes favorable. Furthermore, polymers retaining a cellulosic structure may ultimately have to compete with polymers based on cellulose as a feedstock involving hydrolysis of the cellulose to glucose, fermentation to ethanol and further conversion to ethylene on which many conventional synthetic polymers are based (4). The net energy and material requirements for the overall process will determine the most economical route.

Among the possible new technologies for improving the competitive position of cellulose derivatives are the direct preparation of cellulose ethers and esters from whole wood (8), and the development of new cellulose derivatives which would permit less energy-intensive fiber and film formation than wet spinning.

Wood

Since wood is a composite polymeric material consisting about half of cellulose it warrants consideration in the context of the place of cellulose under energy scarcity. Our total consumption of 250 million tons of wood is approximately equal to the combined production of all metals, cement and plastics. Of this about 150 million tons is used for structural and architectural purposes and manufactured products other than pulp and paper. It is estimated that by the year 2000 the total demand for raw wood will double (2).

Not only is wood a renewable resource, but the energy required for its processing into final products is considerably less than that for competing structural materials. For example, softwood lumber requires a net total energy of about 3 million BTU/ton for extraction, processing and transport compared to 9 million for clay brick, 50 million for steel joists and 200 million for aluminum siding (5).

While the total volume of wood available will be more than adequate to meet our needs the form of this material is changing, requiring new approaches to wood utilization. As large trees from which large solid wood boards can be cut become less available greater reliance will be placed on reconstituted wood products consisting of wood particles or fibers and a resinous binder. Composites such as cellulose or wood fiber reinforced plastics will also become more important. Extruded shapes with more uniform properties in different directions than wood, which is highly anisotropic, will allow the use of smaller members and standardized design.

The replacement of wood by plastics as in furniture manufacture has been reversed with rising petroleum prices with the exception of intricate molded parts. Here too composites of resin and wood fiber will probably prove to provide the best compromise of cost and performance.

Summary

Cellulose is abundant and renewable by direct fixation of solar energy by plants. In its crude form it is cheap. Its continued utilization in many traditional roles is assured, but a greater reliance upon cellulose at the expense of materials which have replaced cellulose in the past will depend on technological improvements in the energy demands and yields of cellulose processing and in the properties of the cellulosic materials themselves.

The use of cotton, paper, cellulose derivatives and wood, all cellulosic materials which are not made from fossil fuels, can be expanded to conserve the fossil energy demands of synthetic polymers and metals by reducing the energy expended in the processing of cellulosics and changing or improving those properties of cellulosics considered disadvantageous in competitive use. The level of research and development on cellulosic materials as a percentage of sales has been only about 10% of that devoted to chemicals and synthetic polymers generally. A redress of this imbalance should provide significant new and improved technology to exploit the full potential of cellulose under energy scarcity.

Literature Cited

1. Lieth, Helmut. Human Ecology (1973) 1 (4) 303-332.
2. USDA Forest Service. Forest Service Report No. 20, Washington, D. C. (1973).
3. Anon. Chem. Eng. News. (1975) 2 June. pp. 31-34.
4. Goldstein, I. S. Science (1975) 189 847-52.
5. National Research Council, National Academy of Sciences. Washington, D. C. (1976).
6. Gatewood, L. B., Jr. National Cotton Council of America, Memphis, Tennessee (1973).
7. Hergert, H. L. Applied Polymer Symposia (1975) 28 pp. 61-69.
8. Durso, D. F. Svensk Papperstidning No. 2 (1976) pp. 50-51.

INDEX

INDEX

D